THE SCIENTIFIC REVOLUTION

THE SCIENTIFIC
REVOLUTION

1500–1800

The Formation
of the Modern
Scientific Attitude

A. RUPERT HALL

BEACON PRESS   BOSTON

First published 1954

Second edition © A. Rupert Hall 1962

Second edition first published as a Beacon Paperback by arrangement
with Longmans, Green and Co Ltd in 1966

Beacon Press books are published under the auspices
of the Unitarian Universalist Association

Printed in the United States of America

International Standard Book Number: 0–8070–5093–8

10   9   8   7

Collegio Christi, suasori meo necnon auctori,
grato animo

PREFACE

To the historian of science the University of Cambridge offers riches in its manuscripts, its libraries, and its associations. Among those who dwell in the places once frequented by Newton, Darwin and Rutherford there are many who, quietly and unostentatiously, are making their contributions to the understanding of this age of science in terms of its long historical evolution. No one who has lived with them, no one moreover who has been fortunate enough to learn from Charles Raven, Herbert Butterfield, and Joseph Needham, can be other than conscious of indebtedness.

To them, and to all those friends who have helped—or tolerated —my endeavours, I offer my grateful thanks. One other only I mention by name, since he is no longer with us, Robert Stewart Whipple, whose life-long enthusiasm for the history of science is commemorated in the collection of historic scientific instruments which he presented to this University.

Above all, this volume could not have been written without the consistent support of my College, which has given generous encouragement to the study of the history of science, and the interest with which the University has fostered the teaching of this subject.

CHRIST'S COLLEGE,   A. R. HALL
CAMBRIDGE
February 1954

NOTE ON THE SECOND EDITION

WITHOUT altering the structure of this book I have revised and corrected it, modifying many passages and adding others, and enlarging the bibliography with some recent publications. I offer my thanks to all those who were kind enough to point out errors in the first edition, and to Messrs. Longmans for this opportunity to amend them.

INDIANA UNIVERSITY   A. R. H.

CONTENTS

INTRODUCTION

I HAVE tried in this book to present something in the nature of a character-study, rather than a biographical outline, of the scientific revolution. Natural science may be defined sufficiently for my purpose as the conscious, systematic investigation of the phenomena revealed in the human environment, and in man himself objectively considered. Such investigation always assumes that there is in nature a regular consistency, so that events are not merely vagarious, and therefore an order or pattern also, to which events conform, capable of being apprehended by the human mind. But science in this sense is not simply the product of one attitude to nature, of one set of methods of inquiry, or of the pursuit of one group of aims. Within it there is room for both economic and religious motives, for a greater or less exactitude in observation (though *some* measure of systematic and repeated observation is essential to science), and for a considerable latitude in theorization.

In many of these respects modern science differs markedly from that of a not very remote past. It demands rigorous standards in observing and experimenting. By insisting that it deals only with material entities in nature, it excludes spirits and occult powers from its province. It distinguishes firmly between theories confirmed by multiple evidence, tentative hypotheses and unsupported speculations. It presents, not a possible or even a plausible picture of nature, but one in which all available facts are given their logical, orderly places. These are the most important characteristics of modern science, which it acquired during the period of transition conveniently known as the scientific revolution, and has since retained. Certainly they were long in gestation, but it is with the period of their coming to fruition and vindication by success that this volume is concerned. Some topics I have chosen to omit: mathematics, because it deals not with the phenomena of nature, but with numbers; medicine, because it was at this time rather an art than a natural science. It has not been my object to attempt a complete narration of events, nor to dwell on biographical and experimental details. Perhaps these omissions may be forgiven in a book designed as an introduction to the study of the historical processes at work in the development of science, and of the major stages in that development.

Science began soon after the birth of civilization. Man's attempt to win an Empire over Nature (in Francis Bacon's phrase) was much older still; he had already learnt to domesticate animals and plants, to shape inorganic materials like clay and metals to his purposes, and even to mitigate his bodily ailments. We do not know how or why he did these things, for his magic and his reasoning are equally concealed. Only with the second millennium B.C. is it possible to discern, dimly, the beginnings of science in the coalescence of these three elements in man's attitude to Nature—empirical practice, magic and rational thinking.

The same three elements continued to exist in science for many thousand years, until the scientific revolution took place in the sixteenth and seventeenth centuries. Reason, in conjunction with observation and experiment, slowly robbed magic of its power, and was gradually better able to anticipate and absorb the chance discoveries of inventive craftsmen, but complete reliance upon a rational scientific method in man's reaction to his natural environment is very recent. Magic and esoteric mystery—the elements of the irrational—were not firmly disassociated from serious science before the seventeenth century, at which time even greater stress than before was being laid on the usefulness to scientists of the craftsman's practical skills. This in turn was not outgrown until the nineteenth century, when it became clear that in the future sheer empiricism and chance would add little to man's natural knowledge, or to his natural power.

Rational science, then, by whose methods alone the phenomena of nature may be rightly understood, and by whose application alone they may be controlled, is the creation of the seventeenth and eighteenth centuries. Since then dramatic achievements in understanding and power have followed successively. In this sense the period 1500–1800 was one of preparation, that since 1800 one of accomplishment. And it is convenient to conclude this history of the scientific revolution with the early years of the nineteenth century for other reasons. Though profound changes in scientific thought have occurred since that time, and though the growth of complexity in both theory and experimental practice has been prodigious, the processes, the tactics and the forms by which modern science evolves have not changed. However great the revision of ideas of matter, time, space and causality enforced during the last half-century, it was a revision of the content, not the structure of science. In its progress since 1800 the later discoveries have always embraced the earlier: Newton was not proved wrong by Einstein, nor Lavoisier by Rutherford. The formulation of a scientific proposition may be modified, and limitations to its applicability recognized, without affecting its propriety in the context to which it was originally found appropriate. We do not need sledge-hammers to crack nuts; we do not need the Principle of Indeterminacy in calculating the future position of the moon: 'the old knowledge, as the very means for coming upon the new, must in its old realm be left intact; only when we have left that realm can it be transcended.' [1]

Despite the progressive accumulation of knowledge and elaboration of theory, only the broader extrapolations of nineteenth-century science would now be described blankly as "wrong," though a larger part of its picture of Nature might be described as "inadequate" or as "true within certain limits." Even in biology, where ancient and extra-scientific notions lingered long, where experiment was most tardy in finding its just deployment, this is still the case. The systematic and descriptive biology of the pre-Darwinian epoch was not rendered futile by the theory of evolution. Earlier microscopists were not exposed to ridicule by the founders of cytology. On the contrary, the revolution in thought about animate nature incorporated, and was founded on, the labours of three or four generations. The same could not be said of science before 1500, or even, without restriction, of the science of the seventeenth and eighteenth centuries. Its progress in these earlier times was not by accretion, for it was now and again necessary to jettison encumbering endowments from the past. Such science *was* on occasion simply wrong, both in fact and in interpretation. Its propositions had to be rejected *in toto*, not merely circumscribed, as the result of experiment and creative

[1] J. Robert Oppenheimer, in his third Reith Lecture (*The Listener*, vol. L, p. 943).

thinking. In this respect the beginning of the nineteenth century seems a useful point of demarcation between the scientific revolution, in the course of which the sciences painfully and in succession acquired their cumulative character, and the recent period during which that character has been successfully maintained.

The cumulative growth of science, arising from the employment of methods of investigation and reasoning which have been justified by their fruits and their resistance to the corrosion of criticism, cannot be reduced to any single theme. We cannot say why men are creative artists, or writers, or scientists; why some men can perceive a truth, or a technical trick, which has eluded others. From the bewildering variety of experience in its social, economic and psychological aspects it is possible to extract only a few factors, here and there, which have had a bearing on the development of science. At present, at least, we can only describe, and begin to analyse, where we should like to understand. The difficulty is the greater because the history of science is not, and cannot be, a tight unity. The different branches of science are themselves unlike in complexity, in techniques, and in their philosophy. They are not all affected equally, or at the same time, by the same historical factors, whether internal or external. It is not even possible to trace the development of a single scientific method, some formulation of principles and rules of operating which might be imagined as applicable to every scientific inquiry, for there is no such thing. Methods of research in each subject are too closely bound up with the content and problems of that subject for the devising of a mechanical method.

Nor can we always exploit any dichotomy between the ideas and the practice of science, though this dichotomy may be a useful tool at times. Interpretations without knowledge and knowledge without interpretations are equally unbalanced and sterile; any branch of science consists of both. Therefore, since knowledge arises from practical operations, the history of science can never be a purely intellectual history (like that of mathematics), nor can it be analysed in a wholly logical manner. For ideas arise from facts, and the facts of science may be revealed in an order conditioned by chance, by technical resourcefulness, or industrial invention. On the other hand, the creative intellect is always playing upon the materials provided by techniques of experiment, observation or measurement; however subtle these become, they can never be said to determine the course of a science, unless in a negative sense.

The dichotomy in science, the fact that its progress requires both conceptual imagination and manipulative ingenuity, is particularly apparent in the scientific revolution, and presents peculiar problems. The reaction against the traditional picture of Nature derived from the Greeks occurred simultaneously on the factual and the interpretative levels. The *theories* of the past were criticized as being inconsistent, speculative, incomprehensible, and as involving spiritual qualities rather than the properties of matter; the *facts* related in the past were challenged as ill-tested, spurious, superstitious, and as casually chosen without care in observation and experiment. The development of each of these phases of criticism may be traced in the later middle ages, when, however, they were rarely associated. Even after 1500 many exponents of a new attitude failed to perceive that doubts of fact and theory were inter-related; that one could not proceed without the other. Hence a great part of the force of the "new philosophy" of the seventeenth century was due to its dual scepticism, its ambition to challenge fact and theory, sometimes without consciousness of making a double attack. This duality reinforced attention to the question of the logical connection between facts and theories, which had also been examined by medieval critics of Aristotle. How were the relevant facts to be brought to light? How, with the facts known, were appropriate propositions concerning them to be arrived at? How could scientific propositions be tested by their usefulness in dealing with facts? The early success of the scientific revolution owed much to the answering of such questions (pragmatically, rather than philosophically) in the scientific method combining observation, hypothesis, experiment and mathematical analysis, a method which demonstrates in itself the essential liaison between the factual and interpretative levels of inquiry.

If the scientific revolution cannot be completely depicted, even in its origins, as an intellectual revolt against the traditional interpretation of Nature, nevertheless such a revolt demands primacy of place in tracing its antecedents. Few men have, or had, the sense of supremacy of pure fact demanded by Francis Bacon. The majority of scientists at all times have sought for facts with an object in view. Consequently, the scientific empiricists of the seventeenth century, including Bacon himself, pointed to the insufficiency of contemporary scientific ideas before they clamoured for more facts on which to frame a sounder doctrine. They did not need new facts to teach them Aristotle's mistakes, for of those they were already aware; they needed observations and experiments to avoid falling into fresh errors of their own. Cartesians, too, were as confident that conventional science was untrustworthy: because it ignored the method of reasoning from indubitable truths, and was therefore philosophically unsubstantial. And does not Galileo himself refute Aristotle with reason, before overwhelming him with observations? Medieval science, and especially the Aristotelean doctrines within it, was not so much swept away by a hurricane of uncomfortable facts, as brought down by its own internal decay. While the mechanics of Greek science, the four elements, the bodily humours, the Ptolemaic system, still commanded allegiance (though becoming less and less relevant to practical affairs) the Hellenistic view of the cosmos was becoming increasingly alien to the European mind. From the fourteenth to the sixteenth century men might teach and work in science as though they were Greeks of antiquity, but they were not so in fact, and the community of thought between Europe and Hellas grew ever more formal. By 1600 it was almost academic. A demonstration in Greek mathematics or a particular piece of reasoning could fire admiration, but the universe could be seen through Aristotle's eyes no more. Even his expositors had transformed him.

In part this was brought about by the rival Christian authority. No Christian could ultimately escape the implications of the fact that Aristotle's cosmos knew no Jehovah. Christianity taught him to see it as a divine artifact, rather than as a self-contained organism. The universe was subject to God's laws; its regularities and harmonies were divinely planned, its uniformity was a result of providential design. The ultimate mystery resided in God rather than in Nature, which could thus, by successive steps, be seen not as a self-sufficient Whole, but as a divinely organized machine in which was transacted the unique drama of the Fall and Redemption. If an omnipresent God was all spirit, it was the more easy to think of the physical universe as all matter; the intelligences, spirits and Forms of Aristotle were first debased, and then abandoned as unnecessary in a universe which contained nothing but God, human souls and matter.

Christianity furnished the scientist with God as the First Cause of things. But if this First Cause had, so to speak, set the universe to run its course and endowed man with free-will to make his own destiny, the phenomena of nature could only be the result of determined processes, manifestations of a mechanistic design, like an infinitely complex automaton or clock. A clock is not explained by saying that the hands have a natural desire to turn, or that the bell has a natural appetite for striking the hours, but by tracing its movements to the interconnections of its parts, and so to the driving force, the weight. If God was the driving force in the universe, were not its motions and other properties also to be ascribed to the interconnections of its parts? The question, slowly compelling attention over three hundred years, received a positive answer in the seventeenth century. The only sort of explanation science could give must be in terms of descriptions of processes, mechanisms, interconnections of parts. Greek animism was dead. Appetites, natural tendencies, sympathies, attractions, were moribund concepts in science, too. The universe of classical physics, in which the only realities were matter and motion, could begin to take shape.

CHAPTER I

SCIENCE IN 1500

EUROPEAN civilization at the beginning of the sixteenth century
was isolated as it had not been since the first revival of learning some four hundred years earlier. Political changes had
severed traditional links between east and west. The heart of the
great area of Islamic culture, from which medieval scholarship
and science had drawn their inspiration, had been over-run by
the Turks; the eastern Christian centre of Byzantium, the last
direct heir of ancient Greece, was destroyed; the learned Moors
and Jews of Spain were expelled by the armies of Castille and
Aragon. The overland route to China, by which many important
inventions had been transmitted to Europe, among them the
new instruments of learning, paper and printing, and over which
Marco Polo had passed, was now barred. Against this loss of a
relatively free interchange of ideas and commodities through the
Mediterranean may be set a developing commercial and scientific
interest in the promise of geographical exploration across the
oceans. The Portuguese had brought spices from the Indies;
Spain had drawn her first golden tribute from Hispaniola. But
the results of the re-orientation of communications southwards
and westwards upon the Atlantic had not yet been assimilated,
and though Europeans were quick to recognize the value of
negroes or American Indians as slaves, they had not yet learnt to
appreciate the magnitude of the civilizations in India and China
with which, for the first time, they came into direct contact.

Fortunately, when these events occurred, European scholars
had already emerged from their tutelage, and Latin literature,
having been enriched with almost everything that Arabic was to
convey to it, had become creative in its own right. The final
dissolution of the Christian empire in the Near East, indeed,
brought to Europe a direct knowledge of classical science with
which it had been less perfectly acquainted through the Islamic
intermediary. At the moment when their intellectual communion
with other peoples was most sharply denied by circumstance, the

humanistic scholars of the renaissance were most engrossed in studying the pure origins of western civilization in Greece and Rome. During the next centuries the trend of the middle ages was to be reversed; Europeans were to teach the East far more than they learned from it.

In 1500 the fruitful cultivation of science was limited to western Europe, to a small region of the whole civilized world stretching from Salamanca to Cracow, from Naples to Edinburgh. Even the peripheries of this region were illuminated rather by the reflected brilliance of the Italian renaissance, than by any great achievements of their own. The northern shift of the focus of learning was beginning—northern Italy had replaced the south and Sicily as the cultural centre of Europe—but the sixteenth is unquestionably the Italian century in science. Scholars, mathematicians, physicians everywhere measured their own attainments by Italian standards; the Italian universities, and the Italian printing-houses, possessed an acknowledged pre-eminence. Beneath inevitable gradations in civilization (no less marked in science than in fine art) lay an essential cultural unity, undisturbed by nationalism or religious schism. Europe had a common tradition; in matters temporal it was consciously the heir of Greece and Rome, shaping the texture and manner of its own life to the model of their golden age; in matters spiritual a single compass of belief had passed from the great Councils of the Church through the early Fathers to the Papacy and the medieval Doctors. If the secular Empire had become a shadow, the spiritual power of the Church seemed still firmly founded upon a single theology and a universal orthodoxy which had triumphed over Albigenses, Wyclif and Hus. Upon this unified Church all learning was in varying degrees dependent; the universities, themselves papal creations, were religious institutions. The study of medicine was the only discipline able to stand, to some extent, detached from learning's prime duty to theology, and even the teaching and practice of medicine were subject to a measure of ecclesiastical control. Human dissection, for instance, was strictly supervised by the Church. Learning had in Latin its own universal language. The system and content of education in every school and university were so similar that a scholar could find himself at home anywhere. In each the student laboured upon the same texts, graduated by debating similar philosophical themes, and (if he progressed as a man of learning) found the stimulus to original thinking in problems familiar to all men of education.

It would be mistaken to suppose that, because the intellectual life of all Europe at the beginning of the sixteenth century was as homogeneous as that of the modern nation state, it was characterized by a dull uniformity. There were, and there had long been, distinct and to some extent rival schools, just as there was also a special development of some type of study at a particular place, theology at Paris, law and medicine at Bologna and Padua. In one place men still clung to the pure Aristotelean tradition, at another they had begun to criticize it; here the Arabic physicians were more highly regarded than the newer Greek texts, there they were despised as poisoning the Galenic well of knowledge. Within the acceptance of a common body of premisses there was vast uncertainty concerning the way in which the premisses should be applied. No one doubted that blood-letting was an essential part of therapy: but there was great dispute over the actual technique. No one doubted that the world was round: but different navigators could not agree on the best way to plot a course. No one doubted that the calendar was out of joint: but the best astronomers could not join in declaring the remedy. It is indeed a misleading view of the early stages of the scientific revolution to see them as involving only the conflict of two types of premiss, as typically in the antithesis between the Ptolemaic and the Copernican world-systems. In fact much more was involved, the practical and successful handling of detailed problems. This is perhaps most clearly seen in the non-physical sciences, in anatomy or natural history in the sixteenth century, but it can also be seen in the practical arts which depended upon the physical sciences. Ultimately, of course, such an art (or science) as navigation advanced very considerably through the substitution of the Copernican for the Ptolemaic astronomy; but in the sixteenth century greater accuracy in cartography and oceanic navigation was independent of the high truths of cosmology.

The unity of learning which the sixteenth century inherited from the middle ages is a strong mark of its diffusion. Scientific knowledge in particular showed little differentiation in pattern, though it varied greatly in quality and method, because it was not the native product of Christian civilization in Europe, but imposed upon it. The primitive Germanic tribes had long submitted to a Romano-Hellenistic culture: first through their contact with the Roman Empire, then through their conversion to Christianity and the various efforts towards a renaissance of learning that had followed; partly assimilated and legitimized (as in witchcraft or folk-medicine), their original attitude to Nature had largely sunk to the level of superstition in comparison with the "literary" science which won its supremacy in the thirteenth century. As a first approximation, it may be said that the cultural history of Europe, since the hegemony of the Roman city-state, is mainly concerned with the diffusion of Hellenistic influences, Greece itself being the heir to the ancient civilizations of the Near East. Perhaps the nadir (by either political, economic or intellectual standards) was reached about the sixth century; by the ninth efforts were being made to assimilate the flotsam surviving from the wreck of Roman Imperialism. In Islam, where the Greek legacy was more direct (much of it had been Hellenized in Alexander's time or later) and where barbaric devastation was slight, study of Greek science and medicine began early. Texts were written that Europe held in esteem centuries later. The Arabic treatise on the astrolabe which Chaucer translated into English c. 1390 had been written in the late eighth century; of the Islamic medical authorities whom he mentions in the *Prologue* to the *Canterbury Tales*, two had lived some four centuries earlier. In the first phase of learning among Franks, Saxons and Lombards, science had been meagrely represented by a few Latin texts—fragments of Plato (the *Timæus*), of Macrobius, and Pliny—and the degeneration of form and thought from these originals was rapid. In the second phase, beginning early in the twelfth century, diffusion was concentrated into a very few channels, for the new philosophical and scientific literature in Latin was created by its very few scholars who were able to translate from Arabic, Hebrew, or more rarely directly from the Greek. The most famous of them, Gerard of Cremona (c. 1114–87) is credited with making at least seventy translations, some of them like Avicenna's *Canon*, or encyclopædia, of medicine, of vast extent. Though the lists certainly exaggerate the personal influence of one individual upon the Latin corpus, it cannot be doubted that the somewhat arbitrary selection exercised by a small group of linguists had a determining effect upon the nature of the material available to the purely Latin scholars of later generations. And it must be further remembered that the Islamic and Hebrew scientific writings, upon which the Latin translators worked, themselves represented a fraction—accidentally or consciously chosen—from the whole original literature.

The intellectual life of medieval Europe was rapidly transformed by this acquisition of a sophisticated method and doctrine in both philosophy and science. Its influence can be traced in such varied directions as the use of siege-engines in war, or the theory of government, as well as in astronomy and physics. At first there was opposition to the new type of study: the sources, being infidel, were suspect, and they seemed to encourage a too presumptuous inquiry into matters beyond human comprehension. But the simplest Christian attitude, that men should not meditate upon this world but the next, and that religion, rather than the works of pagan philosophers, should instil the principles of morality and ethics, had lost its force. The condemnation of the teaching of Aristotle's natural philosophy by a provincial council held at Paris in 1210 proved ephemeral, as did later attempts to limit the influence of Arabist sources.[1] Even before scholasticism developed the study of Hellenistic philosophy upon Christian principles the European mind was awake to the value of the riches laid before it when refined from theological falsities. Robert Grosseteste of Oxford in the first, and Thomas Aquinas in the second half of the thirteenth century were the two foremost exponents of Aristoteleanism, which as a method of reasoning and a fabric of knowledge was thus reconciled with Catholic theology. Inevitably the Hellenistic-Islamic renaissance of the twelfth century emerged with an even greater homogeneity from this further process of moulding. To the extent that Aristotle was seen through the eyes of St. Thomas—and his influence was profound for it was he who made Aristotle the master of medieval thinking—the unity of medieval culture was reinforced by the single Thomist tradition. Averroism, the extreme rationalist appraisal of Aristotle's thought, became heresy. Despite Aquinas' deep study of every aspect of the Aristotelean corpus, he utterly failed to understand the true spirit and methods of natural science, 'and no scientific contribution

[1] George Sarton: *Introduction to the History of Science* (Baltimore, 1927–48), vol. II, p. 568. A reputation for profound acquaintance with Arabic writings was, however, long associated in the popular mind with skill in magical arts, so that a mass of legend was built up round such a figure as Michael Scot.

can be credited to him.'[1] Against this dogmatic tradition, the exponents of experiment (Grosseteste, Roger Bacon) and the mechanistic school of critics (Jordanus Nemorarius, Jean Buridan, Nicole Oresme) of the same and later times were generally unavailing, though they are interesting and important as the precursors of the scientific revolution.

The time of maturation was really astonishingly short. It is a common delusion that medieval intellectual life was stagnant over the many centuries that separate the fall of Rome from the rise of Florence. On a more just assessment the beginnings of intelligent civilization in Europe may be placed no more than four hundred years before the Italian renaissance: a longer interval separates Einstein from Copernicus, than that which intervenes between Copernicus and the earliest introduction of rational astronomy to medieval Europe. Within a century of the first translations Latin contributions to creative thinking were at least equal to those made in any other region of the globe: by the mid-fourteenth century the period of assimilation was over, the emphasis lay on original development and criticism, and the initiative which it has since retained had passed to the West. The renaissance of the twelfth and thirteenth centuries was imitative and synthetic; that of the fourteenth was exploratory, critical, inquisitive. Authorship began to pass from commentary to original composition, although still often in the commentatory form. There are signs of true science: the development of mathematics, now aided by the "Arabic" numerals which were of Indian provenance; anatomical research pursued in actual dissection; the development of a new mechanics employing non-Aristotelean principles and the rudiments of a geometrical analysis. At the same time—and there may be a correlation here which becomes more explicit and conscious in the seventeenth century—there was considerable technological progress. The magnetic compass was first described in Europe by Peter the Stranger in 1269; gunpowder was applied to the propulsion of projectiles about 1320; the third of the great medieval inventions, printing, began its pre-history about 1400. But these are only the most striking examples of a long series: the introduction, or wider exploitation of new materials like rag-paper, of processes like distillation, of new machines like the windmill or

---

[1] George Sarton: *Introduction to the History of Science* (Baltimore, 1927–48), vol. II, p. 914.

the mechanical clock, of new mechanical devices like the stern-rudder for ships, or the use of water-power for industrial purposes. Although philosophers and theologians still regarded slavery as a justifiable social institution, the Latin[1] was the first civilization to evolve without dependence upon servile labour. From the eleventh century onwards the rigours of villeinage were mitigated and slavery disappeared. The phenomenon of the new social fabric was, that while the general wealth increased, the inequalities in its distribution tended to become more moderate. In fact the fourteenth century shows the beginnings of a characteristic of modern European civilization, the utilization of natural resources through means progressively more complex, efficient and economical in the expenditure of human labour. These are the first signs of an attitude to Nature, and to technological proficiency, which was to become overtly conscious in the writings of Francis Bacon but was not to attain its fulfilment before the nineteenth century. A manuscript of 1335, written by Guido da Vigevano, a court physician who was no insignificant investigator of human anatomy, already develops the application of mechanical skills to the arts of war (in this instance a projected crusade to recover the Holy Land) in a manner typical of the "practical science" of the Italian renaissance and later ages. Again, it must be observed that few of the inventions mentioned above originated in the Latin West. The common principles of machine-construction were familiar to the later Hellenistic mechanicians: other discoveries, like the compass and gunpowder, were transmitted to Europe from the Far East through Islam. But it was Latin society which was transformed by them, not that of Eastern peoples: and it was the Latins who, as in the realm of ideas, alone fully realized and extended their possibilities. In the end it was the West that rediscovered the East, arriving in the ports and coasts of India and China with an admitted technical and scientific superiority.

This is the background to what was once regarded without qualification as "*The* Renaissance." The humanists of the sixteenth century, who created the legend of Gothic barbarity,

---

[1] Since, during the middle ages, not all Europeans were members of the Roman Catholic, Latin-writing civilization, it seems convenient to adopt the noun "Latins" and the adjective "Latin" in a general sense as descriptive of those Europeans who were members of this civilization.

participated in a new phase of the diffusion of Hellenistic culture. For them the past was not merely a store of knowledge to nourish the mind, but a model to be directly imitated. Upon them, from its newly discovered texts, and from its extant relics in Italy and France, the pure light of the Hellenistic world seemed to shine directly, and they disdained the reflected splendour of Islam. Medieval scholarship seemed limited, obtuse, pedantic. It had embroidered and perverted what it had not truly understood. Through the perspective of the unhappy and generally, perhaps, retrogressive fifteenth century, the whole medieval period was seen in a jaundiced aspect. The decay of medieval society, and of medieval intellectual traditions, produced a revulsion in the minds of the first exponents of an art, literature and science which were at once more authentically classical, and more modern.

The fathers of the Italian renaissance were Petrarch (1304–74) and Boccaccio (1313–75); they, with other originators of a new literary and artistic movement, were contemporary with the height of medieval science. To its exacting discipline and subtle ratio-cination the new emphasis on taste, elegance and wit was by no means wholly friendly. In one important aspect the renaissance was an academic revolt against the tyranny of expositors and commentators, but the scholars who turned with renewed enthusiasm to original texts came close to surrendering something of value, namely the critical analysis and extension of Hellenistic science due to generations of Islamic and Latin students of philosophy and medicine. These gains, which the superior texts of the scholar-scientists of the early sixteenth century would hardly have recompensed, would have been sacrificed if the renaissance had abandoned the middle ages as completely as the academic revolutionaries wished; but such was not the case, and to a large extent the fourteenth century was assimilated by the sixteenth.

As this suggests, there are important reservations to be made concerning renaissance humanism as a force making for innova-tion in science. For example, it can have had little impact upon the early printers who published large numbers of medieval books, nor upon the readers who presumably desired them. And in their revision of classical authors for the press the scholar-scientists were if anything less prone to comment adversely upon them than their medieval predecessors. Among the humanists intense admiration for the work of antiquity led to the belief that human talent and achievement had consistently deteriorated after the golden age of Hellenistic civilization; that the upward ascent demanded imitation of this remote past, rather than an adventure along strange paths. Thus the boundary between scholarship and archaism was indefinite, and too easily traversed. Some branches of science were indeed directed to new ambitions, others were raised rapidly to a new level of knowledge, but it cannot be said that all benefited from a single influence favouring realism, or originality of thought, or greater scepticism of authority. On the contrary, humanism sometimes made it more difficult to enunciate a new idea, or to criticize the splendid inheritance from antiquity. Against the free range of Leonardo's intellect, or the penetrating accuracy of Vesalius' eye, must be set the conviction that Galen could not err, or the view of Machiavelli that the art of war would be technically advanced by a return to the tactics and weapons of the Roman legion.

Obviously such rigid adherence to the fruits of humanistic scholarship did not yield the renaissance of scientific activity beginning in the late fifteenth century. Rather it sprang from the fertile conjunction of elements in medieval science with others derived from rediscovered antiquity. So in mathematics the revelation of Greek achievements in geometry provoked emulation, yet the algebraic branch of Islamic origin (for which humanism did nothing) made equally rapid progress. Ultimately the union of the two effected a profound revolution. Moreover, to take a more general point emphasized by Collingwood,[1] the renaissance attitude to natural events was so different from that of the ancients that the classical revival inevitably led to some anomalous results. Thus one may account for the vast interest in Lucretius' poetic statement of Greek atomism, *De natura rerum*, rediscovered in 1417, and the eminence granted to Archimedes as the archetype of the physical scientist. Already in the fourteenth century there were clear signs of an approach to a mechanistic philosophy of nature, and the use of mathematical formulations, in the physical sciences where such a novel attitude could do much to liberate the scientist from discussions of the metaphysics of causation and to lead him to concentrate attention upon the actual processes of natural phenomena. In this respect the broader

[1] *The Idea of Nature* (Oxford, 1945).

interests of humanistic scholarship reinforced a tendency which they could not have initiated, a reaction against Aristotle. Although the revival of Greek atomism had an important influence, the "mechanical philosophy" of the seventeenth century represented much more than just this.

A more complete access to the works of Euclid, Archimedes and Hero of Alexandria (among many others) gave a fertilizing inspiration to physics and mathematics, but it was on the biological sciences that the superior information of the ancients, resulting from a more serious attention to observation, had its greatest effect. Encyclopædic study by medical men of the complete works of Hippocrates and Galen, by naturalists of those of Aristotle, Dioscorides and Theophrastus, played a large part in the scientific endeavour of the sixteenth century, only slowly overshadowed by original investigation. Even at the close of the century studies in the structure, reproduction and classification of plants and animals had barely surpassed the level attained by Aristotle. Later still William Harvey could regard him as a principal authority on embryological problems. Undoubtedly a return to purer and more numerous sources of Greek science was a refreshment and stimulus; despite the imperfections in the renaissance scholar's appreciation of the immediate and the remote past, he rightly felt that such sources laid before him new speculations, facts and methods of procedure, suggested many long-forgotten types of inquiry, and disclosed a wider horizon than that known to the medieval philosopher. To a growing catholicity of interest printing added the diffusion of knowledge. The tyranny of major authorities inherent in small libraries was broken for the scholar who could indulge in an ease of compilation and cross-reference formerly unthinkable. But it is not to be supposed that the method of science, or the realization of the problems it was first to solve successfully, were simply fruits of the renaissance intellect. Nascent modern science was more than ancient science revived. Through the revived Hellenistic influence in the sixteenth century, always diverting and sometimes dominating the development of scientific activity, the strong current of tradition from the middle ages is always to be discerned.

In 1500, when Leonardo da Vinci was somewhat past his prime, the scientific renaissance had hardly begun. Scientific humanism, the great critical editions, and the first synthetic achievements under the new influences, belong to the next generation. In the spirit of scientific humanism the astronomers Peurbach (d. 1461) and Regiomontanus (d. 1476) had indeed attempted to recover the original Greek text of Ptolemy's *Almagest*, but they had succeeded in neither translating nor printing it. And though a vast volume of medieval manuscript was left undisturbed and rapidly forgotten, the larger proportion of the scientific books printed before 1500 contained material which was familiar two centuries earlier. The general stock of knowledge had hardly changed, as can be seen from such a compilation as the *Margarita Philosophica* of Gregorius Reisch (d. 1525), published in 1503.[1] This was an attempt at an encyclopædic survey, perhaps rather conservative, certainly immensely simplifying the best knowledge of the age, but useful as a conspectus of scientific knowledge at the opening of the sixteenth century. The pearls of philosophy are divided under nine heads which treat of Grammar, Logic, Rhetoric, Arithmetic, Music (theoretical, i.e. acoustics, and practical), Geometry (pure and applied), Astronomy and Astrology, Natural Philosophy (mechanics, physics, chemistry, biology, medicine, etc.), and Moral Philosophy. There is also (in the later editions) a Mathematical Appendix which deals with applied geometry and the use of such instruments as the astrolabe and torquetum on which medieval astronomy was founded. Reisch's design thus corresponds closely to that of the typical arts course in the universities of the period, which proceeded from the *trivium* (grammar, rhetoric and logic) to the *quadrivium* (arithmetic, geometry, astronomy and music).

Astronomy was the most systematic of the sciences. The phenomena of the heavens had long been subjected to mathematical operations, though opinions differed on the most suitable manner of applying them, and prediction of the future positions of sun, moon and planets, the groundwork of astrology, was possible to a limited degree of accuracy. Mathematical astronomy had been advanced to an elaborate level by the Greeks, and still further perfected by the mathematicians of Islam, who had also been patient and accurate observers. For cosmology in the modern sense there was almost no necessity: the universe had been created

[1] At least nine editions were printed before 1550; others appeared as late as 1583 (Basel) and 1599–1600 (Venice).

at a determined date in the past in the manner described in Holy Writ, and would cease at an undetermined but prophesied date in the future. It could be proved, both by reason and by divine authority, to be finite and immutable, save for the one speck at the centre which was earth, unique, inconstant, alone capable of the recurring cycle of growth and decay. As Man was the climax of creation, so also the earth (though the most insignificant part of the whole in size) was the climax of the universe, the hub about which all turned, and the reason for its existence. It was natural for men who were unhesitant teleologists to believe that the pattern of events was designed to suit their needs, to apportion light and darkness, to mark the seasons, to give warning of God's displeasure with his frail creation. In varying degrees of sophistication such an attitude is universal and primitive. The root of scientific astronomy, however, lay rather in the ancient observation that the alterations of the heavens are cyclic, and can therefore be predicted. But in the tradition of astronomy which prevailed in 1500 there was a third element, of purely Greek origin, which provided a physical doctrine describing the nature of the mechanism bringing about the appearances. The simple form of this mechanism—to which the Greeks were led by their reflection on the Babylonian knowledge of the celestial motions— was described by Aristotle; the complex form, capable of accounting for more involved planetary motions, was due to Ptolemy. The former was a physical rather than a mathematical doctrine, the latter was mathematical rather than physical, corresponding to what would now be described as a scientific model of actuality.

In a debased form, and strongly influenced by Platonic belief in correspondences between the macrocosm (the universe) and the microcosm (man) the simpler Greek theory had never been entirely lost to the middle ages. The twelfth-century translators brought Ptolemy's *Almagest* and many Arabic astronomical texts to Latin science. An abstract of spherical astronomy, which did not treat of the planetary motions in detail, the *Sphere* of John of Holywood, became a common text in the later middle ages and was printed thirty times before 1500. The *Margarita Philosophica* describes the "mechanism of the world" as consisting of eleven concentric spheres; progressing outwards from the sublunary region with the earth as the centre, they are those of the moon, Mercury,

Venus, the sun, Mars, Jupiter, Saturn, the Firmament (Fixed Stars), the Crystalline Heaven, the Primum Mobile, and the Empyræan Heaven, "the abode of God and all the Elect." The spheres were conceived as of equal thickness, and fitting together without vacuities, so that for example the sphere of Saturn was immediately adjacent to that of the fixed stars. Philosophers were less clear on the physical matter of the spheres (Aristotle's perfect quintessence as contrasted with the four mutable elements of the sublunary world) apart from the fact that they were rigid and transparent; having no dynamical theory they were not puzzled by problems of friction, mass and inertia. The tenth sphere, the Primum Mobile, which imparted motion to the whole system, revolved from east to west in exactly twenty-four hours. The eighth, bearing the fixed stars, had a slightly smaller velocity, sufficient to account for the observed precession of the equinoxes. Between these the crystalline heaven was added to account for a supposed variation in the rate of this precession. Then, descending towards the centre, as the spheres became slightly less perfect, more sluggish and inert, each completed its revolution in a slightly longer period, so that the sun required 24 hours 4 minutes between two successive crossings of the meridian, and the moon 24 hours 50 minutes, approximately. This increasing retardation resulted (as we now know) from a combination of the orbital velocity of the earth and the orbital velocity of the planet; the whole heavens revolved round the earth once in 24 hours, but the sun, moon and planets appeared to move also in the opposite direction at much smaller, and varying, velocities.

With these motions established, simple spherical astronomy was taught just as it is today. But this simple theory did not allow of prediction, it did not take into account the motion of the moon's nodes, causing the cycle of eclipses, nor the variations in brightness (i.e. distance) and velocity of the heavenly bodies. It is possible to derive predictions from purely mathematical procedures, as the Babylonians did, without making any hypothesis concerning the mechanism involved. Till modern times the astronomer had nothing to work with other than angles, angular velocities, and apparent variations in diameter and brightness. He was unable to plot the orbit of a planet in space: the best he could do was to fabricate a system which would bring it to the proper point in the zodiac at the right time. The Greeks, however, could not renounce

the system of homocentric spheres altogether, and their geometrical methods based on the circle (in which, it was assumed,
motion was most perfect since it was perpetual and uniform)
facilitated the construction of an analogous, but more complex,
mechanical model.[1] As an example, the system of spheres whose
combined motions are responsible for the observed phenomena of
the planet Saturn is thus described by Reisch. It fills the depth,
represented by the seventh sphere in the simple theory, between
the interior surface of the sphere of fixed stars, and the exterior surface of the sphere of Jupiter, both of which are exactly concentric

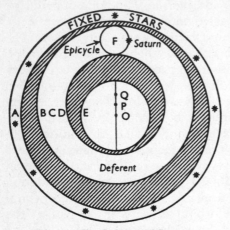

with the earth. Taking a
section through the Saturn-
ian spheres in the plane
of the ecliptic, A (Fig. 1) is
the outer surface of the
largest sphere, concentric
with the earth, B is its interior
surface which is eccentric.
C is a wholly eccentric
sphere, the deferent, carrying
the epicycle F embedded in
it. D is the outer surface of
the third and innermost
sphere, which is eccentric,
and E is its inner surface
which is concentric with the

FIG. 1. The Spheres of Saturn.

earth. Within E follow the remaining planetary systems of spheres
in their proper order. The epicycle F, which actually carries the
planet Saturn, rolls between the concentric surfaces B and D.

This was the mechanical model, which required four spheres,
instead of one, to account for the motions of Saturn. The principles
of Aristotelean physics were observed, since the spheres by completely filling the volume between A and E admit no vacuum. To
represent the phenomena accurately the astronomer had to assign
due sizes, and periods of revolution, to the various spheres, which
were imagined to be perfectly transparent so that their solidity

---

[1] Both the theories could be represented by actual models. Spherical astronomy and the doctrine of multiple homocentric spheres was illustrated by the
Armillary Sphere, of which the *Astrolabe* is a plane stereographic projection, the
Ptolemaic theory of planetary motions by the *Equatory* or computer, which
seems to be a rather late development in both Islamic and Latin astronomy.

offered no obstruction to the passage of light to the earth. The whole system of Saturn (and of every other planet) participated in the daily revolution from east to west; in addition the spheres AB and DE revolved in the opposite direction with a velocity of 1° 14' in a century—this corresponds to the slow rotation of the perihelion of Saturn's orbit in modern astronomy. The deferent C carried round the epicycle in a period of about thirty years, but its motion was not uniform with respect to its own centre P, or to that of the earth O, but to a third point Q so placed that OP = PQ, called the equant. By this means the unequal motion of the planet in its path was made to correspond more closely with observation. Finally, the epicycle itself revolved in a period of one year. The great virtue of the epicycle —which was used by Hipparchus before Ptolemy—was that it enabled periodic fluctuations in the planets' courses through the sky, the so-called "stations" and "retrogressions" to be represented (Fig. 2).

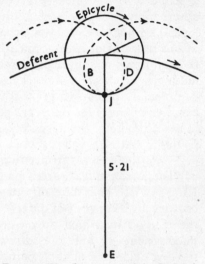

FIG. 2. The Geometry of the Epicycle (Jupiter). E, Earth; J, Jupiter. The dotted line shows the approximate path of the planet, whose "stations" occur about B and D, and whose "retrogressive" motion is from D to B.

The epicycle was essentially a geometrical device for "saving the phenomena," and attempts to make a physical reality of it were late and unsatisfactory. Epicyclic mechanisms similar to that of Saturn were required for all the other members of the solar system, except the sun itself. Those of Jupiter and Mars were identical with that of Saturn, with appropriate changes in the values assigned, while the mechanisms for the inferior planets, Venus and Mercury, and for the moon, were more complex and involved a larger number of spheres.[1] In all, the *machina mundi*

[1] There is an obvious and necessary relationship between the sizes and speeds of rotation assigned by Ptolemy to his circles, and those later adopted in the Copernican system, but this relationship is not consistent. With Venus and

described in the *Margarita Philosophica* required 34 spheres, but at various times much larger numbers had been used in the attempt to represent the phenomena more accurately. Once the constants of each portion of the mechanism had been determined in accordance with observation, it was possible to draw up tables from which the positions of the heavenly bodies against the background of fixed stars in the zodiac could be calculated for any necessary length of time. Unfortunately it was well known by 1500—it was a scandal to learning—that calculations were not verified by observation. Eclipses and conjunctions, matters of great astrological significance, did not occur at the predicted moments. The most notorious of astronomical errors was that of the calendar: the equinoxes no longer occurred on the traditional days, and the failure to celebrate religious festivals on the dates of the events commemorated caused great concern. In fact the Julian calendar assumed a length for the year ($365\frac{1}{4}$ days) which was about eleven minutes too long: the necessary correction was adopted in Roman Catholic states in 1582. But at the beginning of the century the causes of these errors in prediction were by no means clear. The current astronomical tables had been computed at the order of King Alfonso the Wise of Castille at the end of the thirteenth century and were out of date. Were the faults of the tables due to the method by which they were prepared—based on the Ptolemaic system—or were they due to the use of faulty observations in the first place, the method being sound? The fact that Copernicus

Mercury Ptolemy's deferent represents the orbit of the earth in the Copernican system, and the epicycle that of the planet. With the remaining planets Ptolemy's deferent represents the orbit of the planet, and the epicycle that of the earth.

The values of the Ptolemaic constants, and those of their modern equivalents, may be compared thus:

	Ratio of Radii Epicycle/Deferent	Mean Distances (in ast. units)	Angular Velocities (deg. per day) of Epicycle	Sidereal Mean Daily Motion
Mercury	0·375 : 1	0·387	4·09233	4·09234°
Venus	0·719 : 1	0·723	1·60214	1·60213°
Earth	—	1·000	0·98563 [sun]	0·98561°
	Deferent/Epicycle		*of Deferent*	
Mars	1·519 : 1	1·524	0·52406	0·52403°
Jupiter	5·218 : 1	5·203	0·08312	0·08309°
Saturn	9·230 : 1	9·539	0·03349	0·03346°

chose the first alternative, that he accepted Ptolemy's observations and rejected his mechanical system, was to be of great historical significance.

It is important to realize—and the problems became clearer when the first Copernican tables came into use—that the difficulties were not merely conceptual. Tycho Brahe recorded that in predicting the great conjunction of Saturn and Jupiter in 1563 the ancient Alphonsine tables were a whole month in error, while even the Copernican tables were out by a few days. The improvement can be attributed to the greater accuracy of the Copernican calculations, the remaining error to the inaccuracy of the observations on which they were based. Astronomical observations down to the time of Copernicus were liable to an error of at least ten minutes of arc; consequently, tables founded on them could not be more exact and in many computations the error increased cumulatively over the years (as happened with the faulty estimate of the length of the year in the Julian calendar). In small instruments like the astrolabe arcs of circle could not be graduated into lesser parts than the quarter or fifth of a degree, and as yet no means (such as the seventeenth-century vernier scale) existed for reading fractional parts accurately. For these reasons certain oriental observatories had erected huge fixed instruments of masonry to measure the motion of the sun with greater precision, and there was a comparable trend to increase of scale in Europe from the fifteenth century onwards. But it was found that as size increased new errors, caused by distortion, came in, and the expected precision was not attained. A totally new technique of observation was required.

The earth, considering the relative potentialities of human knowledge, was much less known than the skies, where the fixed stars had already been plotted more accurately than any European coastline. It was essential in the prevailing fabric of knowledge that the mutable globe and the unchanging heaven be entirely distinct. It was unthinkable that the same concepts of matter and motion should be transferred from the sublunary to the celestial region, and therefore all transient phenomena—comets, shooting stars and new stars—were regarded as mere disturbances in the upper region between the earth and the moon's sphere. This was, indeed, the region of the element *fire*; below it the element *air* formed a relative shallow layer above the surface of the globe. The

earth was composed predominantly of the remaining two elements, *water* and *earth*, but with air and fire as it were trapped in it,
so that when an opportunity for their release occurred they naturally ascended to their proper regions. Conversely, heavy substances made largely of earth and water sought to descend as far
as possible towards the centre of the cosmos, where alone they
belonged, the water lying upon the earth. It was believed before
the age of geographical discovery that only a sort of imbalance
in the globe enabled a land-mass fit for human habitation to
emerge in the northern hemisphere.

The categories of motion played an extremely important part
in pre-Newtonian science: even Galileo did not succeed in liber-
ating himself from them completely. Perfect circular motion was
an unquestioned cosmological principle, which gave consistency to
the whole theory of astronomy. In the sublunary region natural
motion was invariably rectilinear: away from the earth's centre
in the case of the light elements, and towards it in the case of the
heavy. Since the centre of the earth was a fixed point of reference
the definition of these species of motion occasioned no philosophical difficulties. The earth and its elements were unique, and
as a concept like "heaviness," having no relation to anything but
the sublunary region (to apply it elsewhere would be meaningless)
could only refer to a body's tendency to move away from or
towards the centre of the earth, so also definitions like "up" and
"down" could be quite unambiguous. Violent motion on the
other hand, that is raising that which is naturally heavy, or
lowering that which is naturally light, was unprivileged and could
occur in any direction. Force was required to effect these violent
movements, because they were opposed to the natural order of
the universe, just as force was required to withdraw a piston from
a closed cylinder because nature abhors a vacuum. As soon as
the effective or retaining force ceased, natural motions would
restore the *status quo ante*. This plausible, though limited, doctrine
was of great importance as the foundation of mechanics. Man was
the agent of the great number of violent motions, and one reason
why mechanics played a minor rôle in the tradition of science until
the sixteenth century was perhaps just this fact that so much
mechanical ingenuity was directed towards reversing the natural
order—it did not contribute to the understanding of nature, but
violated it.

A corollary drawn from the classification of motion again distinguished celestial from terrestrial science. Motion was the ordinary state in the former, rest natural in the latter. Wherever the notion of inertia has been perceived, however dimly, it has been seen most clearly exemplified in the motions of the heavenly bodies. While natural motion in the sublunary region exists for Aristotle as a possibility, it is clearly exceptional, except in connection with the displacements effected by living agents. In terrestrial mechanics, therefore, attention was most obviously drawn to the compulsive or violent motions brought about by the action of a force, and the natural motions which follow afterwards in reaction, e.g. the fall of a projectile. It was not difficult, once the question of mechanics was approached in this way, to conceive of force as producing motion from the state of rest, and as the invariable concomitant of violent motion. Within itself inert matter could have no potentiality for any other than its proper natural movement: and though Aristotle never explicitly formulates the proposition that the application of a constant force gives a body a constant velocity, it is implied in the whole of pre-Galilean mechanics. The natural order could not be defied save at the cost of expending some effort, any more than a weight could be suspended without straining the rope by which it is hung. As against this simple principle there was a whole category of phenomena of motion that had to be treated as a special case, of which the motion of projectiles was the leading instance.

Observation could not overlook the fact that no form of motion ceases instantaneously. Effort is required to stop a boat under way, or a rapidly revolved grindstone. What was the source of residual force which could impel an arrow for some hundreds of feet after it had left its mover, the bow-string? The main Aristotelean tradition pronounced that the force resided in the medium, air or water, in which the motion took place, the medium being as it were charged with a capability to move, though it did not move itself. The medium, indeed, plays a most important part in the Aristotelean theory of motion, for it is its resistance to movement which is overcome by the application of a constant force, limiting the velocity which can be attained. If a vacuum in nature were possible (and this Aristotle denied), a moving body could attain an infinite velocity since there would be nothing to limit it. In this special case, then, the medium has a dual function: its resistance brings the moving body to rest, but its charge of motion protracts the movement after the effect of the force has ceased. This apparently contradictory dualism was severely criticized by philosophers who otherwise worked within the general framework of Aristotelean science, and an alternative theory of motion was put into a definitive, and to some extent mathematical, form in the fourteenth century, principally by two masters of the University of Paris, Jean Buridan and Nicole Oresme. The principle they adopted, but did not invent, was that though rest is the normal state of matter, movement is a possible but unstable state. They illustrated this conception by analogy with heat: bodies are usually of the same temperature as their neighbourhood, but if they are heated above that temperature, the unstable state is only gradually corrected. A moving body acquired *impetus*, as a heated body acquired heat, and neither wasted away immediately. The impetus acquired was the cause of the residual motion; and only when the store was exhausted did the body come to rest.[1]

Although the idea of motion as a quality of matter was retrogressive, and the analogy with heat false, the development of the mechanical theory of impetus is of outstanding importance in the history of science. Impetus mechanics was not widely diffused, and the line of investigators who continued discussion of its tenets down to the sixteenth century was somewhat thin. It is not discussed in the *Margarita Philosophica*, but it was well known to Leonardo da Vinci, Tartaglia, Cardano and other Italians of the renaissance period. In its finished form it is absolutely a medieval invention, which was never displaced by the Hellenistic revival. The notion of impetus was not inspired by any new observations or experiments, nor did it suggest any. The facts which it sought to explain were exactly those already accounted for in the original Aristotelean theory. But it was the work of truly creative minds. Before the idea of impetus could be useful, it enforced a revaluation of ideas on the nature of motion and, still more important, of ideas on the natural order. Ideas of motion, even in Aristotelean physics, and still more in that of Plato before him and some medieval philosophers later, had been integrally woven into a

[1] Part of the difficulty of the early mechanicians lay in the disentangling of such concepts as motion, velocity, inertia, kinetic energy, just as much later the physics of heat was obstructed until the concepts of temperature, heat, entropy were defined.

fundamentally animistic philosophy of nature, to the extent that in the extremest form motion and change were denied to matter except in so far as it was pushed, pulled or altered by various animated agencies. The theory of impetus, attributing to inert matter an intrinsic power to move, was a decisive step in the direction of mechanism. In a sense it first conferred a true physical property upon matter, qualifying it as more than mere stuff, the negation of empty space. The impetus theory contained the first tentative outlines of the explanation of all changes in nature in terms solely of matter and motion which was to figure so prominently in the scientific philosophy of the seventeenth century. Indeed, both Buridan and Oresme foreshadow the greatest triumph of seventeenth-century mechanism, Newton's theory of universal gravitation. Since the heavenly spheres were perfectly smooth and frictionless, moving upon each other without resistance and without effort, they saw that when the whole system had been set in motion it would revolve as long as God willed, without its being required that each sphere should be animated by a guiding intelligence.[1]

Although medieval science was ignorant of dynamics in the modern sense, discussions of motion and the displacement of bodies form a very important element in its physical treatises. Medieval accomplishment in the other branch of mechanics, statics, was of somewhat earlier date, since there was little progress beyond the writings of Jordanus de Nemore (c. 1250), based on translations from Greek and Arabic. Archimedes' works on statics and hydrostatics were virtually unknown, though the middle ages were familiar with his Principle and with the distinction between gross and specific weight. In applying the latter, however, even so profound a philosopher as Oresme could go astray.[2] The first use of the principle of virtual velocities is credited to Jordanus in the explanation of the lever; he also dealt successfully with the bent lever and the inclined plane. But the conceptual foundations of medieval statics were shaky, and medieval authors did nothing to develop an experimental tradition. In optics they had a much surer foundation, in the original work of Ptolemy and its extension

[1] A. D. Menut and A. J. Denomy: "Maistre Nicole Oresme, Le Livre du Ciel et du Monde," *Medieval Studies*, vols. III, IV, V (1941-3), esp. IV, pp. 181 et seq.
[2] *Ibid.*, vol. V, pp. 213-14.

by the Arab Ibn al-Haitham (*Latine* Alhazen, , *c.* 965–1039). Alhazen had actually conducted experiments, and made measurements, and in the West the experimental and geometrical study of optics was continued by such men as Robert Grosseteste and Roger Bacon in the thirteenth century. Shortly before the latter's death a great advance was made in practical optics with the invention of spectacles. The magnification of objects by lenses, unrecorded in antiquity, was known to Alhazen; their ophthalmic use brought the glass-grinder's craft into being. As a result the

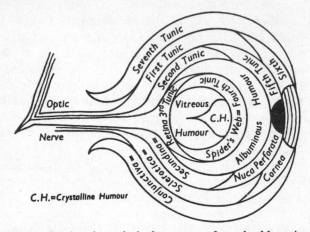

Fig. 3. Section through the human eye, from the *Margarita Philosophica* (translated). *Nuca perforata* = Iris, *Secundina* = Choroid, *Crystalline Humour* = Lens, *Spider's Web* = ? Ciliary processes.

Margarita Philosophica, for example, offers a more rational and complete account of the phenomena of light than of any other part of physics—though it is to be found under a discussion of the powers of the "sensitive soul" not in the section on natural philosophy. Light is defined as a quality in the luminous body having an intrinsic power of movement to the object, which may be either the eye or an opaque body thus illuminated. Colours are somewhat similarly described as potential qualities in the surfaces of bodies which are made actual by the incidence of light. Next the structure of the eye is described, with the aid of a good diagram of a section through the organ, and the functions of the seven tunics and four humours are explained (Fig. 3). The optic nerve is said to conduct the "visual spirit" to the brain, and single

vision with two eyes is simply accounted for by the union of the
two optic nerves into a single channel. The shining of rotten fish
or fireflies is attributed to the element of fire in the composition
of their substance.[1] Reflection and refraction are quite intelligently treated. The fact that the ray of light passing from a rare
medium to a dense (as from air to water) is refracted towards the
perpendicular is explained by the greater difficulty of penetration.
Examples of the effects of refraction, such as the apparent bending
of a stick in water, are elucidated by simple geometrical figures,
and the apparent enlargement or diminution of the object seen in
accordance with the size of the visual angle at the eye is described.
In another section of the book the appearance of the rainbow is
explained: where tiny drops of water in the clouds are most dense
the sunlight is *reflected* as of a purple colour, where they are less
dense the colour is weaker and the light appears green, the
blue is the weakest of all. Refraction is not referred to in this
connection. Acoustics also was comparatively well understood, both theoretically and experimentally. Reisch described
the physiology of the ear, the nature of sound as a vibration, and
the transmission of perception to the brain through the nerves
by means of an "auditory spirit." Music of course had its own
peculiar theory and practice, which are carefully outlined in the
Margarita.

Other aspects of the knowledge of material things, which refer
rather to their composition, structure and generation than to the
phenomena which arise from motion of some kind, were broadly
related to the theory of matter derived from Aristotle. That all
substance accessible to human experience is composed of the four
elements, fire, air, water and earth, in varying proportions, seems
to have been a notion to which the earliest philosophers were
favourably disposed: it was accepted among the Greeks in preference to the single-element theory of Thales, and very similar
ideas prevailed in China and India. It was not required that every
analysis should yield these elements in identical form, nor was
there any means by which such an exact identity could have been
determined; but it is clear that the three ponderable elements
correspond roughly to the three states of matter (solid, liquid,
gaseous), with heat added as a material element. This conception

[1] Phosphorescence was studied with much interest in the later seventeenth
century, e.g. by Robert Boyle.

of heat (and electricity) as material though imponderable fluids—elements in fact if not in name—had not become an anachronism even by the end of the eighteenth century. A great deal of learning was devoted to the question of how mixed bodies are compounded from the elements, and how bodies generate by a synthesis of elements, or corrupt by their dissolution. An important philosophic problem was the relationship between the composition of a substance and its qualities, or properties. Plato's doctrine of "essential forms," or ideal models, exercised its pernicious influence indirectly throughout the middle ages. Aristotle and his followers believed that the elements themselves could be transmuted from one to another, so that water could be condensed into earth, or rarefied into air. Again, in a more strictly chemical form, this conception lingered to the eighteenth century. The four elements were of undefined figure, and matter was thought of as being continuous. Ancient atomistic speculation was known to the middle ages through Aristotle's criticism of it, but that criticism was regarded as wholly convincing. The conjunction of elements in a compound body took place without any conceivable hiatus or division; as Oresme says, they are not mingled like flour and sand, for every fraction of the infinitely divisible compound must contain all its elementary constituents.

Regarding this framework of ideas from the point of view of the chemist or mineralogist, its most important feature was the latitude it offered for an infinite variety of theorizing on the nature of change in substances. The only certain definition of a physical body was that it must have form (i.e. shape and other attributes) and substance (i.e. be material), and possess either gravity or levity. Form and substance are therefore in one sense (as Oresme remarks) the first elements of matter; the four elements proper have a secondary rôle and are transmutable. Form likewise is obviously mutable: only substance could remain constant, since its destruction or creation would require a change in the fixed finite volume of the universe. Consequently, what we should now call physical or chemical changes could be accounted for on any of three hypotheses: (1) variation in the proportion of elements, (2) the generation of elements (fire being the noblest in the series), (3) the degeneration of the elements. *Ex nihilo nihil fit.* The first has in fact formed the main subject of chemistry, but this limitation was only logically established by Lavoisier. In this period the modern distinction between physical and chemical changes would have had no meaning, since this differentiation of properties was not yet established. The analogy, involved in the use of such terms as "generation," between organic and inorganic matter was consciously cultivated, being a natural product of the animistic conception of nature. As the plant grows from earth and water, so equally well could metals, gems and stones.[1] In the seventeenth century it was still believed that veins of ore would grow in a mine if it was left to rest for a time unworked. And by various natural or artificial processes (the calcination of metals, the combustion of organic materials) one or more of the constituents could be recovered. A vital principle, though of a humbler kind, was just as much present in the sand which grew into a pebble, as in the seed that grew into an oak. Natural history has only comparatively recently ceased to signify mineralogy as well as biology, and the apprehension of the problems of formation and change in the organic world as involving problems of a different order from those encountered in the inorganic is a comparatively late product of the scientific revolution. In so far as it demands a differentiation between organic and inorganic chemistry it was never appreciated by Robert Boyle. And it must be remembered that while it was possible to cite as examples of matter "salt" or "flesh" without any doubt that the two were precisely comparable, philosophy imposed an even higher degree of theoretical unity. For the elements figuring in chemical change were the elements of the sublunary world in cosmological theory. The structure of explanation had to be consistent between the macrocosm and the microcosm; the properties of "fire" or "air" as they were stated in the general picture of the universe had to recur precisely when these elements were considered as entering into the composition of organic matter.

Closely allied to the theory of four elements was the doctrine of the four primary qualities—heat, cold, dryness and humidity. A combination of a pair of these qualities was attributed to each element, fire being dry and hot, air hot and moist, water moist and cold, earth cold and dry, and from mixtures of these elementary qualities the secondary material qualities, hardness, softness, coarseness, fineness, etc., were in turn derived, though other qualities

[1] But see below, p. 27.

like colour, taste, smell, were regarded as intangible.[1] Thus the transmutation of elements was accomplished by successive stages through substitution of qualities: dried water becomes earth, heated water becomes air, but to transform water to fire it must be both heated and dried. It was reckoned that with each transformation the volume was multiplied by ten, so that fire had $1/1,000$th part of the density of earth. Meteorological phenomena, for instance, could be accounted for by the application of these ideas in detail. The sun's heat turns water into air; the element air, being light, rises; but high above earth (where it is cold) air is transformed into water, and water being a heavy element falls as rain. Similarly air in the caverns and cracks of hills is cooled till it becomes water, which runs out as a spring. Comets are a hot, fatty exhalation from the earth drawn into the upper air and there ignited. The generation of "mixed substances" such as minerals and metals in the interior of the earth under the action of celestial heat from the primary elements and qualities was considered as a more complex example of the same process. Stone was earth coagulated by moisture. Sal-ammoniac, vitriol and nitre are listed in the *Margarita Philosophica* as examples of salts formed from the coagulation of vapours in different proportions; to mercury, sulphur, orpiment, arsenic etc. a similar origin is attributed. The metals were commonly supposed to result directly from a combination and decoction of mercury and sulphur though, as the alchemists insisted, not the impure, earthy mercury and sulphur extracted from mines and used in various chemical processes. Reisch, while he admits the theoretical possibility of transmuting metals, points out the great difficulty of imitating the natural process of their generation precisely, and seems doubtful of the pretensions of the alchemists. Indeed, the ambition to tincture the base metals and otherwise modify their properties so that they should become indistinguishable from gold is far older than the Greek theory of matter which gave it a *rationale*, and the actual processes or recipes

---

[1] The medieval theory of matter was derived mainly from Aristotle's *De Caelo* (III, 3–8), *De Generatione et Corruptione* (Bk. II), and *Meteorologica*. Cf. *De Gen. et Corr.*, II, 3: 'hence it is evident that the "couplings" of the elementary bodies will be four: hot with dry and moist with hot, and again cold with dry and cold with moist. And these four couples have attached themselves to the *apparently* "simple" bodies (Fire, Air, Water and Earth) in a manner consonant with theory. For Fire is hot and dry, whereas Air is hot and moist (Air being a sort of aqueous vapour); and Water is cold and moist, while Earth is cold and dry.'

traditional in the art of alchemy, some of which can be traced to pre-Hellenic antiquity, had little reference to the philosophic idea of metallic generation. But so long as the organic-inorganic analogy seemed plausible, alchemy could not be dismissed as mere folly.

As for the differentiation between the generation from the elements of the world of inorganic substances like minerals, and the generation of living organisms, it is obvious that this could not be found merely in the principle of growth, or autogenous development from a seed, for this was really common to both. It could not be made material at all, and therefore the two broad kinds of living nature were defined as possessing respectively a vegetative and sensitive "soul," man alone having in addition an intellectual soul which gives him the power of reason. In members of the vegetative class (plants) the organization of matter was more subtle than in minerals: they were not passive creatures of nature, like a stone, for they required to be supplied with water and rich soil if they were to maintain their existence, they showed cyclical variations, and they possessed special structures for reproducing their own kind. The fixing of the margin of life is not simple; our phrase "living rock" is the survival of an ancient mode of thought, and the alchemists used to distinguish between the "dead" metal extracted from the ore and the "live" metal in the veins of the mine. But the need for nourishment, involving some process of "digestion," and the possession of a reproductive system were recognized as distinctive of living creatures. It was certain from theology that these had been created in the beginning, and perpetuated themselves without change ever since. The function of the vegetative soul was to control and indeed *be* the biochemistry of the organism, the whole life of the plant, but only a part of the life of the animal. For the animal is sensitive: it feels pain, it is capable of movement, it can manifest its needs and desires. Therefore it was endowed with a sensitive soul. Man has both a vegetative soul, since he must digest and reproduce, and a sensitive soul because he possesses a nervous system, and he alone being endowed with the power of reflection, of detaching himself from the mechanism of the body, was credited with the third, intellectual soul. The *matter* was everywhere the same, in rock, in tree, or in dog, for it was well known that certain wells and caves had the power of transmuting wood into stone, for instance: the distinction between the animal, vegetable and mineral kingdoms lay in the immaterial organizing principle which brought the elements into conjunction: nature could make pebbles grow in streams, but not grass without seeds.

There was one important exception to this rule, founded (unfortunately) on the authority of Scripture and Aristotle alike. A group of organisms, small as Aristotle originally conceived it, obviously possessing the characteristics of life, seemed to have no special mechanism of generation, but to develop directly from decaying matter. Some insects, lichens, mistletoe, maggots were included in this class of spontaneously generated creatures and to them the middle ages, from sheer ignorance, added others such as the bee, the scorpion, and even the frog. Spontaneous generation provided a fatally easy alternative to investigation, and the belief remained unshaken till the mid-seventeenth century and even later (p. 156). Yet it was less anomalous than it might seem at first sight, for Aristotle was able to imagine the circumstances in which spontaneous generation might occur after a fashion that corresponded closely with his ideas on normal reproduction. Since the whole universe was in a sense, not the full sense, animated, it was not illogical that some humble, borderline creatures should be born of it without parents.

Of the excellent descriptive biology of Aristotle and Theophrastus very little was known at the opening of the sixteenth century. The middle ages had depended only too much on compilers of the later Roman period and on fables both pagan and Christian. Such study was excluded from natural philosophy; it might be just respectable as a form of general knowledge, or moral as it provided material to illustrate a sermon or emphasize the minuteness of divine providence, but it offered none of the sterner food for thought. Botanical or zoological curiosity is of very recent origin; in most periods most men have been content to acquire only the minimum of practical knowledge. A very few in the middle ages, like Albert the Great or the Emperor Frederick II, had an interest and attentiveness far above the ordinary, but they founded no tradition. Observation was blunted, and it has been pointed out that naturalistic representation must be sought not in learned treatises, but in the works of craftsmen, artists, wood-carvers and masons. Botany was cultivated to a minor extent as a necessary adjunct to medicine and medieval herb-gardens included both edible vegetables and medicinal simples. In pharmacology native Germanic lore mingled with the remnants of Hellenistic botany, so causing much confusion. Plants named and described by the Greeks were identified with the different species of western Europe; the nomenclature itself varied widely from place to place; and there was no system by which one kind could be recognized with certainty from its structure and appearance. As the quality of graphic illustration and of verbal description deteriorated, the herbal became little more than a collection of symbols, so that the mandrake which was originally recognizable as a plant becomes a manikin with a tuft sprouting from his head. Again, on the purely empirical side, some progress was made in the selection of strains of corn, just as it is certain that attention was given to the breeding of hawks and hounds, but in none of these things was the plant or animal itself the centre of interest. It existed simply to be cooked, or distilled, or mutilated in man's service, or alternatively to play a part in symbolisms of endless variety. There are, however, signs of a more naturalistic outlook from the earliest beginnings of the renaissance.

Natural history and scientific biology are both modern creations, stimulated indeed by the rediscovery of Greek sources in the sixteenth century. But this was not the only fruit of the Italian renaissance, for naturalism is older in art than in science. It is unnecessary here to discuss at length a change in the artist's spirit and ambition which had such important consequences for biology and medicine. It is at least fairly clear that the medieval draughtsman was not simply incapable of attaining realism, e.g. in matters of perspective, but was not interested in perfecting direct representation. The element of symbolism was as important to him as it has been in the twentieth century. No tradition of "photographic realism" existed which could serve the purposes of science, such as was assiduously cultivated with the aid of perspective instruments, camera lucida, and other devices from the seventeenth century to the perfection of photography, when art and science again parted company. Partly under Hellenistic influence, partly for internal reasons, art moved strongly in the direction of realism and the faithful representation of nature in the fifteenth century. Plants and animals became recognizable as individuals rather than hieroglyphs. The tradition that artists should study the anatomy of the animals and men they depict came into being, reaching its peak in Leonardo da Vinci. This development was powerfully reinforced in its scientific importance by the invention of printing. Biology is peculiarly dependent upon graphic illustration, and it is essential not merely that the illustration should be accurate in the first place, but that it should be capable of being reproduced faithfully. Nothing becomes corrupt more easily than a picture or diagram repeatedly copied by hand; only a mechanical process of reproduction, like the woodcut or the later copper-plate, can maintain faithful accuracy. At the moment when the technique of visual representation was arousing great interest in the draughtsman and artist, printing made it possible for illustrations to be copied in large numbers for teaching purposes, so that the anatomical student and the botanist could recognize the form of the organ or plant described in the text when he saw it in the natural state.

The potentialities of the new interests and skills in biology began to emerge only in the sixteenth century, as living creatures, in their beauty, in the fascination of their habits, in the variety of their species and ecological inter-relations, aroused a curiosity and sentiment which has grown steadily in its extent throughout the modern period. The faithful imitation of nature—the necessary formal basis of all naturalism—was an æsthetic, rather than a scientific revolution; the romantic nature-lover may not be a good biologist. It was not until thought began to play upon the new facts and the faults in the old traditions which passed for natural history that precise observation could reveal consequences important for science, or that science could see in the flower not simply an æsthetically pleasing object, or a symbol of God's mysterious and benevolent ways, but a challenge to man's powers of understanding.

Science is an expanding framework of exploration, not the cultivation of special techniques, or of a pecularly acute appreciation of the wonder and unexpectedness of the universe in which we live. Medieval philosophers, who in this respect had a sharp sense of reality and a just notion of the importance of intellect, had left natural history to such compilers as Bartholomew the Englishman (c. 1220–40) whose book is an uncritical compendium of classical fable and old wives' tale, flavoured with moralizations and somewhat rarely enlivened by a touch of the countryman's lore. They judged that the typical was more significant than the freak, unlike some of the scientists of the later seventeenth century who, in misguided zeal for the principles of Francis Bacon, filled the museum of the Royal Society with the heads of one-eyed calves, internal calculi, an artificial basilisk.[1] None of the great men of the middle ages, with the exception of Albert the Great, showed more than a cool interest in natural history. As a result, bestiaries, herbals and encyclopædias were laborious summaries of the bald notes available in classical sources, and none of them carried more than the slightest indication of personal observation. Heraldic beasts were listed with real animals; the legends of the crocodile's tears, the pelican killing its brood and reviving them with its blood, and the barnacle-goose born of rotten wood added interest to the story: the lion and the fox became popular symbols of bravery and cunning. Creatures were classified according to the element in which they lived. The salamander was the only known inhabitant of fire; birds belonged to air; fishes, whales, mermaids and hippopotami to water; and the rest to land. Trees and herbs formed the two kinds of vegetable life, but only the few species useful to man were noticed. Even common garden flowers were ignored. Such work of compilation was regarded as a purely literary task, and the authorities were quoted in wholly uncritical fashion. Some scholars developed fantastic etymologies, such as Neckham's *Aurifrisius* [Osprey] from "Aurum Frigidam sequens." Roger Bacon pointed out the difficulty of interpreting satisfactorily the more obscure names of creatures mentioned in the Bible— Biblical exegesis was one of the few respectable motives for studying natural history at all.[2] It scarcely existed save as an appendix to some other branch of study. This is not really surprising, since it is a trait of a sophisticated society to be interested in things of apparently no concern to humanity. Natural philosophy had for the middle ages an established place in the House of Wisdom: natural history had yet to establish itself.[3]

Francis Bacon wrote of scholastic philosophy that, from immediate perceptions of nature, it 'takes a flight to the most general axioms, and from these principles and their truth, settled once for all, invents and judges of intermediate axioms.' Once fundamental doctrines, the immovable earth, the four elements,

---

[1] Cf. Nehemiah Grew: *Musæum Regalis Societatis* (London, 1681).
[2] C. E. Raven: *English Naturalists from Neckham to Ray* (Cambridge, 1947).
[3] Cf. Appendix A.

or the souls of living organisms, are accepted as unshakably true, and explanations of the varied phenomena of nature deduced from them, it is an inescapable consequence that the structure of scientific knowledge has great cohesion, a logical unity imposed from above. An empirical science, a science which sees itself as unfinished and progressive, can tolerate inconsistencies—it is for the future to resolve them by some higher-order generalization. Physicists might have debated for a hundred years whether light is undulatory or corpuscular in nature, but the progress of optics did not wait upon a resolution of the argument. And while it is far from being true that medieval science was exclusively deductive or exclusively speculative in every detailed consideration of a particular phenomenon, it is true that its structure was of the form that Bacon described. The character of the structure is not changed by the fact that some philosophers made a few experiments. The structure of modern physics is still experimental, though some physicists do not make experiments. The difference is that the theoretical physicist bases his calculations upon materials obtained by the researches of an experimental physicist and produces a result which is itself capable of confirmation by experiment, whereas the medieval philosopher fitted the results of his experiments into a theory already firm in his mind. He knew what "light" was before making experiments on refraction: he knew the cosmological significance of "weight" before attempting to determine the speeds at which heavy bodies fall. Experiment and induction could only modify the minutiæ of science, and these (for example, the numerical values of astronomical constants) were indeed frequently changed in the later middle ages in accordance with experience. They could not reflect upon the broad lines of the structure. It might indeed have happened that Aristotelean science would have been crushed under the accumulated weight of an adverse mass of experimental testimony. But modern science did not in fact arise in this way.

This too has its unity, of course—a unity derived not from deduction but from the homogeneity of its procedure and the wonderful, unforeseeable interlocking of its branches over three hundred years. Modern science, like medieval science, embraces in this unity statements of fact, concepts and theories. It could not function if it were not free to employ terms like "electron" or "evolution," which are not applicable to crude facts, just as the medieval philosopher could not think about plant-life without introducing the entity "vegetable soul." Within its own context of theory the vegetable soul is not less plausible than the electron, and it cannot be said that one or the other can be disposed of by a straightforward matter-of-fact test. To do so it would be necessary to make a complicated inquiry into the value of the factual information supplied by observation and experiment, to examine the reasoning involved in either case, to trace the relationship between the concept and other elements in the fabric of science, and so forth, and it is because modern science in all such activities differs from medieval that its fruits are different. From such intellectual activities modern science yields a system of ideas, not an unlinked series of factual statements, being in this respect (despite the immense superiority of its factual content) comparable to medieval science. And the latter, as a system of ideas, with all the imperfection of its methods and information, was true science. It offered a system of explanation, closely related to the facts of experience and satisfactory to those who used it, giving them a degree of control over their natural resources and allowing them to make certain predictions about the course of future events. By these standards it was relatively vastly inferior to modern science, just as at each stage in its own development modern science has been inferior to the science of later stages. But to declare that any of its tenets was unscientific is to misuse language. Unscientific is a pejorative term meaning inconsistent with the prevailing framework of the explanation of natural events and the methods used to establish this framework; it can have no other meaning because we do not know that what is scientific now will not be unscientific in the future. There is no absolute standard. All that can rightly be said, when we have understood that medieval men had prejudices, purposes and hopes totally different from our own, is that they were less inquisitive and self-critical than they might have been. They were less interested in natural philosophy, for to them it was but a step forward to higher things. Science was a means, not an end.

CHAPTER II

NEW CURRENTS IN THE
SIXTEENTH CENTURY

The subtlety of nature greatly exceeds that of sense and
understanding; so that those fine meditations, speculations
and fabrications of mankind are unsound, but there is no one
to stand by and point it out. And just as the sciences we now
have are useless for making discoveries of practical use, so the
present logic is useless for the discovery of the sciences.[1]

IN such terms Francis Bacon, in the early seventeenth century,
denounced the existing structure of scientific knowledge as he
knew it. Yet it is clear that the same structure of science had
satisfied the many medieval philosophers of genius who had expounded it; even those who criticized the learning of their day,
like Roger Bacon, could not depart from its strategic concepts.
Clearly the difference between the men who taught the Aristotelean world-system and those who later rejected it was not simply
one of intellectual calibre. Only when the criteria of what constitutes a satisfactory scientific explanation changed, and when fresh
demands were made for the practical application of nature's
hidden powers, could an effective scepticism concerning the
strategic concepts take shape, as distinct from differences of
opinion on matters of detail.[2] When such scepticism arose the
cohesive strength of the science that prevailed in 1500, and on the
whole throughout the sixteenth century, became significant.

In modern science the higher-order generalizations are vulner-
able, but in descending the scale to the substratum of experimental
fact the chances of serious error steadily diminish. In Aristotelean
science the reverse was true: an important dogma, such as the
stability of the earth, might be incapable of experimental proof

[1] *Novum Organum*, Bk. I, x, xi.
[2] An important change also took place in the idea of hidden powers, as the
ambition to force "unnatural" operations on nature by esoteric or magical
means gave way to the belief that man could use processes as yet unknown but
still strictly rational or mechanistic.

or disproof (as Galileo himself confessed), but some of the primary propositions—for example that bodies fall at speeds proportional to their weights—could be exposed as contrary to experience by very simple tests. To modify the accepted scientific opinion of the present time it is usually essential to carry out an intricate investigation verging on the frontiers of knowledge; in attacking the conventional science of the sixteenth century it was possible to outflank the higher-order generalizations altogether by showing that the "facts" deduced from them were simply not true. It could not be a fundamental requirement of the world-order that changes do not happen in the heavens if the new star of 1572 could be shown to be far beyond the sphere of the moon by its lack of parallax. This feature in the structure of Aristotelean science determined a large part of the tactics of the scientific revolution. Further, since the unity and cohesion of science were imposed from above, growing out of its majestic axiomatic truths, it followed that when the results of any one chain of ratiocination were impugned the shock was reflected in similar dependent chains. Once it was known that the liver is not the source of the bloodstream the whole physiology based on this belief was disposed of at one stroke. Admittedly the iconoclasts were not always quick to recognize the necessary extent of their destructive criticism— least of all a Copernicus or a Vesalius—which was often jubilantly pointed out by their opponents. The weakness of conventional science was also its strength; the whole authority of the magnificent interlocking system of thought bore down upon an assault at any one point. How could a mere mathematician assert the earth's motion when a moving earth was absolutely incompatible, not only with sound astronomical doctrine, but with the whole established body of natural philosophy? Logically, to doubt Aristotle on one issue was to doubt him on all, and consequently some problems of the scientific revolution, which may now seem to involve no more than the substitution of one kind of explanation for another, were pregnant with consequence since they implied the annihilation of extant learning.

The sixteenth century shows the tactics of the scientific revolution in two contrasted forms. In the year 1543 were published two volumes which have become classics of the history of science, the *De Humani Corporis Fabrica* of Andreas Vesalius (1514–64) and the *De Revolutionibus Orbium Coelestium* of Nicholas Copernicus (1473–

1543). Neither of these books was "modern" in content—Vesalius was no more successful in escaping the limitations of Galenic physiology than Copernicus in departing from the formal system of perfect circles—but both inspired trains of activity which were to lead to the substitution of other conceptions for their own within two generations. The two books and their authors, however alike in their broad impact upon the scientific movement, are totally dissimilar. *On the Fabric of the Human Body* is a work of descriptive reporting; its value depends upon the trained eye of a great anatomist and the skill of draughtsman and block-maker, while Copernicus' treatise is purely theoretical. Vesalius was a young man whose work, if it was his unaided, shows astonishing pre-cocity. Copernicus was a dying man, of recognized capacity, who had nursed his idea for thirty years. Vesalius' material was taken freshly from the dissecting-table; Copernicus' was the laborious digestion of ancient observations. Vesalius was an ambitious and popular teacher who contributed to the fame of Padua as a centre for the teaching of medicine which lasted till the mid-seventeenth century; Copernicus lived obscurely immersed in his ecclesiastical administration, hesitant to the last over the enunciation of his great hypothesis. The nature of Copernicus' original contribution to science is also quite different from that of Vesalius. The former was the avowed opponent of an *idea*, that the earth is the motion-less core of the universe, but his opposition rested in no way upon his discoveries in practical astronomy, which were negligible, or on the precision of his measurements, which was not remarkable. It sprang from a demonstrable truth, that celestial observations could be equally well accounted for if the earth and planets were assumed to move about a fixed sun, allied to various wholly non-demonstrable considerations—value-judgments—seeming to show that the astronomical system constructed upon this assumption was simpler than the older system and preferable to it. Copernicus criticized the internal logic of prevailing ideas, but to be a Coper-nican did not add one item to a man's factual knowledge of the heavens, whereas it did place him in a position which could be challenged on other grounds. Vesalius, on the other hand, gave vent to no formidably unorthodox opinions, rather indeed his passing comments seem to condemn such innovations. As an editor of Galen's works, and as one who paid Galen the tribute of imitation, Vesalius could not but respect the great master of anatomy and physiology. Though he triumphantly emphasized his detection of mistakes in Galen, he also repeated some of the traditional errors. The depth and precision of knowledge about the structure of the human body contained in the *Fabrica* was the essential foundation for a more rational physiology than Galen's, but of this Vesalius himself had scarcely an inkling. The publication of Harvey's discovery of the circulation of the blood in 1628 aroused the first serious conflict between ancient and modern physiological theories. Yet it may be said that the beginnings of the scientific revolution are to be found as truly in *De Fabrica*, and the series of illustrated anatomies of which it was the outstanding member, as in *De Revolutionibus*. As examples of types of innovation they are complementary.

The twin advance upon the distinct lines of conceptualization and factual discovery constantly occurs in science. The former makes the more interesting history, but it must not be forgotten that each new observation, each quantitative determination accurately made, is adding to the stock of knowledge and playing its part in the genesis of a new idea. Indeed, conceptualization can only progress and rise above the level of speculation through the accumulation of fact by the perfection of techniques of experiment and observation. Anatomy is the crucial instance of this during the early period of the scientific revolution.

In the middle ages it is difficult to distinguish specialized medical sciences from the general practice of the physician and surgeon. There was a research interest of a sort in natural philosophy, even though the research was of a peculiarly narrow kind and its sole instrument formal logic. The medical sciences were even more strongly subject to the limitations of the purpose for which they were cultivated, the training of medical men. Yet some medieval anatomists were imbued with a disinterested love of knowledge, a trait which seems to have been stronger in the fourteenth century than it was about 1500. The tradition in medicine was at least as tightly unified as that in natural philosophy. The principal authority in anatomy, physiology and therapy until the seventeenth century was well advanced was that of Galen (A.D. 129-99). Other writers on medical subjects were of course studied: in clinical matters great respect was accorded to Hippocrates, whose works were made more fully known by the humanists. Aristotle was also followed in these subjects, sometimes in preference to

Galen, but it is not easy to overestimate the latter's power. A humanist physician, Dr. John Caius, as President of the College of Physicians, could order the imprisonment, until he recanted, of a young Oxford doctor who was reported as saying that Galen had made mistakes. When men were educated as logicians and not as observers it was infinitely easier to detect errors in philosophy than in anatomy or physiology. Admiration for Galen was so extravagant that anatomists were more apt to attribute their failure to confirm his descriptions to their own want of skill, than to his. 'I cannot sufficiently marvel at my own stupidity,' wrote Vesalius, 'I who have so laboured in my love for Galen that I have never demonstrated the human head without that of a lamb or ox, to show in the latter what I could not find in the former.' It was only tardily, and hesitantly, that Vesalius admitted to himself the simple truth that the structure then called the *rete mirabile*, described by Galen in the human head, was a feature not of human, but of animal anatomy.[1] Only gradually could anatomists learn to see the body otherwise than as Galen had taught them, and the broader influence of his pathology lingered well into the nineteenth century. Galen was one of the greatest medical scientists who have ever lived, demonstrating in Rome, some six hundred years after Aristotle, the vigour and quality of Hellenistic science. He dissected, he experimented and his work, though dominated by the vitalistic preconceptions of Aristotle, has a strong experiential foundation. He was also an uncritical teleologist, believing that it is possible to discover the purpose of every part of the body and to prove that it could not be more perfectly designed for that purpose. This profound admiration for the divine plan recommended him strongly to Christian writers of the middle ages.

In many ways Galen epitomizes the typical qualities of the Greek tradition in medieval science, itself often far superior to the independent efforts of the later Latins, but so far imperfect that even when it was purified and enriched by renaissance scholars a reaction against it was still necessary before modern science could take shape. In some aspects the Greek intellect was "modern"; but not in relation to medical subjects. Greek medicine never detached itself from teleological arguments, and its anatomy was

---

[1] *De Fabrica* (1543), p. 642. Quoted by Charles Singer and C. Rabin: *A Prelude to Modern Science* (Cambridge, 1946), p. xliv. Doubts on this point had been expressed earlier by Berengario da Carpi.

always firmly subject to *a priori* physiological theories, not lending itself to the reverse and correct process. There was a deep-seated prejudice against human dissection. As a result the study of animal anatomy without sufficient check introduced numerous errors. Galen worked extensively on the Barbary ape, he may possibly have had access to the still-born fœtus, but he never dissected an adult human subject. This lack of experience was scarcely appreciated before the time of Vesalius. Technical nomenclature and classification were very defective in Greek anatomy, and even had description been perfect it would have been useless to a later age in the absence of pictorial illustration. At its best the eye of the Greek anatomist had often been deceived by his preconceived notions of the working of the body. As in astronomy, the Greeks had gone far in the direction of precise and careful research, much of which proved of enduring value, but the observations they made were fitted into a scheme of ideas inherited from primitive times.

The influence of Greek texts upon Islamic physicians had become considerable by the ninth century A.D. Avicenna (980–1037), the greatest scientist of the Arab world and its foremost physician, reproduced in his *Canon* the best features of a Hellenized survey of medicine, as well as original observation. To a considerable extent, with its important debt to Galen, the *Canon* replaced Galen's own writings, even in the West. Gerard of Cremona in the twelfth century translated it and a large part of the Galenic corpus. Other Galenic texts were translated direct from the Greek by William of Moerbeke, half a century later than Gerard. Eventually, too, all the more important works in Arabic on medicine became available, including the *Kitab* of al-Razi (*c.* 850–924), an even larger encyclopædia than Avicenna's *Canon*. In the later middle ages the Islamic commentaries upon and additions to the Greek originals had a decisive influence upon the European knowledge of Hellenistic medicine. But Galen's chief books on anatomy and physiology only became accessible in the early sixteenth century.

Human dissection was discouraged in Islam. In Europe the systematic study of anatomy seems to have begun in the twelfth century, contemporaneously with the rise of the famous medical school of Salerno, though the practice of actual dissection was a north Italian development. Dissection was given countenance, partly by the needs of surgery, and partly by legal recognition (under the influence of the law-school of Bologna) of the value of forensic evidence derived from the post-mortem opening of the body. There has been at all times and in all places a universal revulsion against the dissection of the dead to serve mere curiosity, and certainly it was not extraordinarily strong in the middle ages. At least it is likely that human dissection was comparatively common at Bologna by the early fourteenth century, the legal post-mortem having been transformed into a means of instructing students. Henri de Mondeville, in teaching anatomy at Montpellier in 1304, illustrated his lectures with diagrams probably copied from those used in the school of Bologna.[1]

Shortly afterwards, in Mondino dei Luzzi (c. 1275–1326), medieval anatomy reached its zenith. He dissected for research, and he was probably the first teacher since the third century B.C. to demonstrate publicly on the human body. His *Anathomia*, which was printed many times, long remained a popular text, though it was written with little knowledge of the real accomplishments of Galen's investigations into medical science. Eager to reconcile authorities, he did not venture to assert his own views, and perpetuated many mistakes. Good anatomists succeeded Mondino (though none escaped his influence) but there was a general deterioration in the teaching of the subject. Mondino had expounded while at work upon the body; the standard practice of a later age is familiar from a number of wood-cuts in early printed books. The professor sat in his lofty chair, reading and enlarging upon the Galenic text, while an *ostensor* directed the operations of the humble *demonstrator* who wielded the knife. The body again became merely an illustration to the words of nobler men. Anatomy degenerated into the repetition of phrases and names, physiology into dogmas and disputations. No example of the misleading perspective adopted by the renaissance scholar could be clearer than this: because he knew that the teaching of anatomy had become a literary exercise in the fifteenth century, he assumed that it had never been anything else, and that the only course was to return directly to the works of Galen, now more completely available than before.

The curve moves upwards again towards the end of the fifteenth century. One factor in this seems to have been pressure in the medical schools for more demonstrative teaching; attendance at

[1] Charles Singer: *The Evolution of Anatomy* (London, 1925), p. 73.

dissections had to be restricted so that all might share in the spectacle.[1] Another was the official recognition of human dissection, under clerical licence, by the Papacy. Somewhat later renaissance monarchs permitted their chartered companies of physicians and surgeons to perform anatomies on the bodies of executed criminals. The first printed anatomies with figures appeared in the last decade of the century and the first half of the sixteenth shows a whole group of able practical anatomists at work—Berengario da Carpi, Johannes Dryander, Charles Estienne, Canano of Ferrara, Niccolo Massa, in addition to Vesalius. The humanists applied themselves to the editing of already famous medical texts, and others but recently recovered, most important among these latter being Galen's *Use of the Parts* and *Anatomical Administrations*, and the *De re medica* of Celsus (first century B.C.). There was a powerful reaction against the Arabic authorities, and the Arabic technical nomenclature was gradually replaced by the classical terminology which has established itself. Galen's texts were studied in the original Greek, and re-translated into Latin. Many scholars combined in the edition of Galen's *Opera Omnia* printed at Venice in 1525—the connection between medicine and philosophy was rediscovered. But scholars were not practical anatomists, their enthusiasm reinforced, rather than weakened, the medieval attitude to anatomical study. The pure Galen still held all Galen's faults, for scholarship could discover or purify a text without touching on the errors in observation held within it.

Another and more important source of inspiration was of a different kind, the naturalistic movement in art which has already been mentioned. Italian artists had been engaged in the study of anatomy well before the end of the fifteenth century, and from surviving sketches by, for example, Michelangelo or Raphael, it appears that they occasionally practised illicit dissection. Leonardo da Vinci (1452–1519) certainly did so as his interest in the fabric of the body developed from his first attempts to analyse the structure of the forms he was portraying. Some of the printed anatomical figures of the next generation are indeed reminiscent of Leonardo's famous drawings. The representation of anatomical structures as seen with the artist's eye and recorded by the artist's

[1] Lynn Thorndike: *Science and Thought in the Fifteenth Century* (New York, 1929), p. 69, n. 22.

pencil, which is totally different from the schematized, diagrammatic illustration of an earlier epoch, and even from the merely workmanlike but biologically accurate sketches of later professional anatomists, thus preceded the work of the founders of modern anatomy. The means for duplicating such drawings already existed in the technique of wood-cut printing. From this source the sense of the actual, of the minute, penetrated into academic anatomy and it is significant that the artists were men who had no stake in existing theory, and who, in the case of Leonardo at least, were unschooled in the textual description of organs. But there are limitations to the artistic impulse and to the naturalism which is only interested in superficial forms. The artist's drawing may not be most suitable for the purposes of a text-book; he needs to know something of the configuration and function of the surface musculature, of the run of the visible blood-vessels, and perhaps something of osteology, but he does not normally need to penetrate to the internal organs or the recesses of the skull. The artist at the dissecting table would certainly seek to re-create the appearance of the living body from the structure of the dead but he would not usually be interested in the correlation of function and the ordering of parts which ultimately lead to the discovery of physiological processes. Only when naturalism serves an impulse which is no longer purely artistic and has become the instrument of scientific curiosity can it have significance for medical anatomy. Of course it is abundantly clear that the motive which directed Leonardo to make his perceptive and accurate sketches, to take casts and prepare specimens by the injection of wax, was of a truly scientific order. In him, as in Vesalius, artistic imagination was the servant of science. Yet some even of the most naturalistic of Leonardo's drawings contain ancient errors, perhaps copied from the vernacular anatomical text-books which were already available to him: and to the organization of anatomy as a discipline Leonardo, who had no talent for classification and arrangement, contributed nothing. In any case the influence of his crowded and ill-ordered note-books upon contemporaries is in all respects entirely conjectural.

If the historical situation were that teachers of anatomy and medicine in the universities of France and Italy became themselves the pupils of professional artists unlearned in Galen it would be unique and surprising. But this was not the situation. It was rather that the work of the anatomist reflected, with less artistic merit, the same general cultural trend towards naturalism which affected purely æsthetic representation more profoundly; and that the anatomist made use of the same techniques of draughtsmanship and reproduction as the artist, whose stylistic conventions were impressed upon his illustrations. Naturalism, and the desire to take advantage of the new faculty of the wood-cut print for exposition, forced him along the road to observation. If the teacher of anatomy wished to elucidate the Galenic account of the structure of the human body by pictures of the features described he could do so only by dissecting a body and having drawings made —he could not reconstruct a picture entirely from a verbal text, though the text might influence the instructions he would give to the draughtsman. As, in so doing, anatomists observed discrepancies between the text and the structures themselves they departed with greater confidence from the Galenic model and learnt to rely on observation alone. A few conservative anatomists were well aware of the danger of illustrated texts; too great a reliance upon visual images might lead to contempt for Galen's superior knowledge. So in fact it happened that Vesalius' cuts are sometimes less traditional and more accurate than his text. The practice of making realistic drawings elevated instructional dissection to the level of research, though opportunities for making a recordable dissection were few and hurried. The ambition to make (for example) an accurate map of the venous system need not be taken as evidence of a critical spirit, though it may indicate a more sensitive professional conscience. As the first men to embark on such tasks were firmly convinced of Galen's rectitude, there was no reason why they should not take the liberty to draw what he had already perfectly described. It was thus with Vesalius himself: not until his preparatory work for the *De Fabrica* was well advanced did he realize the extent to which Galen had transferred animal structures into human anatomy. This misleading practice, and Galen's specific mistakes, could not be exposed without an impulse to research which in fact arose out of the needs of teaching and illustration. The errors in text-book anatomy could not be discovered through a desire to amend Galen by reference to nature because no one as yet believed this to be necessary. And none but a skilled anatomist could discover where such a remarkable scientist as Galen had gone astray.

All this has little relation, directly, to æsthetics. The first illustrated anatomies were not indeed beautiful books, though they
contained many new discoveries. It may well be that Vesalius'
superbly produced folio cast an undeserved shadow upon the less
splendid efforts of his contemporaries and immediate predecessors.
Perhaps too much emphasis has been placed upon its interest as
an example of the profitable co-operation between scientist and
artist in the sixteenth century; certainly Vesalius' figures were fine
enough to prompt frequent plagiarism. The preparation of illustrated anatomies demanded such a collaboration, and provided
an incentive for an original research, but Vesalius was not the first
to attempt a complete pictorial survey, which he began with
youthful energy. When in 1537 he set to work on De Fabrica, and
commenced teaching anatomy at Padua, Vesalius was twenty-
three. He could hardly claim to write with mature knowledge,
and though he had studied medicine at Louvain and Paris, so far
as is known his experience of dissection was still very limited. Nor
was he qualified to stand as an arbiter between Galen and Nature,
for at Paris especially, under medical humanists who were unshaken followers of Galen, he had been well grounded in the
renewed Greek tradition. Vesalius' first notable publication was
a revision of Johann Günther's Anatomical Institutions according to
Galen (1538), and his second, the Tabulæ Sex, was a series of wood-
cuts to illustrate the Galenic exposition of human anatomy which,
though based on dissection, was still traditional in character and
repeated many ancient mistakes. It was not, apparently, until the
preparation of De Fabrica was well advanced (about 1539–40) that
Vesalius began to doubt whether an illustrated anatomy founded
on dissection could be reconciled with Galen's descriptions.
Even of Vesalius' finished work it has been said that: 'A few of
his comments reveal an active dissector less experienced than
his contemporaries Berengario da Carpi, Massa and Charles
Estienne.'[1]

Creative scientific ability may run strongly in the direction
either of practical work or of theory. To the former talent the
medieval world offered little opportunity in comparison with that
offered by modern experimental science. As mechanics was in the
early seventeenth century the ideal field for the exercise of the

[1] Charles Singer: Studies and Essays in the History of Science and Learning offered
to George Sarton (New York, 1947), p. 47.

conceptual intellect of Galileo, so anatomy in the sixteenth was a fruitful subject for keenness of observation and, to a less degree, ingenuity of experiment. Unconsciously the new developments in the study of anatomy ran counter to the humanistic endeavour. At the very moment when humanists were rediscovering the philosopher in Galen his scientific authority was being undermined. As with so much original effort in the first stages of the scientific revolution, a more perfect anatomical knowledge contributed not to the integration of science but to its disintegration, in this case to the elaboration of a specialized art of detached description through which alone accuracy could be attained. The anatomists who freed themselves, however partially, from their natural inclination to follow classical masters were framing a new standard of scientific observation. In no real sense was this the moment of the birth of some novel, self-conscious method of observation and experiment in science, but it was the moment when the accepted narrative of fact and theory was first modified effectively and permanently by recourse to nature. A body of original facts relating to a single discipline was for the first time gathered together for comparison with the traditional account. Justified only by practical experience, even opposed to the theories of those who described them, the new discoveries combined to teach the lesson that the whole doctrine of anatomy must be reformed by the use of meticulous observation and independent thinking. While the facts of experience to which appeal had been made in the framing of medieval physical theories had mostly been commonplace, or accidentally revealed, the new anatomy turned to the systematic exploitation of a specialized and laborious technique.

The most complete, and the most striking, use of this technique was certainly in the *De Fabrica* of Vesalius. All authorities on the subject agree that Vesalius' exposition of human anatomy is outstanding in early modern times, basing their judgement on his text as well as on his still more eloquent illustrations. But he was not unique in his originality; rather he was the most successful exponent of a new procedure that was beginning to gain adherents over all Europe, some of whom Vesalius did not scruple to treat unfairly. Among his immediate predecessors Berengario, whose illustrated *Commentary* on the anatomy of Mondino was published as early as 1521, was notable for hesitancy in following Galen.

Another, Charles Estienne, was at work on his book *On Dissection* by 1532, though this remained unprinted till 1545.[1] Estienne was the first anatomist (after Leonardo) to prepare figures illustrating complete systems (venous, arterial, nervous) and he discovered the structures in the veins, later known to be valves, which stimulated William Harvey to the discovery of the circulation of the blood. John Dryander published two illustrated treatises in 1537 and 1541. Canano's incomplete myology also appeared in the latter year (or 1540), and nine years after the publication of *De Fabrica*, a set of copper-plates to illustrate human anatomy was finished by Bartolomeo Eustachio (1520–74), which have been described as more accurate than Vesalius' figures, and almost equally crowded with new observations.[2] Some of the unillustrated treatises on anatomy, such as those of Sylvius and Massa, are not less remarkable for skill in dissection and acuity in criticism. Moreover, in the application of anatomical knowledge to the study of physiological function Vesalius (for the most part cautious and conservative in theorization) did not exceed his contemporaries in originality; soon after 1543, Serveto, Colombo and Fallopio were putting forward notions on the distribution of the blood about the body of which no hint is to be found in *De Fabrica*.

When Vesalius is matched against his not unworthy rivals, his peculiar fame appears all the more enigmatic. He did not lead the way in making discoveries alien to Galenic anatomy, nor did he ever intend an onslaught upon it. He had no greater learning, or more vivid freshness of mind, than his more experienced contemporaries. Even when spoken of by his teacher Günther as 'a young man, by Hercules, of great promise . . . very skilled in dissecting bodies,' the observation for which he was thus praised had already been anticipated in Italy. The task which he set himself would seem to require ample time and mature knowledge, but Vesalius had neither. During the mere three or four years given to *De Fabrica* he was busy with other writing, teaching and travel. His whole study of female anatomy was based on the hasty dis-

[1] *De dissectione partium corporis humani*, a large octavo, has fifty-eight full-page wood-cuts, in which unfortunately the important portions are often too small to be clear.

[2] Charles Singer: *Evolution of Anatomy*, p. 135. Eustachio's figures, after being first printed in the early eighteenth century, were thereafter frequently republished as having contemporary value.

section of three bodies—not surprisingly, it forms one of the weakest portions of his work. Comparison of the *Tabulæ Sex* with *De Fabrica* compresses his development as a scientist into an almost incredibly brief space of time, if it was entirely unaided. Little is known of his personality; in comparison the biography of Copernicus is full. Often abusive, truculent, Vesalius was not a learned man by the standards of his age, and he reveals little in the way of philosophic depth such as was then expected of a scientist. He was ambitious, and apparently ambition prompted his one great work; after leaving Padua in 1543 to become an Imperial physician his career in science ended.

It is impossible to fit Vesalius into the conventional picture of the scientific worthy. But whatever may be the truth concerning the collaboration of artist and anatomist in the production of *De Fabriça*, however serious the charges that may be brought against Vesalius' character as an author, the book itself stands as a monumental achievement of sixteenth-century science, and in its own time it was almost immediately recognized as such. *De Fabrica* was the implementation, in a manner whose total effect is superior to that of any contemporary work, of a single conception of what an anatomical treatise ought to be, though indeed much of this unity derived from following Galen's example. It aimed at a systematic, illustrated survey of the body part by part, layer by layer. The skeleton and the articulation of joints, the muscles, the vascular system, the nervous system, the abdominal organs, the heart and lungs, the brain, were described and depicted with a detailed accuracy never previously attained. The wood-cuts are indeed on occasion better than the text they illustrate, though in one place it is admitted that a drawing had been modified to fit Galen's words. In his section on the heart Vesalius mentions probing the pits in the septum without finding a passage; broadly, however, he accepted the Galenic account of the heart's function. Some whole topics (the female organs, the eye) were less well discussed owing to errors in observation, due in part to Vesalius' inability to free himself completely from traditional ideas. In minutiæ there was much for the anatomists of the later sixteenth century to amend.

For all his insistence on his originality, Vesalius followed closely in the track of his great master. *De Fabrica* is effectively a compilation from Galen's *Use of Parts* and *Anatomical Administrations*, two books wholly or almost wholly unknown before Vesalius's lifetime, made by a man with considerable experience of human dissection (which Galen lacked). Vesalius had the human material from which to correct some of Galen's obvious errors arising from his use of animals and to add many new discoveries; nevertheless much that he wrote can be found, almost word for word, in Galen. The result was embellished and enriched by the splendid illustrations:

> The *Fabrica* is, in effect, Galen with certain highly significant Renaissance additions. The most obvious and the most important is the superb application of the graphic method.[1]

But the graphic method was not invented by Vesalius, and to what extent its quality was due to the draughtsman employed (who, according to the art-historian Vasari, was John Stephen of Calcar)[2] will probably never be known. Although Vesalius hotly criticized Galen for describing human anatomy by analogies drawn from animal dissection, he was himself led into mistakes resulting from this very practice. His boastful claims to originality are not to be trusted without scrutiny; thus it is obviously untrue that human dissection was utterly unknown in Italy before he introduced it at Padua. Vesalius seems to have resented, rather than welcomed, originality in other anatomists. The tradition of anatomical study was not inaugurated in northern Italy by Vesalius during his short residence there, for it had existed for centuries. As for the claim that Vesalius was the first teacher of anatomy in modern times to carry out dissections before the students with his own hands—the idealized scene is represented in the frontispiece to *De Fabrica*—as early as 1528, in the humanistic medical school of Paris, 'the participation of students and doctors in the actual process of dissection was recognized.'[3] It seems, therefore, that if Vesalius introduced any new teaching method to Padua it was but the method he had known at Paris.

In him there was neither the burning passion of a Galileo to refine truth from error, nor the remote vision of a Kepler; yet the work of a man whose spirit is calm, almost plodding, may mark a

[1] Charles Singer: *Studies and Essays in the History of Science and Learning offered to George Sarton* (New York, 1947), p. 81.
[2] This attribution has been disputed and defended; cf. W. M. Ivins in *Three Vesalian Essays* (New York, 1952).
[3] Charles Singer: *Bulletin of the History of Medicine*, vol. XVII (Baltimore, 1945), pp. 435–8.

turning-point in exact science. If Vesalius was pre-eminent among early anatomists he must not be made so at the expense of his contemporaries, who were also men of ability and precision, nor by neglecting his own faults and mistakes. Although the structure of *De Fabrica* was excellent, Vesalius was a poor expositor; nevertheless, his text inculcated the principles of the new mode of anatomical study more effectively than any other work, and brought back the study of anatomy to the dissecting-table. He was responsible for making many useful innovations in the technique of dissection which extended its possibilities. He had the merit of seeing anatomy as more than a catalogue of structures useful to the physician seeking the proper vein from which to let blood or to the surgeon tracing the course of a bullet-wound; it was for him an entrance into knowledge of the living body as a functional organism. True, he was unable himself to progress far towards this desideratum, and therefore he was content to rely upon Galen, but his work was a guide to his successors. *De Fabrica* must be judged as a whole, and it is in the preponderance of its virtues over its defects that the book excels. In a descriptive subject such as anatomy, especially where the relations between the possibilities of research and the necessities of teaching are very close, there seems to exist a critical level of accuracy. Until description has passed the critical point no steady accumulation of knowledge is possible because there is no firm orientation to study, no sound model exists, doubt and duplication lead to wasted effort and repeated surveys of the same elementary facts. Once the critical point is passed and the outline is securely drawn, so that the student can easily identify it with the natural subject, it is possible to press forward to deeper and finer levels of detail. Vesalius' contemporaries had almost attained this point; in particular studies some passed far beyond it; but *De Fabrica* passed it in an extended account of the whole body. The work was far from impeccable, perhaps its originality and the importance of a high æsthetic (which is not the same as a naturalistic) quality in its illustration have been over-emphasized, but it was sufficient.[1] The map was made. It is not enough to say that Vesalius

---

[1] There are some hundreds of figures, not including the initials, in the first edition of the *De Fabrica*. Of these, only about a score enter the "æsthetic" class: they are, of course, the finest wood-cuts in the book, but they do not compose the whole of Vesalius' graphic teaching.

redirected anatomists to the natural facts of bodily structure, for the work of contemporary anatomists was already in the same direction, and in any case it was to be long before the meaning of these facts (like the valves in the veins) could be correctly interpreted. The main point is that the *De Fabrica* was the actual source from which the method of obtaining the facts of human anatomy—by dissection and experiment—was learnt; admittedly it could still have been learnt had the *De Fabrica* never been printed. In the hands of students it was an instrument fit to measure against nature, to serve as a guide to the complexities which Vesalius' successors discovered in their turn, and to create in the exceptional man the critical frame of mind which is the main-spring of research.

In a sense the possession of this quality belongs to the definition of modern science. In many instances the reason for the sterility of conventional science before 1500 was not simply that its exponents were engaged upon a vain endeavour, nor most obviously that they were fondly prejudiced, perverse, or unoriginal of mind. They, like all men, could do no more than accept or modify according to their powers an intellectual inheritance which was a part of the age and society in which they lived. Such an inheritance is never exactly consistent, nor is it wholly dominant over the individual, but the range of possible variation from it is always limited. One characteristic of the intellectual heritage of every medieval philosopher, or physician, was the limited area of contact between the system of ideas in science and the reality of nature; on the one hand was a natural philosophy satisfactory enough to those who knew no other, and on the other the bewildering complexity of the natural world. A juxtaposition of the two was not easily effected. If it seemed that philosophy explained the world, it seemed so because the distinction between the philosophic view of natural phenomena, and the phenomena themselves, was not clearly revealed; and since the *globus intellectualis* and the *globus mundi* were so far distant, constructive, purposeful criticism of the former could hardly emerge from its comparison with the latter. Hence arises the importance of a new field of experience, such as magnetism offered to the middle ages, or of a new attitude to scientific explanation, such as that contained in the mechanistic philosophy of nature. The effect of every major idea in science, of every major observation and experiment, has been to present a new juxtaposition of the realm of thought and the realm of facts, which may in turn demand a far more profound readjustment than the original innovator could foresee.

It is easier to illustrate this from the history of ideas than from the history of factual discoveries, easier to see it in Copernicus, for example, than in Vesalius. It is obvious now that once the central tenet of the old astronomy was rejected, the incidentals which Copernicus had retained must also be challenged. Yet it is important, for the understanding of the process and development of the scientific revolution, to see how the cardinal observations and measurements emerged, as well as the cardinal concepts. The normal development of any established department of science may be, and indeed usually is, conditioned in large part by its conceptual structure and the nature of the theoretical techniques appropriate to it, since these furnish the perspective in which the problems are to be seen. In exceptional situations, however, the disclosure of a new fact—commonly the apprehension of the import of a whole group of facts—may force a crisis. Then existing theory may prove an obstruction. A novel perspective is called for. In this respect stands the work of Vesalius to that of Harvey and later physiologists, who revolutionized ideas by applying experimental procedures to the elucidation of the bare facts yielded by a more exact anatomy. Through the effort of Vesalius and his less famous contemporaries there was introduced into biological science for the first time an acute sense of the importance of minutiæ, of the mastery of special methods, and of the precise and full reporting of observations. Contemporaries admired the *De Fabrica* as the perfection of descriptive art, to this also its claim to mark a turning-point in the growth of scientific knowledge is due. But the book is not less notable as entailing the challenge to Galenic theory which its author scarcely contemplated.

De Revolutionibus Orbium Coelestium has an altogether different character. Its force springs from the other tactic of the scientific revolution, accepting the body of observation on which conventional science relied and even the method by which the facts could be grouped into a theoretical interpretation, but denying that only the conventional interpretation was possible, or even preferable to its alternative. Although his name has become more famous, Copernicus was in many ways less modern than Vesalius, in particular he had a far less acute sense of the reality of nature:

like many medieval men he was far more concerned to devise a
theory which should fit an uncritically collected series of observations than to examine the quality of the observational material.
Tycho Brahe, half a century later, was the Vesalius of astronomy.

If one can discern a fundamental motive for Copernicus'
reconstruction of the cosmos, that too seems to be medieval.
The unity of thought had been the goal of medieval philosophy:
the harmonizing of pagan science and Christian religion, the reconciliation of authorities, the explaining of contradictions and
discrepancies, had been its unending task. And it seems that the
belief that true knowledge must be unified provoked Copernicus'
first doubts. The early sixteenth century was not a time at which
dogmatism on astronomical problems was easy or profitable. The
popular view that the astronomical science of the middle ages
was simply a matter of applying rigid principles in a determined
fashion is mistaken. Of course the main structure of astronomical
theory was firmly established, but the working out of the exact
details involved constant experimentation, and the best authors
were not themselves agreed on the true values to be attached
to all the constants of the Ptolemaic system. In fact medieval
astronomers were well aware that that system was not meticulously
accurate as a calculating instrument and, though they did not
revolt against its fundamentals, they attempted by adjustments
and additions to bring it to perfection. Wherever astronomy was
actively studied, there was a renewed interest in observation,
in modifications to the machinery of the spheres, and renewed
computation of the constants and tables. The greatest problem of
all was of a different order. No man of learning could ignore the
fact that the system of Ptolemy was something more than an
elaboration of that of Aristotle. To describe as *physical* the
Aristotelean picture of the cosmos, and to describe the Ptolemaic
as a *mathematical* hypothesis of planetary motions, is perhaps to
press an analysis of which the middle ages were barely conscious.
Yet, while there was no direct or irresolvable antithesis, it was
always clear that there was something of a hiatus between
cosmological physics and astronomy. There was not merely a
perpetual struggle to reduce the complexities of planetary motion
to mathematical order, but also these complexities were confusing
in physics. The eccentricity of the earth with respect to certain
spheres, and the irregularity of the motion of these spheres about

the earth, were assumptions which practical astronomers were forced to make, without possibility of reconciliation with Aristotle's physical doctrine. Oresme pointed out another difficulty that arose from the introduction of epicycles, concluding that Aristotle was obviously mistaken in his demonstration that the intelligences controlling the spheres do not themselves move.[1] He seems to distinguish, too, between what is true "in philosophy" and true "in astrology," i.e. astronomy. In more practical ways the imperfections of astronomy were recognized. The calendar was out of joint. Even astrologers admitted that their calculations were uncertain because of the inaccuracy of the tables with which they worked. Tables for the cycle of moon and tides, essential to seamen, were unreliable. The observational basis in the celestial coordinates of the fixed stars, and the latitudes of places on the earth, was equally open to doubt. Humanism added its own confusion: the conflict between Arabists and Humanists was less spectacular in astronomy than in medicine, but it was not the less real in the difference of opinion between those who wished to preserve the simpler system of Ptolemy, and those who believed that Islamic elaborations, like the trepidation of Thabit, were necessary.

Already in the fifteenth century Peurbach and Regiomontanus had applied themselves to the technique of observation, and tried like their predecessors to bring astronomy up to date by new determinations and computations. They did not appeal to nature against Ptolemy, rather indeed they looked to the Greek *Almagest* for fresh inspiration. The world of learning upon which Copernicus entered knew that a reform was needful, even though it did not welcome the shape of that reform in his hands. A world expecting some new mathematical formula which would "save the phenomena" along the old lines received instead an unprecedented synthesis of philosophy and mathematical astronomy which attacked much that was orthodox in both. Although united in the higher realm of ideas, the philosopher and astronomer had long been professionally divorced. Those who contemplated the mysteries of the cosmos had done little actual observing or calculating, while the practical astronomer had lost himself in mathematical abstractions. Copernicus was a superbly equipped theoretical astronomer (with some skill in observation, but no great love for it) who shows at the same time a strong sense of

[1] *Medieval Studies*, vol. IV, p. 169.

physical reality. He was affronted by the Aristotle-Ptolemy dualism. He thought that some of the devices used by Ptolemy, especially the equant-point, were cheats counter to sound philosophy. He found that the physical system failed in mathematics, the mathematical in physics. The only solution to the dilemma required two steps: the purging from mathematical astronomy of its needless reduplications, and from physics of those ideas which were obstacles to a true conception of the universe. Thus the reform of astronomy demanded a *limited* reform of physics, and Copernicus was in fact impelled to distinguish between those doctrines of Aristotelean physics which he took to be true, and those which he took to be false. In the end much was to turn upon the justice of his conception of what is "right" in nature.

There is good evidence that Copernicus was highly esteemed by men of learning long before the publication of *De Revolutionibus*, which indeed in the eyes of some contemporaries diminished rather than enhanced his reputation.[1] Rheticus describes him as 'not so much the pupil as the assistant and witness of observations' of Domenico Maria da Novara, himself something of a critic of Ptolemy, at Bologna: and as lecturing on mathematics at Rome in 1500 (when he was twenty-seven) 'before a large audience of students and a throng of great men and experts in this branch of knowledge.'[2] Clearly Copernicus had already become an accomplished student of mathematics and astronomy in his earlier days at Cracow. He was among those replying (in writing) to a call from Rome for ideas on calendar reform. One of his critical writings, the *Letter against Werner* (1524), seems to have circulated in manuscript, as did the first sketch of his new astronomy, the *Little Commentary* which was written, probably, about 1530. There is a record of Copernicus's ideas being explained to Pope Clement VII in 1533. In 1539 his fame attracted the Protestant Georg Rheticus from Wittenberg to Frauenberg, "remotissimo angulo terrae," to become his pupil, and in the following year Rheticus' *First Account*, a review of Copernicus's manuscript of *De Revolutionibus*, was published at Dantzig. Some of his readers, who like Gemma Frisius later became firm advocates of the heliostatic

---

[1] Lynn Thorndike: *History of Magic and Experimental Science*, vol. V (Baltimore, 1941), pp. 408–13.
[2] Edward Rosen: *Three Copernican Treatises* (New York, 1939), p. 111.

system, awaited the publication of the whole work with eager anticipation. In the preface to his book, dedicated to Pope Paul III, Copernicus particularly mentions, as friends who removed his reluctance to publish, Nicholas Schoenberg, Cardinal of Capua and Tiedeman Giese, Bishop of Culm. While his investigations thus received a measure of countenance at Rome, it was Luther who was to speak of 'the fool who would overturn the whole science of astronomy.'[1]

Despite this, the new doctrine failed to secure adherents. Favourable references to it are scarce throughout the sixteenth century: most are couched in terms resembling Luther's.[2] Admittedly the book was difficult, written only for those who were skilled geometers and experienced in astronomical calculation, and it supported a proposition which had been condemned in philosophy for two thousand years, and no astronomical advantages—even if they were genuine—could vindicate it. If the celebrated note by the Lutheran minister Osiander which, unknown to Copernicus, declared that the tenets of the following book were to be taken only as a mathematical hypothesis, a calculating device, and not as truth, had any effect in warding off official censure (which seems doubtful), it certainly did nothing to weaken the general sentiment that speculation on the earth's movement was foolish and futile. Only about half a dozen men are known to have held the opposite view in the years 1543–60: Rheticus, who deserted astronomy; Gemma Frisius of Louvain, one of the first astronomers of the age; Pontus de Tyard, later Bishop of Châlons; and a group of English astronomers and mathematicians, John Dee, Thomas Digges, John Field and Robert Recorde. Later in the sixteenth century the Copernican group grew only slowly, but it finally included such professional scientists as Maestlin, Kepler, and Galileo. To scholars generally, however, Copernicus's theory was still known only as something strange and improbable. Until 1616 its wide spread was restricted not by official condemnation, but by indifference.

To understand this, it is necessary to understand in what measure *De Revolutionibus* was revolutionary, and effected a reform in astronomical thought. In the heliostatic principle, and

---

[1] *De Revolutionibus Orbium Coelestium* (Nuremburg, 1543), sig. iii recto.
[2] Dorothy Stimson: *The Gradual Acceptance of the Copernican Hypothesis* (New York, 1917).

in its necessary physical consequences, Copernicus opposed the common sense of his age: in everything else he was conservative. There was nothing in the method of his madness to arouse contemporary interest. Moreover his madness took the form of a familiar heresy: therefore again the whole theory could be lightly dismissed. A whole stream of references proves that to suppose that the sixteenth century ought to have reacted violently in approval or condemnation of the book when it appeared is to misunderstand a whole facet of medieval astronomy. That certain of the ancients had supposed the earth to move was as well known to the middle ages as Aristotle's powerful reasoning against their contentions. They found the question debated again by their Arabic authorities. As a result, several scholars of the fourteenth century exhibit a marked tolerance in their treatment of the idea when commenting upon Aristotle, among them many members of the "impetus" or experimentalist school, including William of Ockham, Albert of Saxony, Jean Buridan, and Nicole Oresme. Oresme, in his commentary on *De caelo*, denies that the stability of the earth is a logical consequence of the movement of the skies: from the analogy of a revolving wheel he shows that it is only necessary in circular motion that an imaginary mathematical point at the centre be at rest. Further he says that local (i.e. relative) motion need not necessarily be referred to some fixed point or body: rest is the privation of motion, and is in no way involved in its definition. For instance, he says, there is imagined to be outside the universe an infinite motionless space, and it is possible that the whole universe is moving through this space in a straight line. To declare the contrary is an article condemned at Paris, yet this motion could not be considered to be relative to any other fixed body. Suppose that the skies stood still for a day and the earth moved: then everything would be as it was before.[1] In another place he gives his judgement, "under all correction," that it is possible to maintain and support the opinion that it is the earth which is moved with a daily motion (axial rotation) and not the sky, and further that it is impossible to prove the contrary by any experiment ("experience"), or by any reasoning. This was just what Galileo was to declare nearly three hundred years later. Against his proposition Oresme quotes three arguments,

---

[1] *Medieval Studies*, vol. IV, pp. 203-7.

all of which were to be directed against the Copernican hypothesis in the sixteenth and seventeenth centuries. Firstly, the skies are actually observed to revolve about their polar axis; secondly, if the earth turned through the air from west to east, a great wind would be felt constantly from the east; and thirdly a stone thrown vertically upwards would not fall back at the place whence it was thrown, but far to the west. Oresme has replies to all these objections. In answer to the first, he emphasizes the relativity of all appearance of motion: a man seated in a boat gazing at another boat cannot tell whether his own vessel is moving, or the other in the opposite direction, but there is a prejudice in favour of the stability of one's own immediate frame of reference. As for the wind, the earth, water and air of the sublunary world all move together, and so there is no wind other than those to which we are accustomed. To the third objection Oresme replies that the stone thrown up in the air is still carried along in the west-east direction with the air itself, and with 'all the mass of the lower part of the world which is moved in the daily movement.' It is not quite clear what this last phrase meant to Oresme, but it holds an idea of profound consequence. The stone falls back to the place whence it came, as it would do if the earth were still, and Oresme points out (again anticipating Galileo) that all the phenomena of motion appear to be identical in a ship which is moving or a ship which is at rest. As Oresme puts it, in less precise language, a movement which is compounded of motions in two directions is not discernible as such when the observer himself participates in one of them.

Various theoretical or physical arguments against the rotation of the earth anticipated by Oresme turn upon the idea that it would be unnatural and out of place in the texture of natural philosophy. This Oresme also denies. He points out that the Aristotelean system attributes no movement to the earth as a whole, though Aristotle himself declared that a single simple motion was appropriate to each element so that the earth might well turn in its place, as the heavens do, or as the element fire does. Oresme agrees that if the earth turns it must possess a "moving virtue," but this it must have already, since displaced parts of the earth return to it. To the criticism that the motion of a moving earth would falsify astrology, he makes a most important riposte. All conjunctions, oppositions and influences of the sky would take place as before, and the tables of the movements of the heavenly bodies and other books would be as true as they were before, only it would now be recognized that the daily rotation was apparent in the heavens, and real in the earth. It has often been alleged that Copernicus destroyed whatever scientific basis astrology might be supposed to have had. It was not so. The practice of astrology has been entirely unaffected by it, just as it was (and is) unaffected by the fact that the "Houses" of the zodiac no longer correspond with the constellations after which they were named owing to the precession of the equinoxes. The predictions of judicial astrology turn upon the configuration of the skies at any moment: they are unconcerned with the mechanics of the motions by which those configurations are produced. And Oresme's disclaimer was actually re-echoed in the sixteenth century. To quote Thorndike: 'It is a historic fact that the Copernican system was first publicly announced, if not precisely under astrological auspices at least to an astrological accompaniment and that such signifying the future was for long after associated with it in men's minds.'[1] One at least of the leading exponents of the new scheme, Rheticus, had no doubts of the virtues of judicial astrology, and Copernicus himself never declared against it.

The authority of Scripture was constantly brought to bear in favour of the geostatic doctrine, but even this does not silence Oresme. Joshua's miracle can be interpreted in the sense that the sun *apparently* stood still, while the earth was really halted. He seizes upon the point that if the earth be supposed to move from west to east, all the celestial motions will take place in the same sense, which he thinks increases the harmony of the system. Also it would place the habitable part of the globe on the right or noble side of the earth. Again, it will be found that in this way the celestial bodies which are farthest from the centre will make their circuits more slowly (instead of more quickly as in the geostatic theory) which seems reasonable; and the principle that God and Nature do nothing in vain will be more faithfully observed; for example, there will be no necessity for a ninth sphere. After all this potent reasoning has been marshalled against the conventional doctrine, it comes as an anticlimax to find Oresme returning to it. 'Nevertheless,' he concludes, 'all maintain, and I believe, that the heavens are thus moved, and the earth not: "for God fixed the orb

[1] *Op. cit.*, vol. V, p. 414.

of the earth, which shall not be moved,"[1] notwithstanding the reasons to the contrary, for these are not conclusive arguments. But considering all that has been said, one might believe from this that the earth is moved, and the sky not, and there is nothing obviously to the contrary, which seems *prima facie* as much, or more, against natural reason as are the articles of our faith.'[2]

One might think that the famous cosmological debate of the seventeenth century had been rehearsed in the fourteenth![3] There were, indeed, subtle changes in the background in which Galileo set the same, or similar, arguments; but if Oresme had been diligently followed the stability of the earth could hardly have been defended save as an act of faith, not by reason and observation. The diurnal revolution was again discussed in the fifteenth century, when it was upheld by Nicholas of Cusa, and in the early sixteenth before the appearance of *De Revolutionibus*. This diurnal revolution, however, and the long argument in favour of it by Oresme, must be clearly distinguished from the theory of the annual motion of the earth which was developed by Copernicus, not to speak of the third motion which he attributed to it to account for the parallelism of its axis in space. A heliostatic theory imposed a far more severe strain on intellectual adaptability than a geocentric theory which admitted the diurnal rotation. To have conceived the *annual* motion is Copernicus' chief claim to originality, and it was the annual motion which condemned the Copernican hypothesis both in the eyes of the majority, and also in the eyes of the Church because of its heretical consequences.

It can hardly be imagined that the earlier discussion of the diurnal motion was unknown to Copernicus, and he must have considered its adoption as the first step to a reform of astronomy—a step which unfortunately did little to solve the problem of planetary motions and the multiplicity of spheres. It is not unlikely that his conception of the earth as of the same kind as the heavenly bodies, and therefore equally suited "by nature" to planetary motion (being spherical for instance), is an extrapolation from the reasoning used (by Oresme among others) to support the idea that the earth is "by nature" suited to axial rotation. Certainly

[1] Psalm XCIII, 1: 'The world also is stablished, that it cannot be moved.'
[2] *Medieval Studies*, vol. V, pp. 271-9.
[3] As was long ago pointed out, with too great emphasis and some misunderstanding, by Duhem in *Revue Générale des Sciences Pures et Appliquées*, vol. XX (1909).

Copernicus had to abandon the idea of "one element, one motion," but this had been virtually abandoned by Oresme. Not, however, until Copernicus had begun to consider this second, annual motion could he have begun to see the possibility of results important to mathematical astronomy—in this the middle ages had done little or nothing to help him. Even Oresme's very correct remarks on the illusions of relative motion only refer to a geocentric system, though they were equally valid as Copernicus applied them to the heliocentric. Unfortunately, the steps by which Copernicus proceeded from the first to the second motion, if that was his course, are totally unrecorded. He relates simply, in the Preface to his book,

> Nothing urged me to think out some other way of calculating the motions of the spheres of the world but the fact that, as I knew, mathematicians did not agree among themselves in these researches. For in the first place they are so far uncertain of the motion of the Sun and Moon that they cannot observe and demonstrate the constant magnitude of the tropical year.

Then he goes on to say that there was no consistency of principle in the devices that had been used—some had employed the simple homocentric spheres, others eccentric spheres, and yet others epicyclic systems. They had passed over the symmetry of the universe, as though one should put together a body from different members in no way corresponding to one another, so that the result would be rather a monster than a man. This uncertainty, he thought, must prove that some mistake had been made, otherwise everything would have been verified beyond doubt.

> Then when I pondered over this uncertainty of traditional mathematics in the ordering of the motions of the spheres of the orb, I was disappointed to find that no more reliable explanation of the mechanism of the universe, founded on our account by the best and most regular Artificer of all, was established by the philosophers who have so exquisitely investigated other minute details concerning the orb. For this reason I took up the task of re-reading the books of all the philosophers which I could procure, exploring whether any one had supposed the motion of the spheres of the world to be different from those adopted by the academic mathematicians.

Coming across Greek theories which attributed both an axial and a progressive rotation to the earth, and following the example of his predecessors who had not scrupled to imagine the circles they required to "save the phenomena," he thought he might himself be allowed to try whether, by allowing the earth to move, more conclusive demonstrations of the rotation of the spheres might be found. And he discovered that if, as he puts it, the motions of the planets were calculated according to their own revolutions, with allowance for the circulatory motion of the earth, then the phenomena of each worked out duly. Even more important, the orders and sizes of all the celestial bodies, spheres and even the heaven itself, were so harmonized that nothing could be transposed in any detail without causing confusion throughout the universe (Fig. 4).[1]

Copernicus was a theoretical astronomer of genius and originality. He must have noted the suspicious reversal of the constants of epicycle and deferent between the upper and lower planets: he must have perceived the unaccountable intervention of the sun's period of revolution in the calculations for each of the five planets. Believing that the assumption of an annual motion in the earth was no more wild than the assumption of a diurnal revolution, he was capable of taking the celestial machinery to pieces and re-assembling it in accordance with a different pattern. Working from ancient and respected observations, he proved that the new pattern gave as good results as the old.

It could hardly give much *better* results because the parts of the machine of the world as revised by Copernicus were essentially of the same dimensions as before, though arranged in a different order. His determining observations were those of the Greek astronomers, Timocharis, Hipparchus, Ptolemy, supplemented by the work of their Moslem successors Arzachel, al-Battani and Thabit, and his own measurements, few in number and mostly made about 1515. Sometimes, as in his determination of the length of the sidereal year, his result was less accurate than an earlier one (that of Thabit). Copernicus admitted most of the variations or anomalies which had been introduced to account for the apparent changes in some constants (for example, the variation in the obliquity of the ecliptic causing an oscillation which was correctly explained by Copernicus for the first time; and various complexities in the motion of the sun, which he transferred, of course,

[1] *De Revolutionibus*, Preface. Cf. T. S. Kuhn, *The Copernican Revolution* (Cambridge, Mass., 1957), 136–8, 140–2.

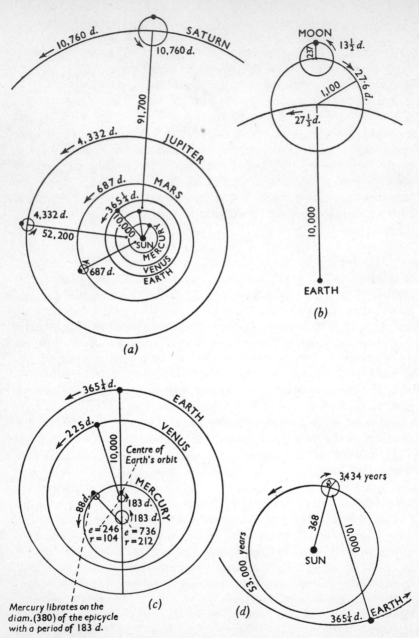

FIG. 4. The Copernican System of the Universe. (a) General arrangement, (b) Lunar theory, (c) The inferior planets, (d) The Earth. The dimensions are to scale unless otherwise indicated. (d=days, e=eccentricity, r=radius.)

to the earth). In addition he introduced new elaborations; some, like the motion of the apse-line of the earth's orbit already suspected by al-Battani, were real and necessary, others, like the "third motion" of the earth were superfluous or due to the collation of inaccurate observations, such as the variation which he supposed to occur in the eccentricity of the earth's circle about the sun. In fact, mathematically speaking, Copernicus did not create a new astronomy in the sense of isolating and measuring every celestial motion afresh; what he did do was to reinterpret the Ptolemaic structure from the heliocentric point of view, adding such refinements as he believed to be necessary. His remark to Rheticus, "If only I can be correct to ten minutes of arc, I shall be no less elated than Pythagoras is said to have been when he discovered the law of the right-angled triangle," shows that his ambition in the matter of precision was very limited. And when the first astronomical tables were calculated on the new theory— by Erasmus Rheinhold who yet denied its physical truth—they proved little superior to their predecessors.[1]

Astronomy remained the doctrine of the sphere, and the imagery drawn from it which is so prominent in sixteenth-century imaginative writing was hardly less appropriate to the new theory than to the old. Having rejected the prime assumption of cosmology, Copernicus had no occasion to challenge others. His geometry of the heavens is still that of rolling orbs, save that the earth has replaced the sun in the third sphere. He preserved the fundamental division between the sublunary region and the celestial, between the natural laws of earth and sky. He insisted even more severely than Ptolemy upon the inviolable perfection of circular motion, repeated again the proof that the alternative methods of eccentric sphere and deferent-with-epicycle are identical in their results, and reduced all motions to combinations of these two forms. The reality of the crystalline spheres was unquestioned. Apart from his one great innovation, all Copernicus' astronomical thought is thoroughly medieval. Truly he reformed the medieval universe, because he brought its pattern into a new order, but he introduced no new doctrine concerning its composition or the deeper logic of its various appearances.

What, therefore, were the merits of the new astronomical system? What arguments could be adduced to show that in this

[1] Cf. Appendix B.

complicated dance of relative movements the stability of the sun was more real than the stability of the earth? Copernicus dealt with these two questions as one, but as will be seen they require separate answers. In the first place, the pattern of celestial motion described by Copernicus and the methods used to work it out in detail had several distinct advantages. On the whole, the devices were rather more economically displayed, though this is a feature whose importance has been exaggerated and, since the observer's point of view is geocentric in any case, the geocentric conception had some advantages in the ease of handling. The fixed stars were fixed indeed, and with the whole system no more than a point in comparison with the size of their sphere—as Copernicus realized from the fact that they revealed no annual parallax—the problem of predictive astronomy was properly limited to the planetary bodies. These now recovered their independence without the intervention of any extraneous factor in their revolutions, and the main pattern of the planetary mechanisms could now be made uniform for all five. Moreover, Copernicus was able to declare, for the first time, the relative sizes of the planetary spheres, though because he followed Ptolemy's estimate of the earth's distance from the sun the whole system was far too small.[1] As the lower planets, Venus and Mercury, were now placed between the earth and the centre of the universe the peculiarities of their motions and their special relation to the sun were explained: clearly they could never be observed at a greater distance from it than the lengths of the radii of their spheres. The "stations" and retrograde motion of the planets, which had favoured the epicycle theory, could now be seen as pure illusions caused by the addition and subtraction of the earth's movement and the planets' relative to the unchanging background of the starry sphere. The equants and uneven revolutions of the Ptolemaic system were removed and the principle of perfect circular motion observed more purely. Again, the fluctuations in the apparent size of the heavenly bodies (as the earth approaches or recedes from them) required by the Copernican geometry corresponded more closely to observation than those required by the Ptolemaic, particularly in the orbit of the moon, which Ptolemy supposed to vary by a factor of nearly two. Most of the advantages that can be obtained by supposing

[1] About 5 per cent. of the true value: the dimensions are derived from the Ptolemaic planetary constants (see above, p. 16).

the sun to lie near the centre of all the orbits were actually worked out by Copernicus in his calculations, which thus removed some redundancies, and threw light on many obscure points.

On the other hand, it could not logically be claimed that these advantages proved the sun to be at rest and the earth moving. An interesting variant of the conventional geocentric theory, described by Martianus Capella in the fifth century A.D. made the lower planets revolve about the sun, and there are references to the same idea to prove it was not forgotten in the middle ages.[1] A natural step beyond this was to suppose all the planets to revolve around the sun, while the sun turns about the fixed earth, and it was taken by the great Danish practical astronomer Tycho Brahe towards the end of the sixteenth century. Mathematically and observationally the Tychonic system and the Copernican are indistinguishable, in fact the Tychonic is an exact representation (in Copernican terms) of what the observer actually sees with his instruments. Whatever advantages the Copernican pattern of motion had over the Ptolemaic could equally well be claimed for Tycho's defence of geocentricity. History has not been kind to Tycho's hypothesis, and it has seen (with Galileo) the great cosmological debate as being conducted between Ptolemy and Copernicus. It is a misleading view. Tycho had enormous authority, and his system was quoted as the third possible hypothesis until late in the seventeenth century. It reconciled the combatants without compromising any essential issue; it took all that was scientifically sound from Copernicus, and for the rest clung to common sense. If we would understand the power of the geocentric notion over men's minds we must give it its best defence, not its worst, and this Galileo naturally did not do. To dismiss Tycho Brahe's contribution as a rather pointless and unnecessary obstacle to the advance of truth, or as a conservative aberration on the part of the creator of modern positional astronomy, is to misrepresent entirely the scientific situation and the scientific mind of the sixteenth century. Its importance is that Tycho pointed out, with absolute justice, that there was no particle of evidence to be derived from astronomy which could decide whether the earth moved or not, whereas there was much good evidence of another kind to lead one to conclude that the

[1] Heraclides of Pontus taught this system, with the addition of the diurnal rotation of the earth; it is mentioned approvingly by Copernicus.

earth does not move. Therefore, while the ingenuity of the Copernican geometry of the heavens is to be admired, and was in fact admired by many who opposed his chief premiss, it must be recognized that for the sixteenth century this was irrelevant to the main issue, which turned upon different considerations concerning the earth's mobility.

Taken by themselves, and omitting the mathematical improvements on Ptolemy which are quite neutral in their effect, Copernicus' arguments in favour of the moving earth were scarcely compelling in their contemporary scientific setting. Nor were they very novel. He answered the traditional objections much as Oresme did, developing the thesis that the globe is naturally fitted for circular motion. The earth is absolutely round; it is the property of a sphere to revolve in a circle, expressing its form in its motion; *ergo* the earth revolves. By this reasoning the movement of the earth is natural, not violent: why then should it fly to pieces, as Ptolemy supposed? It would rather be much more likely that the almost infinitely distant sphere of the fixed stars should disrupt under its own velocity, if it were forced to turn round once in twenty-four hours, than that the earth should do so. Air and water move with the earth, so terrestrial phenomena are unaffected. The absence of measurable stellar parallax is explained by the immense distance of the stars, and it is argued that it is easier to believe that the earth moves, than that the whole heaven moves about it. If, as Aristotle seemed to maintain, the condition of rest and permanence is more noble and divine than that of change and instability, Copernicus replies that rest should in that case be attributed to the whole cosmos, and not to the earth alone. In discussing the effect of terrestrial motion upon the doctrine of gravity and levity, he enunciates an idea of surpassing fertility:

> Gravity is nothing but a certain natural appetite conferred upon the parts by the divine providence of the maker of all things, so that they should come together in a unity and as a whole in adopting a spherical form. It is credible that this property exists even in the Sun, Moon, and other planets, ensuring the unchanging sphericity which we see, and nonetheless they perform their revolutions in many ways.

Once more Copernicus has emphasized the fact that the earth is of the same kind as the heavenly bodies; and if it is, as he thinks, a planet itself, then it must have a progressive circulatory motion as well as axial rotation. Finally Copernicus has this glowing passage upon the sun:

> In the very centre of all the Sun resides. For who would place this lamp in another or better place within this most beautiful temple, than where it can illuminate the whole at once? Even so, not inaptly, some have called it the light, mind, or the ruler of the universe. Thus indeed, as though seated on a throne, the Sun governs the circumgyrating family of planets.

If an argument has to be upheld by appeal to principles other than those usually accepted, it has a clear weakness. Copernicus's arguments in favour of the moving earth suffered from it. His aim was to prove that the movement of the earth did not conflict with the principles of physics, which of course meant Aristotelean physics. But in so doing he interprets these principles in a sense different from that of Aristotle, and that understood by most of his own contemporaries. He is forced to allege that gravity, a tendency to cohere, is a universal attribute of spherical bodies. He questions whether rest is inevitable to the elemental earth, and motion to the weightless heaven. Contemporaries can hardly be blamed if it seemed to them that physics had been distorted to fit a newly imagined astronomical theory. Copernicus was not a natural philosopher but an applied mathematician, and in historical perspective it would have been no weakness in him if he had failed to subscribe to contemporary physical notions. In the discussion of cosmological systems two possibilities were open: either a heliostatic system could be adopted, in which case Aristotelean physics was palpably false, and it would be necessary to replace it by a new intellectual framework: or alternatively the traditional physics might be held to be true, in which case a geostatic cosmology was enforced. To attempt, as Copernicus did, to reconcile traditional physics and a heliostatic cosmology was to choose a course open to fatal criticisms. It was easy enough for the opponents of the Copernican hypothesis to show that the reconciliation could not be effected, and that its author's justification of it in terms of contemporary doctrine was wholly spurious. The new astronomy demanded a new physics; and this was ultimately to prove of great advantage to science. But a new physics was not Copernicus' creation, and therefore his appeal to natural philosophy for tolerance of terrestrial motion was an appeal to the unknown; his arguments could carry no conviction because they were not derived from the physics in which men believed, but to an unsubstantiated adaptation of it. In *De Caelo et Mundo* Aristotle had welded a unity of explanation and, so long as its fundamental concepts of motion remained unchanged, this unity was not to be broken.

Ultimately the decision in controversies upon cosmological matters which early astronomy was incompetent to decide unaided turned upon physical considerations, and here Copernicus, with his thoughts still modelled on Aristotle, barely hinted at a new approach. Two things make it difficult to view his work dispassionately. In the sheer majesty of its mathematical achievement *De Revolutionibus* is traditional, but it is a grandly conceived and meticulously executed demonstration of the comprehensive powers of a new hypothesis. To recalculate every motion and every anomaly from the crude observations in accordance with an entirely original pattern was a task never previously attempted. Secondly, it is impossible to escape the compelling power of Copernicus' intuition. Like many other original thinkers, he uttered the truth for the wrong reasons. His work did not form the basis of modern positional astronomy, and within a hundred years the doctrine of the spheres no longer played a part in serious science; and yet his major premiss was essential to the development of both terrestrial and celestial mechanics. His generalship was medieval, but the fruition of his victory lay in the future. Lesser men might debate the logic of solar and terrestrial motions while an imaginative mind could fasten upon the harmony, the irresistible neatness and dexterity of the Copernican pattern. Galileo relates that he first relented towards Copernicus when he found that the Copernicans were usually well informed in their science, in contrast to their opponents who knew only stock arguments. He read the book and was converted. Men of power and vision could learn that the new system, though incapable of rigorous proof in detail, contained a transforming conception. The constitution of the fertile line of advance at any particular moment is not always clear in scientific investigation; Galileo and Kepler found it in the Copernican hypothesis. In their work Copernicus' intuition that the earth is a planet—it can hardly be called a reasoned judgement—was justified.

Even a brief survey of the total scientific activity of the sixteenth century would require a volume to itself. Here it must be characterized in a few sentences. The scientific renaissance caused no sudden break in the course of academic studies, nor did it suddenly enable a "scientific method" of investigation to prove its value. Most of the problems that were discussed belong clearly in a medieval context, and with a few exceptions the procedure and style of argument conformed to familiar patterns. Experimental science was not born in the sixteenth century. On the other hand, the study of pure mathematics flourished exceedingly, and the use of mathematical methods in science was successfully extended.[1] Arithmetics were published in the vernacular languages and printing also helped to spread and standardize mathematical symbols. The Greek geometers were edited and their work thoroughly assimilated. Great advances were made in the formulation and solution of algebraic equations, and in trigonometry, which in its modern form was wholly unknown to the middle ages. The calculations involved in astronomy and in practical arts such as navigation, cartography, mining, surveying, and "shooting with great ordnance," became easily practicable, and books instructing in these various forms of applied geometry appeared in considerable numbers. When these arts were mathematized, the practitioner who had given up rule-of-thumb methods required instruments for the measurement of angles and distances, and the rise of an instrument-making craft of importance can be traced somewhat earlier than the middle of the century. In many places it was closely allied with the domestic clock- and watch-manufacture which began at about the same time.

Opinions concerning the pseudo-sciences, astrology and alchemy, show no remarkable change during the sixteenth century. As in earlier times, there were disputes over the merits and sanction of judicial astrology, but as the general sentiment was strongly favourable there was no sign yet that the dependence of astronomy on its mother-science was almost concluded. There were no sixteenth-century alchemists whose authority stood so high as that of the medieval Latin writer Geber, or the early seventeenth-century adept who called himself Basil Valentine,[2] but the mystical

---

[1] D. E. Smith. *History of Mathematics* (Boston, 1923), vol. 1, pp. 292–350.

[2] The former of these was traditionally an Arabic author, but it now seems that the works ascribed to him were not composed by the real Jābir (an eighth-century physician), though they were put together in Latin from Arabist sources, probably in the late thirteenth century. To the latter also a false antiquity was traditionally credited.

attitude to chemical operations was powerfully reinforced by the
personality of Paracelsus, though he was not an alchemist in the
conventional sense. Under his influence, strengthened by empirical discoveries such as that of the specific action of mercury
against the "new" disease syphilis (first reported *c.* 1492–1500),
inorganic chemical remedies were gradually introduced into
medical practice, against strong opposition from the Galenists.[1]
The use of chemical compounds in medicine stimulated a more
rational interest in their preparation and properties than that of
the alchemists, but even more important, perhaps, in promoting
a purely empirical attitude to material transformations among
educated men was a new kind of book describing industrial operations in a practical manner. Works on smelting and assaying were
circulating early in the century; the early masterpieces of technological description, Biringuccio's *Pirotechnia* and Agricola's *De
re Metallica*, appeared in 1540 and 1556 respectively. Mining and
mineralogy, smelting and casting, the extraction of saltpetre and
the manufacture of gunpowder, the making of glass and mineral
acids, the purification of mercury and the precious metals, were
here treated in detail, systematically, and from a thorough knowledge of actual practice. Somewhat later similar works on machinery for lifting, pumping, sawing, textile manufacture etc., in a
slightly less realistic vein, likewise did much to place engineering
on a sound basis of description.

In medicine, it is probable that the education of physicians and
surgeons improved considerably, owing to the new accessibility
of the Greek authorities, and to the new anatomical tradition
founded by Vesalius and his contemporaries. There were many
serious problems: war, which has always stimulated the progress
of surgery, presented the new horror of gunshot wounds, and the
rapid growth of towns favoured the spread of disease. Public
health was less a matter of public concern in the sixteenth-century
city and state than it had been in the medieval. While, broadly
speaking, there was no great revolution in the theory and practice
of medicine, there were many advances in detail. Jesuit's bark
introduced from Peru contained quinine as a specific against the
recurring fever of malaria. The pharmacopœia was standardized—

---

[1] The origin of syphilis has been exhaustively investigated. Some recent
opinion seems to be against the traditional (and sixteenth century) view that it
was introduced into Europe from the Americas.

first in Italy and Germany, not in England until the London Pharmacopœia was issued in 1618. Most dramatic of all perhaps was the influence of the French surgeon, Ambroise Paré, who led the way in abandoning the use of the fiery cautery, previously applied to *all* gunshot wounds (which were believed to be envenomed) as well as in cases of amputation. Although Paré still modelled his practice broadly on that of the *Chirurgia* of his great fourteenth-century countryman, Guy de Chauliac, (which was itself often reprinted in the sixteenth century) his writings did much to raise the prestige and skill of the surgeon. Guy, who himself leaned heavily upon Galen and Avicenna, had taught that the 'surgeon who is ignorant of anatomy carves the human body as a blind man carves wood,' and Paré reinforced this truth by making free use of Vesalius' *De Fabrica*.

The sixteenth century also saw the current of realism at work in natural history. Of this something more must be said later (Chapter X). The best work was still compilatory, and encyclopædic on the vastest scale. It was devoted entirely to the superficial characteristics and habits of plants and animals, and botany remained an adjunct of medicine rather than a discipline in its own right, but much rubbish was purged from the medieval garner of fact and legend. The humanistic naturalists used a scissors-and-paste technique upon the authentic texts of Aristotle or Pliny, and if the range of their reading and collation was wide and discursive, some of them show for the first time a real eye for genuine observation, and painstaking endeavour to confirm at least the external features of their subjects. The first monographs in natural history have an authentic realism and attention to specific detail which are entirely new.

On the whole, however, the scientific spirit of the century developed naturally from the work and progress of the later middle ages. A hasty reference to the output of the printing-press may act as a guide to the nature of contemporary taste and estimations; the most sought after and respected books were still those composed long before the invention of printing.[1] Neither the library, nor the academic training, of the medieval world were suddenly outmoded by a vast efflux of novel aspirations and methods. The mythical "renaissance man" of the early sixteenth century, though his tastes might be more hedonistic than those of his

[1] See Appendix C.

forefathers, though he might be more enamoured of the powers and potentialities of this world and less regardful of the next, was still largely limited in his science to the achievements of the medieval renaissance: his very classicism only attached him more deeply to the same roots of western learning. The cosmos of Shakespeare is the cosmos of Dante, save that the former was a far less philosophical poet: the Fables of Bartholomew, the complex vocabulary of astrology and alchemy, and the doctrine of the four humours still enclosed the framework of science which most men knew. It was not the experimentally minded Dr. William Gilbert, with his glass rods, magnetic needles and other trivia, who was most honoured at the court of Elizabeth, but Dr. John Dee, the mysteries. To the sixteenth century the story of Faust was not quite a fable.

CHAPTER III

THE ATTACK ON TRADITION: MECHANICS

U NTIL the end of the sixteenth century scientific innovations were put forward with deference and almost a sense of humility. A great deal of the best work of this period was entirely non-polemical: to this class belong the first stages of the Vesalian tradition in anatomy, and much purely descriptive writing on natural history, mineralogy and chemistry. The works of Agricola or Rondelet were excellent contributions to science as positive knowledge, but they created no ferment of new ideas. It is true that Paracelsus is supposed to have burnt the books of the masters before his inaugural lecture at Basel in 1527, and that he declaimed against official medicine:

> I will not defend my monarchy with empty talk but with arcana. And I do not take my medicines from the apothecaries. Their shops are but foul kitchens from which comes nothing but foul broths. . . . Every little hair on my neck knows more than you and all your scribes, and my shoebuckles are more learned than your Galen and Avicenna, and my beard has more experience than all your high colleagues.[1]

Paracelsus, whatever his other merits, was a picturesque ranter and it is futile to portray him as a herald of the scientific revolution. The texture of his thought—in which it is difficult to see any precise pattern—was woven upon a mystical conception of nature entirely alien to that of natural science.[2] Medicine was indeed torn by faction: the Arabists and the Humanists, the Paracelsians and the anti-Paracelsians, the cauterizers and the non-cauterizers sharpened their wits in vituperative debate as they contended

---

[1] Paracelsus: *Selected Writings* (edited with an introduction by Jolande Jacobi), (London, 1951), p. 79.
[2] As, for instance, the perennial fallacy that the virtues of herbs are indi cated by their structure. 'Behold the Satyrian root, is it not formed like the male organs? Accordingly magic discovered it and revealed that it can restore a man's virility. . . . And then we have the thistle: do not its leaves prick like needles? Thanks to this sign, the art of magic discovered that there is no better herb against internal pricking . . .' etc. (*ibid.*, p. 186).

73

for supremacy within the profession, without aiding the advancement of knowledge. Natural philosophy and natural history were not similarly divided by quarrels arousing professional passion, though inevitably personal jealousies were not lacking. More typical of the relations between old and new were Copernicus taking leave to speculate afresh on the revolving spheres, Vesalius adapting his text to follow Galen. As yet, content if they could show that new ideas were no worse than old ones, men were far from asserting that the House of Learning was a crazy, rambling warren that needed to be pulled down and reconstructed. With the exception of Paracelsus, no scandalous defiance of authority had been noised abroad; and this was partly because the shape of authority, the policy and content of conservatism, were themselves unsettled at this nascent stage of modern science.

When the famous crisis was reached in 1615–16, it had already been foreshadowed in the tragedy of Giordano Bruno's life which must be mentioned in its proper place. Bruno was no scientist, and his impact, his historical importance, his ultimate influence upon the development of non-scientific attitudes to science, were all the more startling for that reason. It must not be imagined that the ways of the ordinary, common-sensible religious man and of the critical natural scientist were bound to deviate at the first novelty, or that the sixteenth century was afflicted by the same opposition of loyalties and criteria of truth that troubled the nineteenth. On the contrary, every flaw in the conventional account of nature was examined not in the hope that it would profit one philosophy against another, but in the belief that its examination would advance truth, and that in truth all matters of importance were ultimately reconcilable. The view that a man's access to truth might be measured by the nature of his ideas on celestial mathematics was not known to the sixteenth century. Copernicus' narrow vision had embraced no more than the validity of a single hypothesis: it was with the wider philosophical perspective of Bruno, and the wider scientific range of Galileo, that iconoclasm assumed a massive, threatening character. Disputes among mathematicians, astrologers or physicians could be tolerated (these things were not suddenly born at the time of the renaissance) but criticisms shaking the roots of philosophy had to be repelled. It was not simply a question of the liaison between Aristotle and religious doctrine, nor was Catholicism the only opponent of innovation.[1] The new scientific philosophy of the early seventeenth century, speaking with a confident voice that demanded credence, met sheer mental inertia and the weight of academic omniscience. Men resent admitting that they have given their lives blindly to the defence of absurdities, and it was a new generation, not similarly committed to tradition, that devoted itself to the prosecution of Galilean science.

Galileo's greatest fame is as an astronomer, yet in intellectual quality and weight his one treatise on mechanics almost outweighs all the rest of his writings. Although his book on cosmology became notorious, and had a more general public influence, it had no comparable effect upon the future development of scientific astronomy, for its polemics were suited only to its own age. The contradiction here is more apparent than genuine. Though formally divided between two branches, Galileo's creative activity in science was a unity, not twofold: it was a unity in laying down revised principles of procedure in science, and again in its specific exemplification of these principles, since Galileo saw that the science of motion and the just appraisal of the results of observational astronomy were the twin keys to an understanding of the universe. This is not to say that his comparatively minor (though in themselves major) researches, into thermometry, the properties of pendulums, the grinding of lenses, the strength of materials, the theory of the tides or the measurement of longitude at sea were not conducted with the same zest for knowledge, or that they were either irrelevant or auxiliary to his main achievement. Each one of his minor discoveries would have been notable in a lesser man. Each has its important place close to the origins of modern physical science. But Galileo's physical experimentation, like his turning of the telescope to the heavens, was original and fecund only in the sense that it was initiatory. Other men took up his experiments, and even drew more valuable conclusions from them than Galileo had done, by following and expanding his methods; and while the initiation of a new kind of scientific activity is itself an achievement of the first order, it is not of the same dignity as the induction of fundamental principles which henceforth

[1] In the first half of the seventeenth century the "new philosophy" advanced most rapidly in two Catholic countries, Italy and France. Kepler, a Protestant, was tolerated in Catholic Austria, but Descartes, a Catholic, sought the intellectual freedom of Holland. Both Galilean and Cartesian scientific ideas were very rapidly adopted by men of orthodox religion in France.

dominate a whole field of science. Besides fulfilling this function in dynamics, Galileo demonstrated the connection between its principles, when properly understood, and the disputed points of cosmology.

Unlike his predecessors Galileo consciously assumed the attitude of a publicist and a partisan. Writing more often in his native language than in Latin (for Galileo was one of those who led the way in abandoning for science the official language of philosophy), he shaped his arguments to a broad audience. His dialogues were lively, his irony biting, and he did not scruple to make a merely debating point. Zealously he magnified the weaknesses of conventional science in order to turn it to ridicule. Almost alone among the ancients the experimenter and mathematician Archimedes was singled out for Galileo's commendation; Aristotle he treated almost as an ignoramus, as though the subtlety and intricacy of that intellect, preventing vision of the simple truth, had composed fantastic webs of improbability and artificiality. He had no doubt at all that the modern way was infinitely superior to the ancient, and hardly mentioned any contention of the Peripatetics that he was not prepared to deny. This habit of opposing conventional ideas was not, apparently, a product of Galileo's maturity nor of his great discoveries. Rather the critical attitude which stands out even in his juvenilia was the source of his original ideas. Born in 1564, by 1589 Galileo was already a teaching member of the University of Pisa, where he attracted the attention of the Grand Duke of Tuscany. Two of the most famous stories belong to these Pisan years: here he observed the equality in time of the swinging cathedral lamps, and carried out (as his first biographer relates) the famous experiment of dropping weights from the Leaning Tower.[1] Unlike most academics of the time, Galileo remained a layman. The resentment aroused among his colleagues by his criticisms of Aristotle prompted his removal in 1592 to Padua, within the anti-clerical Venetian state, where there was a long tradition of freedom in scientific thought and where anti-scholastic opinions, if not exactly encouraged, were at least tolerated. At Padua Galileo studied mechanics, constructed

[1] The experiment from the Leaning Tower, told in glorification of his master by Viviani, has been rejected by many scholars as lacking support in contemporary documents. At any rate Galileo was not the first to subject Aristotle's dynamics to this particular test.

his first telescope, and began his long battle on behalf of the Copernican system. After his growing fame had brought about his recall to Florence, under the special patronage of the Grand Duke and with an appointment at Pisa (1610), Galileo was particularly warned to avoid utterances having a theological implication. From 1609 to 1633 he was immersed in his astronomical observations and the polemics they provoked; then after the publication of the *Dialogue on the Two Chief Systems of the World* (1632) there followed his trial and recantation: yet having been condemned to silence, Galileo published his greatest work at Leiden in 1638. The significance of his summons to Rome is easily exaggerated. Galileo did not compel the Roman Church to condemn Copernican astronomy: this intervention in scientific discussion had already taken place decisively with the decree of 1616, and was a natural consequence of the condemnation of Bruno. The only interest in the trial concerns the nature of the judicial process, which is not impeccably transparent; but though Galileo broke no private pledge, he had certainly contravened the spirit of a general order.[1] Good Catholics were placed in a false position for two centuries by the decree and Galileo's trial under it, without these preventing excellent work in practical astronomy in Italy, or the uninhibited development of other new studies there. Even the Copernican Alfonso Borelli was able to make his suggestive contribution to celestial mechanics (1660) by means of an ingenious quibble.[2]

A reading of the *Mathematical Discourses and Demonstrations concerning Two New Sciences* (1638) does not, however, give a true picture of the revolution in dynamics as Galileo effected it, any more than the earlier series of *Dialogues* can now be considered as a balanced statement of the respective merits of the two cosmologies. In his book Galileo the polemicist is in full cry after the absurdities of Aristotelean physics, but in fact it is very

---

[1] However obvious this may now seem, it was not so to the ecclesiastical censors who affixed the *imprimatur* to the *Dialogue* in 1632.

[2] The decree distinguished between the annual revolution of the earth, 'utterly heretical because contrary to Holy Scripture,' and the diurnal rotation, 'philosophically foolish.' One of its more curious results, as I learn from Dr. Joseph Needham, was that Chinese astronomers, being instructed by Jesuit missionaries, remained ignorant of the Copernican theory until the late eighteenth century. Borelli observed its letter by overtly limiting to Jupiter and its satellites a discussion of planetary motions which he obviously intended to apply to the earth and sun.

likely that Galileo was never an Aristotelean physicist in the true sense, and that the original account of motion was never urged upon him as a true and satisfactory explanation. Much patient scholarship has been devoted to the origins and development of Galileo's mechanics; at one stage it seemed as though its godfather was Leonardo da Vinci, until it was found that Leonardo was only re-echoing a current of medieval thought. At first these medieval discussions were regarded as no more than imperfect gropings at a truth which Galileo apprehended perfectly. Today the theory of impetus appears as a theory of motion in its own right, not an uncertain anticipation of Galileo, but a coherent doctrine which provided Galileo with a firm starting-point. The theory of impetus could not develop the modern conceptions of inertia and acceleration by a smooth process of transition and expansion, but it did provide a more convenient, and more adaptable, starting point than the ideas expressed in Aristotle's *Physics*. Galileo's achievements are more properly measured by comparing his own science of dynamics not with the absurdities he discovered in Aristotle, but with the theory of impetus which was already well formed.

Since it had been handled by Oresme in the fourteenth century this theory had made little progress up to the time of Galileo. Remaining a somewhat specialist complexity, it had not sunk to the popular level of exposition, but it was taught by respectable mathematicians and philosophers, including Cardano (1501–76), Tartaglia, Benedetti and Bonamico who was Galileo's own master. Tartaglia (1500–57), moreover, succeeded in making a useful application of the impetus theory to ballistics, and was the first writer to aim at computing the ranges of cannon by means of tables derived from a dynamical theory, in which task Galileo later was to believe himself successful. After Tartaglia, if not before, it was at least clear that in this particular respect a dynamical theory ought to be quantitative, that is it should be capable of making exact numerical predictions. Many writers on gunnery, with varying degrees of imagination, continued to search for this desideratum within the framework of the impetus theory until well after Galileo had offered a far better one. This essential sterility of the impetus theory in the sixteenth century is an important point. Able men failed to derive from it a mathematical description of the phenomena of motion, yet failed also to develop concepts which could take its place. It is not, therefore, the case either that Galileo's ambition to mathematize dynamics was something unusual in the contemporary scientific *milieu*, or that, once this ambition had been framed, it was easy to mould the necessary modern dynamical concepts out of their crude impetus forebears. Modern dynamics did not spring from a modified version of the impetus theory: Galileo was compelled to return once more to fundamental ideas of motion.

Towards a new Kinematics

The theory of impetus, the living tradition of dynamics in the sixteenth century, utilized by Galileo himself in his early treatise *On Motion* (1592), had the basic function of *explaining* certain phenomena. It had not changed the language in which they were *described*. While it gave reasons for the continuation of motion by a body after the moving agent was withdrawn, and in particular accounted for the properties of motion shown by falling bodies and projectiles, it had not produced more accurate or fertile definitions of force, velocity, acceleration. In the technical terms of philosophy it dealt with the "accidents" of motion—the decelerating ascent of a projectile was caused by an "accidental levity," its accelerating descent by an "accidental gravity," demonstrated by its having effects on impact equal to those caused by a body at rest of much greater weight. Impetus as a causal factor, responsible for the accidents of free motion, being thus associated with the categories of Aristotelean dynamics, no mechanician before Galileo attempted to deny the validity of the substantial distinction between the two kinds of motion, natural and violent. Hence great difficulties arose when the attempt was made to describe the path of a projectile, for example, since the two parts of the problem—the ascent and the descent—were qualitatively distinct and subject to different considerations. Further, progress towards a kinematics through this type of problem was obstructed by the limitation of the impetus theory to strictly rectilinear motion. Though both Leonardo and Tartaglia in the sixteenth century represented the trajectory of a projectile as a continuous curve, this was no more than a pictorial device, since their common theory of motion allowed no compounding of natural and violent. In theory the categories of motion were exclusive, and rectilinear-natural motion could only occur after the rectilinear-violent was completed. Again, if the relations between the impetus theory and the later concept of inertia are examined, it is clear that while in isolated statements exponents of this theory seem to anticipate the rigorous definition of inertia, in the full physical context their interpretation of the phenomena was different. For it was not supposed that a body, having acquired an impetus to motion, would continue to move at a uniform velocity. The resistance of the medium ensured that it would slow down and come to rest; and motion in a vacuous space was inconceivable when the universe was regarded as a plenum. Impetus was like heat, an evanescent accident of matter, of its own nature wasting away. As before, thought on the application of these ideas to the special problem of projectiles was misled by utter deficiency of observation and description. The impetus mechanicians universally believed that the velocity of a projectile increased for a space after it had parted company with the propellant, so that in some strange way its impetus was actually increased during a part of its free motion. Here also the theory, instead of leading to universally valid concepts of inertia and acceleration, required the formation of a special case.

If the scientific tradition inherited by Galileo did not offer any simple, universal axioms in dynamics, and in its preoccupation with causation had neglected accuracy in description, it did provide a suitable mathematical analysis for dealing with changing quantities, such as the velocity of an accelerated body. It is a matter of historical record that the geometrization of motion, the establishment of a formula connecting velocity, time and distance appropriate to the motion of an accelerated body such as a freely falling mass, was more easily accomplished than the development of a framework of kinematics within which such a formula would be not merely possible but inevitable; the formulation of a true dynamics involved a still greater effort which was hardly completed before the eighteenth century. The effect of supposing a uniform (linear) variation in the intensity of a quality, such as heat, had been studied by a number of philosophers in the fourteenth century. They found that the second variable (e.g. heat) could be related either arithmetically or geometrically to the first variable (e.g. time). Oresme, for example, had demonstrated that if a quality be supposed to vary uniformly from the magnitude represented by the line AB to zero at C (Fig. 5), the first variable being represented by the line AC, this varying quality is equal in effect to a uniform quality of the magnitude ½AB, since the area of the triangle ABC is equal to ½AB.AC. He expressly stated that if the varying quality were velocity, the equivalent uniform velocity would be that attained at the mid-point in time (i.e., if AC represents time, of magnitude ½AB) and deduced that any uniformly varying quality (or velocity) could be

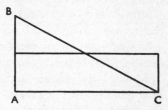

FIG. 5. Geometrical analysis of uniformly varying qualities.

equated to a uniform quality (or velocity). Somewhat earlier another master of the University of Paris, Albert of Saxony, had tried to prove that the motion of a freely falling body could be fitted into this theoretical pattern; he had taught that (i) such a body is uniformly accelerated, and (ii) that in uniform acceleration the instantaneous velocity is directly proportional to a first variable which is either the time elapsed *or* the distance traversed.

The first of these propositions was supported, from Albert's time to that of Galileo, by reference to the theory of impetus. It was asserted that the continually increasing speed of a falling body was caused by the addition, to the impetus already acquired by it at any point in its descent, of the constant endeavour (*conatus*) towards the attainment of its natural place which it always possessed. If this *conatus* were suddenly abolished while the body was actually falling, its impetus would nevertheless bring it down to the ground at a nearly constant speed, but as this *conatus* acted as a cause of motion on a body already in motion, it made it move ever faster. By an intuitive process, rather than by strict reasoning, it was argued that the increase of velocity is linear;[1] an argument not approved by all philosophers. It was not accepted by Galileo in 1592, for it might be that the rate of increase would slacken as the falling body moved more quickly.

The second proposition, or rather pair of alternative propositions, involved even more obvious difficulties. In the first place,

[1] Intuitive, because Albert's proposition really implies the true law of inertia, which he did not enunciate, and also because it is untrue when the descent occurs in a resisting medium, such as the air. Nor was it logically certain that the *conatus* would act in the same way on a body in motion, as on one at rest.

the notion of instantaneous velocity was very imperfectly grasped —no term existed to describe it—for while the calculus of varying qualities could equate these, over a given range, with uniform qualities, it did not deal with the instantaneous magnitude of the changing quantity. Moreover even problems on simple linear relations involve integration, and it was by no means easy to decide what the integral (the area ABC in Oresme's demonstration) actually represented. To Oresme it was the total quantity of a quality over a given range, expressed as a simple product.[1] But what was the "total quantity of a velocity"? Oresme identified it with distance—by intuition, for he could not have proved that an area gave a measure of distance. This was equivalent to saying that if the speed of a body increases uniformly from rest to v, the distance passed over is the same as that traversed in the same time at the constant speed $\frac{1}{2}v$. That result was not forgotten. But no one knew how to interpret it. No one before Galileo succeeded in deducing from it the general relations between time, velocity and distance. The difficulty was not a mathematical one—Galileo's geometry was identical with Oresme's—it lay rather in attaching the correct conceptual significance to mathematical quantities.[2] No one, moreover, before the sixteenth century suggested that the discussion of uniform acceleration could be applied to the free fall of physical bodies. Oresme's result, a special case of the "Merton Rule" defining uniform variation of things, belonged to pure theory, not to the world of real things.

Finally, the philosophers who followed Albert of Saxony had to decide between the alternatives in his second proposition. Was the speed to be taken to vary with the time elapsed, or the distance traversed? This was the principal obstacle to the progress of kinematics in the sixteenth century, though Oresme had defined velocity as a variable of time. And it was so, not because a choice could not be made, but because it was not even clear that a choice was necessary. Leonardo typified the confusion of thought in his

---

[1] This assumes, for example, that a hot iron having a temperature t_1 falling to t_2 over time T melts as much ice in that time as an identical iron of constant temperature $\frac{1}{2}(t_1 + t_2)$. It was only much later that Black cleared up the confusion between temperature and quantity of heat, analogous to the confusion in mechanics here discussed.

[2] Galileo's geometry was equivalent to the derivation of the integral ($\frac{1}{2}at^2$) from the differential equation $\frac{ds}{dt} = at$. He realized, correctly, that this represented the distance traversed.

completely clear, and completely incompatible, statements in different passages of his note-books that the velocity of a falling body at any instant is proportional to the distance fallen, and to the time elapsed. For the writers on kinematics who preceded Galileo, what he was to call the supreme affinity between time and motion was inevitably obscured. They could not measure small time-intervals. They most naturally stated the problem whose solution they sought in a form which made time a function of distance, that is: "If a stone falls x feet in one second, how long will it take to fall 100 feet?" Even in making experiments on dropping bodies from different heights, it was more natural to think of the greater velocity as a function of the greater elevation. It seemed that the lifting of the stone to a greater height was the direct cause of its greater velocity on reaching the ground. This confusion of hypotheses was indeed to trouble Galileo himself, and some of his brilliant contemporaries. Only one philosopher of the sixteenth century steered his way unambiguously through it. This was a Spanish theologian, Dominico Soto, whose commentary on Aristotle's *Physics*, fully in the tradition of Oresme and Albert of Saxony, was published in 1555. After defining "uniform difform" motion (i.e. uniformly accelerated motion)—not velocity—as proportional to time, he declared that this kind of motion was proper to freely falling bodies and to projectiles.[1] He did not, however, prove these propositions nor did he suggest a formula relating time, velocity and distance.[2] Though he gave the definition and application of acceleration correctly, he was still far from a true kinematics, and his propositions were still dubiously derived from the concept of impetus. If they were significant, but limited, anticipations of Galileo's theories on motion, they were also nothing more than one version of Albert of Saxony's propositions. And there was no further advance. It was left to Galileo to perform two functions of highest importance: to formulate new, clear concepts of motion, and to derive from them the complete elements of kinematics, using the fragments provided by those who had shaped his intellectual inheritance.

[1] Marshall Clagett: *The Science of Mechanics in the Middle Ages* (Madison, 1959), pp. 555–6.

[2] i.e., though Soto loosely gives the equivalent of $\frac{ds}{dt} = at$, he did not attempt to integrate this equation. As will be seen, this was a task that defeated Descartes, and (at first) Galileo.

The Law of Acceleration

During more than two centuries before Galileo's birth the application of the calculus of varying qualities to the concept of impetus promised the discovery of the law of acceleration—or rather of two such laws.[1] Dominico Soto had decided correctly that acceleration was a rate of change of motion (velocity) in time. He had even struck a blow at the dichotomy of natural and violent motion by deducing that in the violent motion of ascent the acceleration is negative. Although Galileo's great achievement was to be in the mathematical analysis of motion, it was not his first preoccupation. Instead, in his treatise *On Motion* of 1592, he examined the physical nature of acceleration, and criticized the opinions commonly derived from the concept of impetus. In the physical sense, he maintained, acceleration was a mere illusion. His argument at this stage denied the proposition which became one of the axioms of modern dynamics—a constant force gives a body a constant acceleration—for he argued that since the cause of the natural motion of a body is its weight, each must have a natural velocity proportional to its weight. He explained the *appearance* of changing speed in this way: suppose a heavy mass projected upwards, the impetus conferred being greater than the natural *conatus* to descend. It will rise until the tendency to fall and the impetus are of equal strength. At this point the body still possesses a certain degree of impetus and consequently as it begins to fall back it will increase its speed until all the impetus has disappeared; after this its motion will have the constant velocity proper to its weight. In the case of a body falling from rest, he declared, the impetus acquired by its displacement from the centre was preserved and the same explanation held.[2]

Certainly this novel modification of impetus mechanics—one enforced by allowing notions of the causal functions of impetus to prevail over its usefulness in kinematics—introduced no remarkable clarity. It may be that Galileo's somewhat unfruitful speculations along these lines had the effect of turning his thoughts in

[1] The law relating instantaneous velocity to distance traversed $\left(\frac{ds}{dt} = as\right)$ makes s an exponential function. It was therefore beyond the mathematical competence of Galileo's age.

[2] This theory is discussed in detail by A. Koyré: *Études Galiléennes* (Paris, 1939), pp. 59–64.

a new direction. The followers of Oresme and Albert had long abandoned the idea that the speed of a falling body is proportional to its natural weight, though they did not conclude, as did Galileo later, that in a vacuum all bodies would fall at the same velocity. Oresme, for example, opposing Aristotle, held that the uniform velocity of a body is not proportional directly to the "puissance" applied (e.g. its weight), but to the ratio between the "puissance" and the resistance to be overcome. He recognized also that the natural weight of a body is unaffected by its motion; the only change is in apparent or effective weight, and it is to this (at any instant) that the velocity is proportional.[1] Moreover, Oresme accepted the fact that acceleration may continue indefinitely, and thus differed in principle from Galileo who assumed in 1592 that the velocity of a falling body tends towards a uniform value.[2] But it is true that the identification of impetus with accidental gravity involved conceptual inconveniences which Galileo had correctly appreciated, though he set them into an outmoded Aristotelean pattern.

The first evidence of a major success already shows that Galileo had turned from a physical-causal to a kinematical approach. In 1604 he wrote, in a famous letter to Paolo Sarpi, that on the basis of an axiom sufficiently obvious and natural he had proved that the spaces passed over by a falling body are as the squares of the times. The axiom adopted was that the instantaneous velocity is proportional to the distance traversed; an axiom already rejected by Dominico Soto. The demonstration of this impossible conclusion (which happens to survive)[3] made use of the medieval calculus of varying qualities. Galileo decided that the integral (area ABC in Fig. 5, Oresme's "quantity of velocity") represented the space traversed; but the process by which he derived this integral from his axiomatic function was entirely false. His reasoning assumed, in fact, that velocity was plotted against time, not against distance.[4] Thus the familiar theorem, $s = \frac{1}{2}at^2$, was first derived by Galileo in a process mathematically correct but vitiated by a conceptual error. Perhaps the theorem was first tested about this time by the experiment of rolling a brass ball

[1] Oresme, *Medieval Studies*, vol. III, pp. 230–1; vol. V, pp. 179–80.
[2] In the *Discourses* Galileo explains that the resistance of the air tends to limit the velocity of a falling body to a maximum value.
[3] Cf. *Opere* (Edizione Nazionale), vol. VIII, p. 373.
[4] Duhem: *Études sur Léonard de Vinci*, vol. III, pp. 565–6; Koyré, *op. cit.*, p. 98.

down an inclined plane described in the *Discourses*.[1] It is certain, at least, that from this time Galileo was convinced of its accuracy as a mathematical description of the motion of freely falling bodies.

In Galileo's scientific method experiments were designed to give ocular confirmation of reasoning; therefore he could not be satisfied with his newly discovered theorem as a merely empirical fact. At this point he was most concerned to prove that his axiom —the false law of acceleration—followed logically from an analysis of the nature of motion, and to establish it as a reasoned premiss.[2] Could it be justified in philosophy? In tackling this question he must have been aided by the progress of his thought since 1592, of which unfortunately little is known. Probably he had already gone far in renouncing that concern for the causation of phenomena shown in his early writing after realizing the confusion into which it plunged dynamics. In the later phases of his thought impetus was no longer appealed to as a causal factor but became a mathematical quantity—the product of weight and velocity. As Galileo's problem became more purely kinematical, he accepted the facts of gravitation and the fall of heavy bodies without trying to impose an explanation, although, from his favourable references to William Gilbert after 1600, it may be presumed that he approved the notion of gravity as a quasi-magnetic attraction.[3] Already in 1604 he saw acceleration as a fact to be defined, not explained; but it was not until afterwards that he was satisfied that the constant effect produced by the constant cause, a force, is not a velocity but a rate of change of velocity and so resolved the paradox he had treated very differently in 1592.

With the abandonment of the impetus causation of acceleration is involved the transformation of this vague conception into the law of inertia. Mach insisted, logically, that this law is the special case of the law of acceleration where the force applied is nil, and therefore no separate statement of it is strictly necessary.[4] His argument is just, but not historical. Historically the special case was more readily understood than the general law. It was derived

---

[1] *Dialogues concerning Two New Sciences*, trans. by H. Crew and A. de Salvio (New York, 1914), pp. 178–9.

[2] The empiricist would have derived the law of acceleration mathematically from the law of distances verified by experiment; but this was not Galileo's method.

[3] *Dialogue on the Great World Systems* (ed. G. de Santillana), pp. 409 ff.

[4] Ernst Mach: *Science of Mechanics* (Chicago, 1907), pp. 142–3.

from the consideration of motion in a resisting medium: if the impetus of a body is expended in overcoming the resistance, then in the absence of resistance its impetus and velocity will remain constant for ever. It is in this way that Galileo explains the law of inertia in his *Discourses*, and the non-resisted motion of the celestial spheres had long been described as a peculiar instance of undiminished impetus, or inertial rotation.

In Aristotelean physics, rationalizing experience of the real world, rest was normal and motion a state requiring special explanation; in Galileo's physics where space was a vacuous geometrical framework—a notion partially anticipated by Benedetti—only changes in the state of motion or rest required explanation. Here the concept of impetus as the "impressed force" overcoming resistance to motion becomes redundant, or worse, erroneous. Its abandonment was already foreshadowed in the treatise *On Motion*, where Galileo discussed the motion of a sphere rolling on an infinite horizontal plane, a motion which was neither violent nor natural and could therefore be produced by an infinitely small force; or where he showed that when the resistance was nul, as in a vacuum, it did not follow that movement would be infinitely swift. In Galileo's space there could be no more "force" impressed upon a moving body than upon an unmoved one; what bodies demonstrated was a sluggishness (inertia) to respond to changes of their uniform state of motion or rest. Galileo never gave a quite clear and accurate definition of inertia. The *Dialogue* defended the concept of impetus in the language of the fourteenth century; nevertheless, Galileo certainly understood that only by its changes is motion in the real world differentiated in kind from motion in geometrical space, and consequently that such changes alone required physical explanation.[1]

Once the vague concept of impetus had been supplanted by the complementary concepts of inertia and momentum, however tentative and imperfect, the potentialities of the geometric method in kinematics were vastly extended. With the further addition of an arbitrary law of acceleration the fundamental theoretical structure of kinematics was almost complete, and at once the distinction between "natural" and "violent" motion appeared as an unnecessary hindrance. The terms were still used by Galileo, but purely

[1] Koyré, *op. cit.*, pp. 71, 93.

for classification, without any causal or dynamical significance. Natural motion had become, by definition, accelerated motion in accordance with the normal law: the violent, a motion retarded in accordance with the same law of opposite sign. Gravitation had become a force like any other, which might be greater or less than other forces opposed to it, and the effect of a force was to accelerate or retard a body whose only physical properties were weight (or as Newton said, more properly, mass) and inertia. Privileged directions with respect to the centre of the universe, intrinsic lightness and heaviness, causal distinctions between enforced and unenforced movements, all disappeared once the aptitude of the law of inertia in perfectly geometrical vacuous space, where all planes are infinite, all perpendiculars are parallel, and only simple forces operate, was realized.

It need not be supposed that Galileo had, by 1604, reached the stage where the distinction between the essence and the accidents of motion for which he had sought so long, and which was essential for the elucidation of the true laws of kinematics, was perfectly clear in his thought. Rather, his erroneous definition of acceleration at that time, together with the imperfect conception of inertia which he was never to amend, prove that the steps in the process of reasoning he had followed still lacked clarity and rigour. Having abandoned the traditional theory of impetus Galileo's intuition had brought the laws of kinematics almost within his grasp, at a time when his analysis of the nature of motion was still far from impeccable. The more adequate reasoning of the *Discourses* was to be developed over the next thirty years, yet even so the ultimate confutation of the false law of acceleration adopted in 1604 rests upon a paralogism.[1] Powerful and original as Galileo's thinking already was, and close as it came to the essential idea of motion, his kinematics of Euclidean space was still a no more complete theory than impetus dynamics. The false law of acceleration stated by impetus mechanicians still seemed plausible, and Galileo was still ignorant of the true law, explicitly stated by Dominico Soto. Galileo had not yet appreciated the crucial importance of the distinction between the two possible

[1] *Discourses* (Crew and Salvio, *op. cit.*, p. 161), Galileo here argues that if the velocity of a falling body is always proportional to the space fallen, it will fall 8 feet in the same time as 4 feet, which is impossible. But his reasoning clearly contains the hidden assumption that velocity is proportional to time, contrary to the initial hypothesis.

hypotheses; an ancient train of thought (strengthened by his geometrical proclivities) bound him to spatial relations. Was his statement of the distance fallen as a function of time thus a happy accident which owes nothing, logically, to the progress of his ideas after 1592, since the axiom of 1604 was available to him then? Only in a qualified sense. It is true that the geometric method of Galileo in 1604 was an attempt to integrate instantaneous velocities in Oresme's graphic representation, but Galileo's calculation is purely kinematical. It is not bound in terms of explanation to the impetus theory. The quantitative result could have been obtained in 1592, but it would have belonged to a different pattern. Secondly, the error which enabled Galileo to derive a correct function from a false axiom was not an accidental error. It was probably inevitable that it would be made by anyone at that stage of thought attempting the same calculation.

Actually, the identical error was made in precisely the same fashion independently by Descartes and the Dutch physicist Isaac Beeckman.[1] Their acquaintance began in 1618, when Beeckman was already convinced of the perfect conservation of motion: 'quod semel movetur semper eo modo movetur dum ab extrinseco impediatur.' He also believed (following Gilbert) that gravitation was the result of the earth's attractive force. According to Descartes, Beeckman proposed the following problem: 'A stone falls from A to B in one hour; it is perpetually attracted by the earth with the same force, without losing any of the velocity impressed upon it by the previous attraction. He is of the opinion that in a vacuum that which moves will move always; and asks what space will be traversed in any given time.' Offering certain dynamical axioms, therefore, Beeckman asked Descartes to furnish him with a mathematical function relating time and distance. And Descartes replied with a geometrical construction from which he asserted that if the spaces fallen are as

s, $2s$, $4s$, $8s$, etc., the times of fall are as t, $\dfrac{4t}{3}$, $\left(\dfrac{4}{3}\right)^2 t$, $\left(\dfrac{4}{3}\right)^3 t$, etc.[2]

When his demonstration of this function is examined it is

---

[1] The story of their collaboration is told by Duhem in *Études sur Léonard de Vinci*, vol. III, by G. Milhaud in *Descartes Savant* (Paris, 1920) and by A. Koyré, *op. cit.*, pp. 99–128.

[2] By Galileo's theorem they are of course t, $\sqrt{2}\,t$, $2t$, $2\sqrt{2}\,t$, etc.

apparent that Descartes has done exactly as Galileo did—taken the instantaneous velocity as proportional to the distance fallen, and arrived at a law of acceleration by a process of integration of these instantaneous velocities which is as mistaken as Galileo's in 1604. Beeckman's interpretation of Descartes' geometry is even more curious. For Beeckman wrote out a perfect demonstration that, from his hypotheses of motion, the instantaneous velocity is proportional to the time, and the distance fallen to the square of the time, without ever perceiving that it was different from that given to him by Descartes, in fact he noted this proof as devised by Descartes. Evidently neither Beeckman nor Descartes was able to make a clear distinction between the two laws of acceleration: neither was capable of seeing that the true function relating distance and time can only be deduced from the one hypothesis. Thus the investigations of Beeckman, Descartes and Galileo show the same intellectual difficulty: even when the essence of motion was justly apprehended in physical terms, its geometrical expression defeated their initial efforts. The disentanglement of the velocity-time and the velocity-distance relationship was still hazardous and the sheer mathematical task of integrating changing quantities could not be attempted with any assurance of success.

There is evidence to suggest that Galileo had developed the correct formulation of acceleration by 1609, but the steps he followed are unknown. It may have been that first he discovered the error of his calculation, and so was led to substitute the true axiom for the false: but it would seem more likely that it was in meditating further upon the foundation of the law of acceleration in his essential idea of motion that he realized the 'intimate relationship between time and motion.'[1] It is doubtful if the purely mathematical error would have been apparent to him with any clarity (since he repeated the same kind of error in his own confutation of his first axiom), whereas he may have reflected that velocity and rate of change of velocity may be conceived as changes in time irrespective of spatial considerations. Yet if Galileo had worked backwards from the relation $s = \frac{1}{2}at^2$, as he represented it geometrically (Fig. 6), he may have seen that the areas ABC, ADE, can only represent distances ($\frac{1}{2}vt$) because of their relative dimensions, and that therefore the dimensions AB,

[1] Crew and Salvio, *op. cit.*, p. 161.

AD must be in time $\left(\text{being such moreover that } \left(\dfrac{AD}{AB}\right)^2 = \dfrac{ADE}{ABC}\right).$

It may be, indeed, that if the function $s = \tfrac{1}{2}at^2$ is assumed to be true, it is easier to deduce the true conception of acceleration, than it is to perform the inverse and more logical operation (in which Descartes and Beeckman failed). At all events, Galileo inserted into the *Discourses* a long passage to the intent that the "natural" idea of velocity is a rate of change in time, and acceleration therefore a rate of change of velocity in time, which was written, perhaps, in 1609, and followed this by the exposure of the *reductio ad absurdum* in the alternative proposition. What he did *not* do was to derive the definition of uniform acceleration from the action of a constant force: though again it may be conjectured that Galileo's, and Beeckman's, progress towards this definition may have been influenced by Gilbert's suggestion of gravitation as attraction—a suggestion which from the causal point of view liberates natural acceleration from spatial considerations.

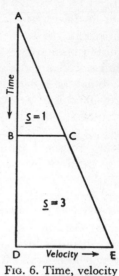

Fig. 6. Time, velocity and distance.

The law of acceleration and the theorem $s = \tfrac{1}{2}at^2$ derived from it are the foundations of dynamics, and dynamics exercised a preponderant influence in the evolution of scientific method during the seventeenth century. It is no exaggeration to describe this double discovery, with all the new structure of thought of which it was the crowning achievement, as the justification of the new philosophy, as the beginnings of exact science which consciously set itself to proceed by other ways than those of the past, and which hardly doubted that all the past philosophy of nature was vain. Yet the law of acceleration had been discussed since the fourteenth century, and was defined, admittedly not impeccably, by da Vinci and Soto among the more immediate predecessors of Galileo. Was the extent of his achievement, then, no more than to effect the proper integration which would give the law of distances? If this had been all, it would have been a great feat, for the perplexities of Descartes and Beeckman show that the law of acceleration did not prove a ready key even to the most acute and resourceful of his contemporaries. But this was not all; Galileo, a less ingenious mathematician, excelled Descartes in a mathematical problem because he understood the conceptual structure into which the key would fit. In fact the single law of acceleration, set in the framework of impetus theory, had hardly proved more useful in the sixteenth century than the crude observation that falling bodies travel more quickly as they approach the earth. Its full significance was revealed only when Galileo applied it in a context of dynamical theory in which the law of inertia had replaced the idea of impetus, a theory so highly abstracted that causation, friction, resistance, were no longer considered as relevant. To Galileo the law was no longer a descriptive deduction from physical principles (as it was to Soto), or an arbitrary assumption (as it was to Descartes) but a primary fact, inevitable in theory and confirmed by experiment. There are many similar instances in the history of science of the isolated statement of an anterior phase becoming the core of a comprehensive generalization; so comprehensive, in this case, that Galileo obtained from it knowledge of a whole class of mechanical theories. It is the combined effect of this rapid evolution in thought, requiring clarity of definition and elaboration of mathematical expression, the re-thinking of the nature of motion and the statement of functions making quantitative calculations possible, that determines the magnitude of Galileo's achievement. The commentary upon Aristotle's physics had been replaced by the mathematical scaffolding of a new branch of science.

Galileo and Descartes

The strategic lines of the *Discourses on Two New Sciences* were probably sketched out about 1609. The origins of the earlier dialogues, in which Galileo discussed cohesion and disputed the doctrine that "nature abhors a vacuum," attacked the Aristotelean view of the accidentals of motion (that there are absolutely light bodies, that velocity is proportional to weight, etc.) and began the study of the strength of materials (the other "new science") may be traced to a period more than ten years earlier, when Galileo, in his most Archimedean manner, was introducing into wider fields the principles of the sciences of statics and hydrostatics founded by his favourite Greek author. Certainly the secret of the trajectory of a projectile was known to Galileo by 1609, of which he wrote later, 'truly my first purpose, which moved me to speculate on motion, was the discovery of such a line (which most certainly when found is of little difficulty in demonstration); nevertheless I, who have proved it, know what pains I had in discovering that conclusion.'[1] The theory of projectiles, the propositions relating to the oscillations of a pendulum by which Galileo established its isochronous property, and the various theorems depending on the principle of the conservation of *vis viva* (momentum) are all straightforward deductions from the complementary laws of acceleration and inertia. Fragments of this reasoning had been used by Galileo in earlier years; for example, the experiment of rolling a ball down an inclined plane assumes the conservation of motion or *vis viva*. Now the whole was welded into a single coherent mathematical structure. It was not a faultless structure, for a number of Galileo's minor theorems were later found to be false, but its foundations were sound.

With one important exception, the theory of impact and the partition of kinetic energy between colliding bodies, the whole of seventeenth-century kinematics springs from Galileo's *Discourses*. When, finally, this became the instrument by which Gilbert's conception of attractive forces could be interpreted mathematically, the classical theory of dynamics was created. But if the lines of descent are direct, the historical process was complex. Galileo's intellectual evolution was by no means unique, though it has chronological priority. An independent "modern" tradition in dynamics can be traced to the fertile conjunction of Descartes and Isaac Beeckman. Other writers also, less strikingly, demonstrate a general tendency for the idea of the conservation of impetus to be transformed into the conservation of motion. As has so often happened in science, the decisive advance was made by one man in accordance with a broad progressive movement. Moreover, the reaction to the publication of Galileo's discoveries in 1632 was not simply favourable or adverse.[2] In the later seventeenth century the mathematico-mechanical group of scientists, the true heirs of Galileo, confided entirely in the essential conception of motion developed in the *Discourses*; in England,

[1] That is, the trajectory according to the usual assumptions of Galilean kinematics, in a vacuous, perfectly Euclidean space. Cf. *Opere*, Ediz. Naz., vol. X, p. 229, vol. XIV, p. 386.

[2] The elements of kinematics were indicated in the *Dialogues*, though a full discussion waited for the *Discourses* of 1638. See below, p. 110.

however, where this group was very strong and produced the dominating figure of Newton, the Galilean tradition was partly modified by the companion influence of Francis Bacon in the direction of more forthright empiricism. Thus Newton completed the mathematization of the distinction between the *ideal* laws of motion, and the real motions of terrestrial bodies. The representatives of conservative, anti-Copernican science, such as Giovanbattista Riccioli, after a vain attempt to dispose of the *Discourses* altogether, ultimately accepted the law of acceleration as a rough empirical truth, a mathematical hypothesis approximately agreeing with actual experiments, but continued to oppose the philosophy of motion from which this law was derived. This attitude was not dissimilar to that adopted by Tycho Brahe with regard to Copernicus: the innovations could not be philosophically true, but they could, by twisting them a little, be taken as quantitatively reliable. Far more important for the development of science was the position of the Cartesians, who likewise accepted Galileo's law of acceleration as a quantitative statement, while opposing Galileo's ratiocination in philosophy.

Descartes himself would never allow any supreme merit to Galileo as a scientist, and in his later utterances, after the crisis of 1619–20 in Descartes' intellectual development, he felt himself far superior because he alone possessed the true method of philosophy. In general the reception in France of Galileo's new science of motion was not uncritical: along with Descartes, neither Mersenne, nor Fermat, nor Roberval would give complete credence to a natural philosophy which applied such a violent process of abstraction to the complicated world of experience. For Descartes this objection was insuperable, and he finally came to regard Galileo as a mere phenomenalist who, lacking insight into the total mechanism of the universe, had merely been successful in isolated feats of mathematical description. Of the *Dialogue on the Two Chief Systems of the World* he wrote in 1634 that Galileo philosophized well enough on the subject of motion, but that very little of his doctrine was wholly true. He admitted that Galileo was more correct when he opposed current notions, and had indeed expressed some of Descartes' own ideas—it is strange to find him identifying Galileo's theorem, $s = \frac{1}{2}at^2$, with that which he had himself devised in 1618. Four years later Descartes' judgement was more harsh. Although he approved Galileo's introduction of mathematical reasoning into physical questions and his criticism of scholastic errors, he added that Galileo had built without foundation because he did not at all proceed by order in his investigations, nor consider the first causes of natural phenomena. The abstraction of Galilean science which has been its merit in the eyes of many subsequent historians was for Descartes its cardinal defect. He had now conceived his own model of the universe, and he found the Galilean model too remote from reality to be worthy of serious consideration. Its prime requisites, such as vacuous space and the constancy of gravitational forces, were assumptions contradicted by a true insight into nature.[1]

Descartes was no less convinced of the errors of the past than Galileo. He was equally a pioneer of the new natural philosophy. But whereas Galileo had arrived at a new method of procedure by patient inquiry into particular problems so that his essential idea of motion was the product of the endeavour to resolve the inconsistencies in the prevailing conception, Descartes' method taught him that it was essential to settle the simplest and most general ideas first, and then (as in the three treatises appended to the *Discourse on Method*) particular applications could be made to specific problems. What, for instance, is the simplest idea of matter? Extension, Descartes replied: matter is that which occupies space. To suppose that space could exist without a material content, that dimension could be conceived without any physical consideration, was meaningless. God could formally create a vacuum, but so interchangeable were the ideas of space and matter for Descartes that if a vessel was imagined to be divinely exhausted of matter, it meant only that its walls had been brought together so that there was no space between. Of course he admitted that vessels could be sensibly exhausted of grosser matter, such as water or air, when they would remain filled with a purer species capable of passing through the pores in the coarse material of the vessel. Similarly with motion: the simplest idea was the translation of a body from one situation among other bodies to another situation among a different set of bodies, and motion and rest were simply states of matter. This definition enabled Descartes to declare formally that the earth does not move, since despite its

---

[1] *Œuvres de Descartes*, publiées par Charles Adam et Paul Tannery, vol. I (Paris, 1897), pp. 303–5; vol. II (Paris, 1898), pp. 380–8. Also J. F. Scott, *The Scientific Work of Descartes* (London, 1952), pp. 161–6.

revolution around the sun in the vortex of the solar system, its environment in surrounding matter is unchanged. Movement, properly, was displacement; as a body moves, it is forced to displace other matter from its path, and a corresponding quantity must occupy the volume which the motion of the body would otherwise leave vacant. From these considerations Descartes deduced more specific laws. In the first Beeckman's conservation of motion became the perpetuation of the state of matter unless it is acted upon in some way. In Descartes' second law this principle of inertia was perfected by the statement that all bodies in motion continue to move *in a straight line*, again unless they are acted upon. Thus the necessity for some centripetal lien or force in circular motion was explicitly recognized by Descartes—a fundamental contribution to mechanics. The third law of motion was the fundamental rule for determining the partition of motion between impacting bodies. It was only later that the mistake in Descartes' formulation of the laws of impact became significant; they were, however, essential to the development of his corpuscular philosophy of matter. The notion of an intangible, incorporeal force, like Gilbert's attraction, was to Descartes barbarous and unthinkable. The state of matter could only be altered by the direct action of other matter, that is, by contact between bodies.[1] Hence what appears as the action of an incorporeal force is in reality the action of a stream of impalpable particles.

Action at a distance—such as the gravitational pull of masses across empty space—was to remain for Cartesians the bitter pill of Newtonian mechanics. Implicitly its impossibility conditioned Descartes' reaction to Galilean mechanics. When he declared that Galileo had not inquired what weight is, when he maintained that Galileo's description of how bodies fall was vitiated by his ignorance of why they fall, he meant that in nature there are no constant forces producing constant accelerations. 'That is repugnant to the laws of Nature, for all natural forces (*puissances*) act more or less, as the subject is more or less disposed to receive their action; and it is certain that a stone is not equally disposed to receive a new motion or an increase of velocity when it is

[1] Thus Descartes was led into an insoluble problem—the nature of the contact between the soul (spirit) and the matter of the human body. (Cf. M. H. Pirenne: "Descartes and the Body-Mind Problem in Physiology," *Brit. J. for the Philosophy of Science*, vol. I, 1950.)

already moving very quickly, and when it is moving very slowly.'[1] Descartes' reasoning is quite clear. Terrestrial gravity he believed to be the effect of a stream of corpuscles of impalpable matter sweeping towards the centre of the earth. As this stream swept through bodies of ordinary coarse matter, it pressed them likewise towards the centre. Clearly this pressure would diminish with the increasing velocity of the falling body, which would tend towards a limit, the velocity of the stream itself.

In 1618, under the influence of Beeckman, Descartes had treated motion as a geometer, and had made the same sort of mistakes as Galileo in 1604. In later years, at the time when Galileo's treatises were published, he thought of motion in the context of physical or cosmological theory. Unfortunately, as he penetrated from the level of description to the deeper level of explanation, Descartes denied himself the possibility of mathematizing the law of acceleration: the mechanism, as he framed it, contained too many unknown variables.[2] The work in which his system of nature is expounded, *The Principles of Philosophy* (1644), is in fact almost completely non-mathematical, although Descartes is one of the greatest figures in the history of pure mathematics. For all the originality of its starting point, this was a synthetic system, blending harmoniously elements derived from the animistic science of Aristotle with others taken from the mechanistic philosophy of Democritus and Epicurus; and, like these early deductive philosophers, Descartes was inevitably preoccupied with the uncovering of chains of causation, starting from his initial postulates concerning the nature of the physical world. When by deductive reasoning the end of the chain was reached, it was impossible to adopt the reverse process of analysis and abstraction which alone makes a mathematical science possible. Therefore Cartesian mechanics could never consist of more than general principles and the false laws of impact. Therefore also Descartes could never do more than admit that Galilean kinematics gave a rough approximation to the real dynamics of the world of experience.

In Descartes the dichotomy between the mathematical and the physical attitudes to nature is almost complete, since he could not

[1] Adam and Tannery, *op. cit.*, vol. I, p. 380.
[2] Koyré, *op. cit.*, pp. 126–7; Paul Tannery: *Mémoires Scientifiques*, vol. VI (Paris, 1926), pp. 305–19.

sacrifice causation to computation. The Cartesians of later generations were more eclectic and less rigorous. They could not resist
the intellectual charms of Descartes' discoveries both in pure
mathematics and in natural philosophy. In attempting to combine
them they became the illegitimate heirs of Galileo. Having
realized the futility of seeking directly for the descriptive dynamical laws of the actual universe, they sought instead for the correction which ought to be applied to the ideal laws of Galileo in order
to apply them to a world where bodies move in resisting mediums
and there are no pure forces. No tradition is perpetuated un-
adulterated and the later Cartesians compromised with the phenomenalism which Descartes had condemned. While they firmly
maintained his method of philosophizing, his theory of the origin
of the universe, of the corpuscular mechanisms involved in the
phenomena of light and vision, magnetism and electricity, gravity
and the planetary revolutions (none of which, from its very
nature, could be subjected to direct empirical verification), in
discussing any problem of mechanical science descriptively they
turned to the mathematical model of Galileo, to which there was
no alternative. There was a subtle change of perspective: instead
of proceeding from indubitable first principles to the details of
nature and the ultimate descriptive mathematical understanding,
as Descartes had sought to do, later Cartesian scientists followed
a different path, that of working out the modifications required
in transposing a problem from a mathematical model to the
natural scene.

This is most true of the Dutch physicist, Christiaan Huygens,
for many years the mainstay of the French Académie des Sciences.
Certainly Huygens, when still a boy, had worked out a purely
mathematical proof of Galileo's law of acceleration, and certainly
in full age he confessed that his thoughts had been too greatly
influenced by Cartesian fictions. But it is also clear that he could
never have become an entire phenomenalist: he was too great
a Cartesian to be a Newtonian. He has been appropriately considered in relation to the Cartesian tradition.[1] Much of Huygens'
work was suggested by Descartes' own original thoughts (for example, his study of centripetal forces and the laws of impact),
much of his basic theory was Cartesian. Always denying the reality

[1] e.g. by P. Mouy: *La Développement de la Physique Cartesienne* (Paris, 1934),
chap. II.

of a physical vacuum, he believed in impalpable matter or æther, and his theory of gravity was wholly in agreement with Descartes'. He was never converted to "action at a distance." Yet his particular investigations are in the highest class of seventeenth-century mathematical physics, and as an experimenter also Huygens, with his strong affiliation to the Royal Society of London, was not negligible. In the actual conduct of his scientific career no one was more justly an exponent of the Galilean tradition in science than Huygens who, like Galileo, allied astronomy and physics. Only on the widest issues did he fall below the example which Galileo had set, when as it seems, measuring the mathematical model against the physical model of Descartes, he found his attachment to the latter unbreakable. The Cartesian mechanical theory of causation in nature was inevitably the frame of reference for each of his researches, deeply though his spirit as an inquirer was akin to that of Galileo.

The example of Huygens sufficiently demonstrates the complexity of intellectual inheritance in the second half of the century. There was no simple antithesis, uniform at all points, between conventional science and the new philosophy of the scientific revolution. There was no pure line of descent in mechanics from Galileo to Newton, nor among the scientists of the French school who broadly accepted the framework of Cartesian natural philosophy. The conflict between the Newtonian and the Cartesian system of the universe which lasted till about 1740 was indeed absolute and irreconcilable, but it should not be supposed from this that the Cartesians had borrowed nothing from Galileo, nor that the Newtonians had learnt nothing from Descartes. Much of the application of mathematics to mechanical problems which had been such a fertile and active field of science during the half-century between 1638 and 1687 was in fact the work of men who did not subscribe without reservation to Galileo's philosophy of science.

In addition, the many lesser contributions to mechanics cannot be overlooked. Admittedly the main source-books for the later seventeenth century were the *Discourses* of Galileo and the *Principles of Philosophy* of Descartes, but other works of the nascent age of modern science held fruitful suggestion, among them the practical machine-books and even such a curious confection as Baptista Porta's *Natural Magic*. Inventive interest in the development of

new mechanical devices was a prominent feature in the activities
of the early scientific societies, the Académie Royale des Sciences
enjoying the privilege of examining them and officially approving
those which it found sound and beneficial. Again, though the
progress in statics during the late sixteenth and early seventeenth
centuries was less pregnant with consequence than that in dynamics—there was a profound change in degree rather than a
change in kind—this progress too contributed to the methodology
of science as well as to the formation of the whole body of experimental learning. Simon Stevin, whose investigations provided
another part of the foundations of seventeenth-century mechanics,
was a man of wide competence little short of the first rank. In
the history of mathematics he has an important place as, among
other things, one of the leading early exponents of decimal arithmetic. With another Fleming, De Groot, he carried out, about
1590, an early experiment on the fall of heavy bodies to test
Aristotle's theory of motion. In statics his work on parallel forces
utilized the important principle of virtual velocities or displacement ; he recognized (like Galileo) that the equilibrium of a system
of pulleys or levers depends upon the constancy of the product
of the weight and the distance moved in each of the balanced
members. Stevin extended the same principle to the action of
non-parallel forces. He showed that unless perpetual motion is
assumed to be possible (and Stevin was one of the first to consider
this ancient fallacy in the light of plain mechanical reason) two
weights resting upon a pair of inversely inclined planes must
balance when they are proportional to the lengths of the
planes. His demonstrations contain the principle of the triangle of
forces, and therefore the first implication of vector-quantities in
statics, just as Galileo's treatment of the trajectory of a projectile
contains the first use of vectors in dynamics. In hydrostatics he
examined the conditions necessary for the stability of a floating
body, and the distribution of pressure in liquids, involving a problem of integration which he solved successfully. He was the first
discoverer of the so-called "hydrostatic" or "Pascal's paradox,"
that the pressure of a liquid upon a surface varies only with the
area of the surface and the height of the column of liquid above
it, irrespective of its cross-section. This, along with his other discoveries in mechanics, Stevin described in Flemish in 1586. Isaac
Beeckman noted it there, and drew the attention of Descartes to

this strange phenomenon. Descartes—without acknowledging his source—discussed its theory at some length. From Stevin too the paradox passed to Pascal, who analysed it very completely in 1648.[1]

The essence of mechanistic philosophy in the seventeenth century was the axiom that all natural phenomena could be reduced, by a sufficiently prolonged process of abstraction, to one single kind of change, the motion of matter. This axiom was the foundation of Cartesian science, and with limitations and qualifications it was generally shared by scientific men. Descartes' great contribution to dynamics, in a sense, is that he made it the primary science. In so doing he caused research to enter some false paths, but his idea was to be vindicated in the kinetic theory of the nineteenth century. Galileo provided the elements of kinematics, the basic conceptualization and the mathematical procedures— again further elaborated by Descartes in coordinate geometry— which rendered definable the properties of matter in motion. For the next century a major concern of science was the extent to which nature could be explained broadly in terms of Cartesian mechanism interpreted with the aid of Galileo's descriptive analysis of motion.

[1] Milhaud, *op. cit.*, p. 34.

CHAPTER IV

THE ATTACK ON TRADITION: ASTRONOMY

T HE period of silence in which comparatively little comment, favourable or unfavourable, was made on the new celestial system proposed by Copernicus lasted for about a generation after the publication of *De Revolutionibus*. From the later years of the sixteenth century to the middle of the seventeenth there was a noisy and not invariably elevated dispute between the adherents of the new and the old opinions in natural philosophy in which astronomy was the touch-stone of faith. The triumph of the innovators did not come rapidly, for it is always easy to exaggerate the adaptability of the scientific intellect. Books were written assuming the truth of the geostatic system, astronomical clocks and armillary spheres were constructed to show a motionless earth, until at least the end of the seventeenth century. And it is well to remember that though the ascendancy of the "new philosophy" made modern science possible, the material point so often at issue was not of long-term importance. It is now recognized that before using words like "motion" and "rest" in relation to the solar system the frame of reference must be explicitly defined. It was by no means logically essential to the progress of astronomy that men should believe the earth to have an annual rotation around the sun. No phenomenon known to the seventeenth century required such a motion for its explanation; in practical astronomy the relative movement of earth and sun could be equally well interpreted by supposing either to be at the centre of the other's orbit. Before the celestial mechanics of Newton's *Principia* was developed, there was no positive, demonstrative argument that can be called conclusive either way; only in the sense that the progress of science demands the liberty to theorize, to extrapolate beyond the available positive knowledge, is it true that the Copernicans had enlightenment on their side.

The first conflict between the innovators in philosophy and authoritarian learning, which was full of consequence in framing not only the attitude adopted by the Roman Catholic Church towards the great astronomical question but also (in part) the spirit of the scientific movement, at once defensive against suspected allegations of irreligion and hostile to the older kind of literary scholarship, had strictly no connection whatever with natural science. This conflict was exhibited in the famous condemnation of Giordano Bruno. Bruno was burnt in 1600 because he taught the plurality of worlds; he was moreover technically an apostate from a religious order. He believed that beyond the universe we observe there are other universes similar to our own, equally of divine creation, equally inhabited by immortal souls. Speculation of this kind has a natural fascination for some minds; it had occurred very early in the history of systematic thought, and had received the strong disapproval of Aristotle. It had always been regarded as theologically dangerous. Yet Nicole Oresme, in his *Livre du Ciel et du Monde* gave considerable attention to it. He envisaged the possibility of a plurality of worlds in time or in space so that there might be one world enclosed in another or separate worlds scattered in space, for example beyond "our" universe, all the work of one creator. Aristotle's argument that the earth of another world would be drawn to a natural place at the centre of this he confuted by rejoining that the natural place for such earth would be at the centre of its own world. The restriction of the divine creative power to the fabrication of one universe he regarded as a denial of omnipotence. Further, he was bold enough to declare that it is natural to human understanding to believe that there is space beyond "our" finite universe; in this God could bring other universes into existence. From this discussion Oresme concluded that reason alone could not eliminate the possibility of a plurality of worlds, but that in fact there never had been more than one, and probably there never would be.[1] In the seventeenth century a similar speculation produced the first scientific fantasies, such as John Wilkins' *Discovery of a World in the Moon* (1638).

The allied belief in the infinity and eternity of the whole cosmos was also ancient. Expounded by Lucretius, who had them from Democritus, these doctrines were further developed by an important group of Muslim and Hebrew philosophers of the middle ages, though little favoured by Christians. Oresme seems to hint

[1] *Medieval Studies*, vol. III, pp. 233, 242, 244.

at the infinity of space. Nearer to Bruno's own time the impossibility of conceiving a boundary to space was asserted by Nicholas of Cusa in the fifteenth century: and it was from him that Bruno received his inspiration. In his own time the idea was specifically related to the Copernican hypothesis by the English mathematician Thomas Digges. Digges believed the fixed stars to be infinitely remote from the earth, a notion he was free to adopt since it was no longer necessary to suppose them fastened to a revolving sphere. All that the Copernican hypothesis required, however, was that the ratio of the distance of the fixed stars to the earth's distance from the sun should be a large number; had this number been far smaller than it actually is, the sixteenth-century astronomer would still have failed to detect any evidence of the earth's motion in his celestial observations. As for the infinity of space outside "our" universe, it was perfectly reconcilable with Ptolemaic astronomy and not at all a Copernican innovation. However great the interest of Bruno's ideas in themselves, the idea of the plurality of worlds was not inspired by the scientific renaissance, they were not a logical deduction from heliocentric astronomy, and they were totally irrelevant to the progress of science. Bruno was not a scientist, and his dispute with Rome turned on a purely metaphysical problem.

It is well to attempt to define the contemporary appreciation of a situation such as this. That the introduction of religious considerations into a question of quasi-scientific speculation is quite distinct from a similar intervention in the interpretation of observations or experiments was perfectly clear to philosophers of the middle ages and the early modern period alike. A very high proportion of scientists up to the mid-seventeenth century were men of unusually profound religious conviction, and none used science as a lever against religion. Thomas Hobbes won no countenance from the Royal Society. The furtherance of science and religion were commonly regarded as inseparable objectives. The English scientists of the seventeenth century, especially, were far more complacent than medieval scholastics in their belief that reason and research properly conducted could never conflict with religious dogma. The attitude of the middle ages had been that where reason was incompetent to decide, faith should pronounce: and that in many instances faith must even prevail against reason. In the period of the scientific revolution natural theology was still distrusted, and divine modification of the laws of nature in rare events was a commonplace. A general antithesis between science and religion was consequently out of the question; a particular antithesis over a single point could only be due to some misinterpretation either of nature or of religious truth. Even for the most empirical of scientists, like Boyle and Newton, the introduction of religious considerations into the pattern of scientific investigation was natural and inevitable. They would not have dreamed of denying the validity of a universal moral or religious law—a concept to them more binding than scientific law. To the Protestant mind of the seventeenth century the judgements upon Bruno, and later Galileo, would seem full of human error and expressive of the bigotedly narrow outlook of the unreformed Church, but they would agree that it was the duty of the responsible officers for the time being to enforce the moral law according to their own enlightenment.[1] No sympathizer with Bruno or Galileo believed in complete freedom of thought and expression; none would have asserted that scientific activity and theorizing are completely outside the range of universal moral law. For the seventeenth century the burning of Bruno could not be wholly wrong in principle as it was to the liberals of the nineteenth. Galileo, though he never surrendered his inner conviction of scientific rectitude, apparently admitted the right of the ecclesiastical authorities to pass religious censure upon his arguments. He lived and died in the Catholic faith; and when compelled to make his formal recantation, it would seem more reasonable to attribute his compliance not to lack of courage, but to recognition of the then universal belief that moral and religious truths are of a higher order than the scientific. Galileo could only lament that in his case the moral law had been misapplied.

Although the pronouncements of the Holy Office were not opposed to any positive scientific knowledge of the time, and in the case of Bruno only condemned a form of quasi-scientific speculation, they had a deep effect upon the scientific movement. They were widely interpreted as a final declaration against the Copernican system, and there is evidence that some (like Descartes) who were disposed to favour a heliostatic model were impelled to express their ideas in veiled and guarded terms. It seemed as though innovations in natural philosophy must lead to

[1] As the reformed Church of Geneva had upon the person of Serveto.

outbreaks of heretical opinion, as reactionaries had long predicted. Even in England, critics of the newly founded Royal Society did not fail to assert that its methods were subversive of the Church of England. An odium, which had not existed in the early sixteenth century, was for a time attached to any originality in astronomical thought. But the challenge provoked a powerful reaction in men like Galileo and Kepler. It created a situation in which the new doctrines had to be effectively vindicated; it was no longer possible for the two systems of the world to exist peacefully side by side.

The pre-Newtonian development of heliostatic astronomy may be analysed as involving four principal steps. Firstly, the dissolution of the prevailing prejudice against allowing any motion to the earth, which involved a careful criticism of all existing cosmological ideas in order to create a new pattern in which such a motion would no longer seem implausible, and a broad discrediting of Aristotle's authority. A necessary auxiliary to this was the most important second step, in which physical theories were revised to show the invalidity of objections against the Copernican hypothesis arising out of terrestrial mechanical phenomena. Thirdly, the new astronomy was greatly enriched by qualitative observation, which suggested that the old teaching was very inadequate. Fourthly, exact quantitative observation provided new materials for recalculating the planetary orbits, thereby leading to the abandonment of the ancient preconception in favour of perfect circular motion, and to the enunciation of new mathematical laws. Kepler's discoveries might have been expressed in the terms of the geostatic system; but as Kepler was a staunch Copernican, his whole discussion of the solar system was constructed in such a way as to give further force to the heliostatic hypothesis. The accomplishment of these four steps extended over a period of roughly half a century (c. 1580–1630), while their assimilation and particularly the gradual recognition of Kepler's laws of planetary motion occupied another generation. During the middle period of the seventeenth century the nature of the problem was again slightly changed under the influence of Descartes' natural philosophy, which tended to enlarge the disparity between the natural-philosophical and the mathematical-astronomical approach which the discoveries of the first third of the century seemed rather to diminish.

The contributions of Galileo to the first three of these steps were of major importance: to the fourth, quantitative observation, he brought almost nothing. Galileo was not an astronomer as the word had previously been understood, he was never interested in the traditional procedures of positional astronomy, but as a philosopher he applied new astronomical techniques, mostly of his own invention, to the examination of cosmological problems. Before 1609 his studies were very largely, if not wholly, devoted to physics and mechanics. He must, of course, have been thoroughly acquainted with elementary astronomy, and it appears that he had already read Copernicus and become converted to his doctrine, which he had at first distrusted. In 1609 Galileo learnt of the Dutch optical device which made distant objects seem near, and with this hint and considering the invention as a problem in optics he constructed his own telescope. During succeeding years he endeavoured to increase the magnification of this instrument and to effect improvements in lens-grinding, besides carrying out celestial observations which were recorded in a series of treatises.[1] Galileo was the first to appreciate the usefulness of the recently invented telescope in astronomy, and was therefore rewarded by many discoveries, but within a few years a considerable group had taken up qualitative investigation in astronomy. The old and the new branches of the subject remained practically distinct until about 1670, when the application of the telescope to measuring instruments became effective; during Galileo's lifetime astronomy with the telescope brought into being a completely new branch of scientific investigation. Its first results were striking, and provided powerful arguments against Aristotle. In January 1610 Galileo saw four of the satellites of Jupiter, which he called the Medicean stars and regarded as forming a visible model of the whole solar system. Later he observed the phases of Venus, from which it could be deduced that the planet revolves around the sun, and not between the sun's sphere and the moon's as Ptolemy's hypothesis supposed. He observed the moon itself, and confirmed his conjecture that it was a body resembling the earth, with valleys and mountains whose heights he estimated from the lengths of their shadows. His telescope resolved part of the Milky Way into a dense cluster of stars. The discovery of sun-spots was made by

[1] *Sidereus Nuncius* (1610); *Istoria e dimostrazioni intorna alle macchie solari* (1613); *Il Saggiatore* (1623), and the *Dialogue* of 1632.

several observers of whom Galileo was one. They are, of course, sometimes visible to the naked eye; Kepler had failed to recognize one in 1607 when seeking for a transit of Mercury. Fabricius probably made the earliest discovery of these *maculæ*, and Scheiner certainly wrote the largest book on them, but it was Galileo who realized their importance for astronomical theory.

To anyone who was prepared to think, the discoveries that followed upon the invention of the telescope would suggest a single train of thought which was certainly anti-classical but equally departed very far from the ideas of Copernicus. It would seem that celestial phenomena were much more complex than any extant astronomical system allowed. It would seem that the stars were not finitely or infinitely remote, but distributed through space. It would appear also that the heavens, far from being incorruptible and unchanging, were undergoing regular and irregular mutations, as Tycho Brahe had suggested as early as 1572. The planets, notably Saturn, whose ring-satellite was not yet fully understood, altered their aspects, and the sun itself was stained by spots that enabled its revolution upon its axis to be traced. The image of the moon could be conceived as comparable to that of the earth seen at the same distance. All this tended towards one general conclusion, that the universe was a physical structure, not composed of light and a matter totally different from the matter of the terrestrial region, but rather of two types of physical body. The first, the stars, were incandescent sources of light and plainly physical since they were not invariable. The second, of which more could be learnt, the planets, were physical bodies practically indistinguishable at this stage from the earth itself, which could now be placed without hesitation in the class of solar satellites on physical grounds as well as by reason of its motion. Physical astronomy was thus a creation of the telescope, for in the past the sole subject of astronomical science had been the analysis of the positions and motions of the heavenly bodies without consideration of their nature, which had been resigned to the speculative discussions of philosophers, while astrology had embraced the supposed influences of these bodies upon the terrestrial region. The concept either of physical astronomy or of celestial mechanics (which came later) was completely irreconcilable with the philosophic background of sixteenth-century practical astronomy, and it was necessary to reinterpret the Copernican heliostatic universe in this new light.

This reinterpretation was achieved in Galileo's *Dialogue on the Two Chief Systems of the World* (1632) which was, therefore, far more than a mere defence of the heliostatic principle. At the same time, it will be seen that Galileo's treatise did not advance beyond *De Revolutionibus* in one respect—it retained perfect circular motion —and in another respect it was even far inferior, for as a guide to positional astronomy the *Dialogue* is worthless. Galileo wholly neglected the complexities of planetary motion with which astronomical theoreticians had struggled for two thousand years, and on which his contemporary Kepler was to spend his life. In astronomy, Copernicus and Kepler set themselves against Ptolemy; Galileo's opponent was Aristotle, the philosopher. And, oddly enough, only a combination of errors made Galileo's view of the universe consistent: his belief in circular inertial motion, and in circular planetary orbits.

Within a few years after 1609 Galileo's teaching at Pisa as modified by his discoveries with the telescope had become unconventional enough to occasion his first encounter with the Holy Office. The *Dialogue* was conscious propaganda for the new philosophy, though the opinions expressed were put in the mouths of imaginary characters. Galileo did not scruple to indulge in a certain buffoonery with the Aristotelean Simplicius, whose feeble defences are mercilessly attacked. The first pages of the *Dialogue* contain a delightful battle of wits in which of course the Copernican is made to score heavily. Galileo examines the reasoning by which it is argued that the heavens and the terrestrial region are distinct, both in their motions and their natures. He concedes that the motions of the heavenly bodies are perfectly circular, since only by such motion could the pattern of the heavens be preserved without change, and that rectilinear motion 'at the most that can be said for it, is assigned by nature to its bodies, and their parts, at such time as they shall be out of their proper places, constituted in a depraved disposition, and for that cause needing to be reduced by the shortest way to their natural state.' But he denies that terrestrial bodies do, in fact, move along straight lines, and so the antithesis is not a true one. As for the Aristotelean contention that the elements move directly towards and away from the centre of the universe along straight lines, Galileo replies:

If another should say that the *parts* of the Earth, go not in their motion towards the Centre of the World, but to unite with its *Whole*, and that for that reason they naturally incline towards the centre of the Terrestrial Globe [a notion distinctly reminiscent of William Gilbert], by which inclination they conspire to form and preserve it, what other *All*, or what other Centre would you find for the World, to which the whole Terrene Globe, being thence removed, would seek to return, that so the reason of the *Whole* might be like to that of its *parts*? It may be added, that neither *Aristotle* nor you can ever prove, that the Earth *de facto* is in the centre of the Universe; but if any Centre may be assigned to the Universe, we shall rather find the Sun placed in it.

A number of propositions in mechanics are carefully elucidated, and it is made apparent that the opposition of the Copernican to the traditional world-picture will depend upon his completely different analysis of the properties of moving things. Mechanics in fact is the foundation of cosmology:

... none of the conditions whereby Aristotle distinguisheth the Coelestial Bodies from Elementary [i.e. terrestrial], hath other foundation than what he deduceth from the diversity of the natural motion of those and these; insomuch that it being denied, that the circular motion is peculiar to Coelestial Bodies, and affirmed, that it is agreeable to all Bodies naturally moveable, it is behooful upon necessary consequence to say, either that the attributes of generable or ingenerable, alterable or unalterable . . . equally and commonly agree with all worldly bodies, namely, as well to the Coelestial as to the Elementary; or that Aristotle hath badly and erroneously deduced those from the circular motion, which he hath assigned to Coelestial Bodies.[1]

That is, if the earth moves, all Aristotle's physical theory of cosmology is baseless. Soon after this, Galileo makes the general discussion of the "architectonics" of the world break down on the argument that hot and cold are not qualities proper to the heavenly bodies. That sort of dictum (he remarks) leads to 'a bottomless ocean, where there is no getting to shore; for this is a Navigation without Compass, Stars, Oars or Rudder.' Accordingly the debate shifts to the evidence for or against the changelessness of the heavens, in which the new observations made with the telescope are fully discussed, and a detailed comparison is made

[1] I retain the language of Galileo's seventeenth century translator, Thomas Salusbury. The passages may be found in the revision of his work by Giorgio de Santillana, pp. 37, 40, 45.

between the optical properties of the earth and the moon. From this their physical similarity is deduced. Incidentally Galileo points out the futility of the notion that changes in the celestial region would be impossible because they would be functionless in the context of human life, the purpose of the heavenly bodies being sufficiently served by their light-giving and regular motion. Nor does he pass over the curious judgement which made sterile immutability a mark of perfection: rather if the earth had continued 'an immense Globe of Christal, wherein nothing had ever grown, altered or changed, I should have esteemed it a lump of no great benefit to the World, full of idlenesse, and in a word, superfluous.' By such asides as these, in a strictly scientific argument, the values of conventional thought were challenged in turn, its texture made to seem weak and strained.

The second Dialogue opens with caustic mockery of foolish deference to Aristotle's authority: 'What is this but to make an Oracle of a Log, and to run to that for answers, to fear that, to reverence and adore that?' Those who use such methods are not philosophers but Historians or Doctors of Memory: our disputes, says Galileo, are about the sensible world, not one of paper. As for the motion of the earth, it must be altogether imperceptible to its inhabitants, 'and as it were not at all, so long as we have regard onely to terrestrial things,' but it must be made known by some common appearance of motion in the heavens; and such there is.[1] But in so far as motion is relative, the science of motion cannot decide whether earth or heaven really moves.[2] Consequently opinion turns on what is "credible and reasonable." It is more reasonable that the earth should revolve than the whole heaven; that the celestial orbs should not have contradictory movements; that the greatest sphere should not rotate in the shortest time; that the stars should not be compelled to move at different speeds with the variation of the Poles. On all these points Galileo's view of what is "reasonable" supposes a different perspective from that of anti-Copernican philosophers; but Simplicius is not allowed to argue the matter, and merely remarks that 'The business is, to make the Earth move without a thousand inconveniences.' The

[1] Santillana, *op. cit.*, p. 125 ff.
[2] 'Motion is so far motion, and as Motion operateth, by how far it hath relation to things which want Motion: but in those things which all equally partake thereof it hath nothing to do, and is as if it never were.'

first group of "inconveniences" to be dealt with includes the usual mechanical phenomena, the stone falling vertically, the cannon-ball ranging equally far east and west, which it was thought would not occur if the earth moved beneath. They naturally lead Galileo into a long exposition of his new ideas on mechanics, in which a partial version of the law of inertia is enunciated. Much of this reasoning against the Aristotelean doctrine of motion had already been exactly anticipated by the impetus philosophers of the middle ages. There is a long discussion of the so-called devia-tion of falling bodies—a problem which attracted attention inter-mittently throughout the seventeenth century as affording a possible proof of the earth's rotation. It had been alleged, for example by Tycho Brahe, that if a stone was allowed to fall freely from the top of the mast of a moving ship, it would not fall at the foot of the mast, but well aft. Galileo shows that this is inconsistent with the true principles of mechanics. In spite of Simplicius' cry 'How is this? You have not made an hundred, no not one proof thereof, and do you so confidently affirm it for true?', Galileo places his faith in *a priori* reasoning to predict the result of an ex-periment he has never made, though previously he has taken pains to point out (against Aristotle) that experiment is always to be preferred to ratiocination. Therefore, since all heavy bodies have inertia: 'we onely see the simple motion of descent; since that other circular one common to the Earth, the Tower and our selves, remains imperceptible, and as if it never were, and there remaineth perceptible to us that of the [falling] stone, onely not participated in by us, and for this, sense demonstrateth that it is by a right line, ever parallel to the said Tower.'[1]

In his attempt to analyse the true path followed by a falling body in space—the resultant of its double motion about and towards the centre of the earth—Galileo committed a serious error caused by his imperfect definition of inertial motion. Though familiar with the common effects of centrifugal force, he over-looked the fact that the inertial motion of the falling stone is along a tangent to the earth's surface, so that in the absence of gravity it would never describe a circle about the earth's centre unless it was held to it in some way, but would continue along a straight line into space. Galileo thought, on the contrary, that its inertia, impetus or *virtus impressa* could cause a free body to revolve in a

[1] Santillana, *op. cit.*, p. 177.

circle, and so he declared that if a stone fell at a uniform velocity towards the centre, its compound path would be an Archimedean spiral. Since the motion of descent is accelerated, however, he devised a demonstration showing that the true path in space is probably an arc of circle. Marin Mersenne, who obtained the accurate law of inertia from Descartes, described this curve correctly in 1644 as a paraboloid. It is possible that Galileo was induced to make this error, which he never corrected despite its incongruity with the parabolic theory of projectiles worked out in the *Discourses*, because of his preconception in favour of perfect circular motion. Eager to prove that even the local motions of terrestrial bodies are truly circular, as further testimony against the Aristotelean antithesis between circular-celestial and rectilinear-terrestrial displacements, he also noted that, on his theory, the absolute velocity in space of the falling body is uniform, acceleration relative to the earth's surface being only apparent. This being so, the question of finding the cause of acceleration, which had made Galileo doubtful of the original impetus theory many years before, could be shown to be spurious; the problem is solved by denying absolute acceleration in falling bodies altogether, and visualizing the phenomenon of relative acceleration as the product of the uniform movement of observer and object along intersecting circular paths.

With one exception the mechanical objections that could be raised against the earth's diurnal revolution are disposed of by appeal to the principles of inertia and of the relativity of motion which Galileo illustrates with a variety of ingenious examples. Other problems in mechanics auxiliary to the main argument are touched upon, such as the conception of static moment, the isochronism of the pendulum, and the law of falling bodies, here quoted by Galileo without proof. In replying to the objection that the rotation of the earth would hurl down buildings, etc., Galileo makes the first investigation into centrifugal forces. Using the idea of virtual forces, enunciated in connection with static moment, he proves that with equal peripheral velocities the force is inversely proportional to the radius. He points out that the angular velocity of the earth is very small, and its radius very large: therefore the force set up would not be sufficient to overcome a body's natural gravity. From Galileo's argument ('thus we may conclude that the earth's revolution would be no more able to extrude stones, than any little wheel that goeth so slowly, as that it maketh but one turn in twenty-four hours') it is clear that he did not realize that when *angular* velocities are equal, the centrifugal force is *directly* proportional to the radius. It was not indeed till much later that the rotatory stresses in equatorial regions were detected.

The extensive illustration of the perfect agreement between the heliocentric theory and a rational natural philosophy is certainly Galileo's most important contribution to the cosmological debate. In this there was no conflict of principle between the new philosophy and the old: they were agreed that the science of motion was the foundation of physics, and that physics and astronomy must speak the same language. Galilean mechanics was thus the necessary complement to Copernican astronomy, and though it is true (as Professor Heisenberg has remarked)[1] that nothing could have been more surprising to the scientists of the seventeenth century than their discovery that the same mechanical laws were appropriate for celestial and terrestrial motions alike, on the smallest and largest scale, the coincidence was not fortuitous, for it followed from Galileo's conscious endeavour to interpret Copernicus' mathematical model in terms of natural philosophy. There could be no question—as Galileo frequently emphasizes in the course of the *Dialogue*—of proving that the Copernican hypothesis was necessarily true; but with the readjustment of physical ideas effected by him it could be shown to be at least as plausible as the Ptolemaic. Aristotle's physical theory of the cosmos, terrestrial and celestial, had been an integral whole; for Galileo astronomy and physics were so far independent that he acknowledged the incompetence of purely physical observations to determine the system of the world, but he had no doubt that the same laws of motion were universally applicable, to celestial and terrestrial bodies alike, even to the point of satisfying himself that if the planets had fallen freely towards the sun from the same determinable point, they would have acquired, when they had attained their actual orbital distances from it, the velocities with which they actually revolve about it. A true mechanical theory, besides wholly destroying all physical objections to the heliocentric system, actually created a preponderance of belief in its favour.

[1] *Philosophic Problems of Nuclear Science* (London, 1952), p. 35.

In the Third Dialogue Galileo takes up the arguments for and against the annual motion of the earth. Beginning with purely astronomical considerations, he makes plain his distrust of the quantitative measurements of his own time, from which, however, he confirms that the new star of 1572 was truly celestial. The irradiation of light, exaggerating the stars' and planets' apparent diameters, is next explained, and the observations are proved to verify the Copernican arrangement. Galileo then discusses, in an eloquent and lucid exposition, the problem posed by the absence of a detectable stellar parallax. He was not of the opinion that the stars are infinitely remote, but he does argue that the size of the universe is such that its dimensions are beyond human standards of magnitude. If its immensity can be grasped, then it cannot be beyond the power of God to make it so immense; if its immensity is beyond comprehension, it is none the less presumptuous to suppose that God could not create what the mind cannot comprehend. Simplicius objects that a vast region of empty space between the orbit of Saturn and the fixed stars would be superfluous and purposeless, so that Galileo can again condemn the introduction of teleological reasoning into science.[1] He appears to think that the remoteness of the fixed stars, though vast, is not to be exaggerated, and calculates that even if the radius of the stellar sphere bore the same proportion to the semi-diameter of the earth's orbit, as that bears to the radius of the earth, a star of the sixth magnitude would still be no larger than the sun, which is, according to Galileo's reckoning, five and a half times as big as the earth. The assiduity and skill of astronomers in making observations of stellar parallax is in any case doubtful, since these would demand 'exactnesse very difficult to obtain, as well by reason of the deficiency of Astronomical Instruments, subject to many alterations, as also through the fault of those that manage them with less diligence than is requisite. . . . Who can in a Quadrant, or Sextant, that at most shall have its side 3 or 4 *braccia* long, ascertain himself . . . in the direction of the sights, not to erre two or three minutes?'[2]

Galileo's general explanation of the manner in which the heliocentric theory "saves the phenomena" is modelled on that of Copernicus, save that he denies the reality of the third motion

[1] A teleological argument is, however, used by Galileo himself later.
[2] Santillana, *op. cit.*, p. 398.

which Copernicus had ascribed to the earth in order to account for the parallelism of its axis. Thus, for instance, the principle of the relativity of motion solves the appearance of stations and retrogressions in the planets. But Galileo nowhere indicates that their orbits, which Copernicus had made eccentric, are other than purely circular about the sun, nor does he attempt to justify particular orbits from the records of positional astronomy. He further differs from Copernicus in making the centres of the orbits coincident with the body of the sun. It cannot be said, therefore, that Galileo improved the Copernican argument in terms of technical astronomy in the *Dialogues*, except through his use of the new qualitative evidence derived from the telescope, which had already been commented upon in his earlier writings. Indeed, the extremely simple astronomical model described is flatly incompatible with precise observation. Galileo, it is clear, was far more confident of the truth of the mechanical principle that bodies possess the property of inertial rotation in a perfect circle than of the accuracy of astronomical measurements. It was a characteristic of his scientific method of abstraction that it could more easily analyse, and describe in mathematico-mechanical language, a version or model of the real phenomena which was less complex than the phenomena themselves; and Galileo was not always sufficiently conscious of the genuine significance of the greater complexity. In this instance Galileo was deceived, partly by his imperfect definition of inertia (only implicitly rectified in his discussion of centrifugal force) and partly by lingering cosmological ideas. It is noteworthy that, while he does not debate whether the spheres are real or not, the word itself he uses naturally and without comment. No more than Tycho Brahe long before did Galileo believe that the heavenly bodies are supported by solid crystalline spheres, yet it seems that Galileo, whose approach to the problem was kinematic rather than truly dynamic, did not sufficiently reflect upon the consequences of taking away the heavenly spheres, and leaving the stars and planets as free bodies in space. Unlike Newton, Galileo never compared the motion of a planet to that of a projectile; unlike Kepler, he did not know that the geometry of planetary orbits vitiated any kind of spherical model. Unlike both of them, he rejected the idea that the sun affects the planetary motions.

The scientific work of Johann Kepler (1571–1630) was of an

utterly different quality from that of Galileo. The *Dialogues*, and to a less extent the *Discourses*, are popular books. In them Galileo does not fear to explain many elementary matters with which the expert would already be well acquainted. Kepler's writings, on the other hand, are so highly abstruse that the spread of his ideas was retarded by the difficulty of discovering and understanding them. They required that the reader be fully trained in the elaborate mathematics of positional astronomy. As Galileo was complementary to Copernicus, so the mathematician Kepler was complementary to Galileo, and there is perhaps no more remarkable example of the way in which two cognate, yet unlike, minds can follow parallel paths without interaction. Both Galileo and Kepler sought to strengthen the Copernican doctrine; they corresponded, and referred favourably to each other; but there is no shred of evidence that either was to the slightest degree deflected from his own course, or impelled to modify his own ideas, by the other's work. The synthesis of their two distinct points of view was only effected a generation later; until then Kepler and Galileo had as little in common as Ptolemy and Aristotle. As in the traditional, so also in the new science of the early seventeenth century this failure to overlap was not due to fortuitous causes, nor to any lack of sympathy between the two modes of approach. It was simply that the key by which a synthesis could be effected was not yet found. A cosmological dichotomy existed in Hellenistic science because physics and positional astronomy required similar, but not identical models; a comparable dichotomy existed from about 1630 to 1687 because there was no way in which the observed celestial motions could be interpreted in accordance with the kinematical principles of Galileo. The two lines of progress could only be brought together by a higher generalization—for which dynamics was essential and kinematics inadequate.

Kepler was the discoverer of the new descriptive laws of planetary motion, but he achieved more than this, for he made the first suggestions towards a physical theory of the universe adapted to the necessities of the new description. Although his life was given over to mathematical drudgery—in which he was aided by the much improved trigonometry of his day, and the invention of logarithms—Kepler was a man of vigorous and original scientific imagination. The mathematical operations he set himself could hardly have proved creative without it. In his first work he sought for the divine canon in celestial architecture, and this pursuit ran through all his later computations. But Kepler was no empty theorist: the divine art of proportion, the harmony of the grand design of nature, was to be elucidated from the most precise mathematical observation of the universe. Hence the turning-point in his career, the necessary foundation for his work, was Kepler's encounter with the Danish astronomer, Tycho Brahe (1546–1601), who, after a quarrel with King Christian IV, had deserted his royally endowed observatory at Hveen to enter the service of the eccentric Emperor Rudolph II, patron of alchemists and astrologers, at Prague. Thither Kepler was drawn from Graz in Styria by the undigested mass of observations, of unparalleled accuracy, which Tycho had brought with him. Ultimately the accumulated result of thirty years' labour yielded up the three laws of planetary motion which, as Kepler framed them, contained the decisive argument against the geostatic hypothesis so warmly defended by Tycho.

In many ways the Danish astronomer's rôle in the early history of modern astronomy is analogous to that of Vesalius in anatomy. Perhaps he was even more fully than the anatomist the first modern exponent of the art of disinterested observation and description. For if Tycho imported into his astronomical theory dominant factors which were physical in nature, it cannot be said (as it may of Vesalius' physiological preconceptions) that the evidence to confute them lay before his eyes. The problem of attaining precision was no less real for him than for Vesalius, and the methods he devised to solve it were probably more original. And certainly Tycho was unique among early modern scientists in his insistence upon the crucial importance of accurate quantitative measurement; always a desideratum in astronomy, certainly, but never previously handled with the analytical and inventive powers of Tycho, who first consciously studied methods of estimating and correcting errors of observation in order to determine their limits of accuracy. The most accurate predecessors of Tycho were not Europeans, but the astronomers who worked in the observatory founded at Samarkand by Ulugh Beigh, about 1420. Their results were correct to about ten minutes of arc (i.e. roughly twice as good as Hipparchus'); Tycho's observations were about twice as good again, falling systematically within about four minutes of modern values.[1] This result was achieved by patient attention to detail. The instruments at Hveen were fixed, of different types for the various kinds of angular measurement, and much larger than those commonly used in the past, so that their scales could be more finely divided. They were the work of the most skilful German craftsmen, whom Tycho encouraged by his patronage and direction. He devised a new form of sight, and a sort of diagonal scale for reading fractions of a degree. In measuring either the longitude of a star, or its right ascension, it is most convenient to proceed by a way that requires an instrument measuring time accurately, and Tycho studied the improvement of clocks for this purpose: but he found that a new technique of his own by which observations were referred to the position of the sun was more trustworthy. He was the first astronomer in Europe to use the modern celestial coordinates, reckoning star-positions with reference to the celestial equator, not (as formerly) to the ecliptic. Another innovation in his practice was the observation of planetary positions not at a few isolated points in the orbit (especially when in opposition to the sun), but at frequent intervals so that the whole orbit could be plotted.

The techniques and standards of precision in astronomy, of which Tycho was the real founder, were evolved slowly over a period of thirty years to fulfil a very simple ambition. When he made his first observations with a home-made quadrant, he found that the places given in star-catalogues were false, and that events such as eclipses occurred as much as two or three days from the predicted times. As the Copernican "Prutenic Tables" were computed from old observations they had brought in no significant improvement. The task Tycho set himself, therefore, was very simple: to plot afresh the positions of the brightest stars, and with the fundamental map of the sky established to observe the motions of sun, moon and planets so that the elements of their orbits could be recalculated without mistake. It does not seem that he undertook this with any violently partisan intent, but it is likely that he wished to show the falsity of the Copernicans' claim to have

---

[1] As was first pointed out by Robert Hooke, the unaided human eye is incapable of resolving points whose angular separation is less than about two minutes of arc, so that Tycho's work approximately attains the minimum theoretical limits of accuracy for instruments such as he used.

increased the accuracy of celestial mathematics, and to bolster up the geostatic doctrine by publishing unimpugnable tables and ephemerides calculated upon that assumption. To vindicate the "Tychonic" geostatic system was the last ambition of his life, which he charged Kepler to fulfil. However, he was no slavish adherent to conventional ideas. He did not believe that apparent changes in the sky were due to meteors in the earth's atmosphere; he proved that comets were celestial bodies, and that the spheres could have no real existence as physical bodies since comets pass through them; and his description of the planetary motions is relativistically identical with that of Copernicus.[1] As an astronomer, indeed, Tycho in no way belonged to the past: it was as a good Aristotelean natural-philosopher that he believed the earth incapable of movement.

Tycho's observations, including his catalogue of 1,000 star-places, have not proved of enduring value. The earliest observations that have an other than historical interest are those of the English astronomer, Flamsteed, early in the eighteenth century (error c. ten *seconds* of arc), for within about sixty years of Tycho's death the optically-unaided measuring instrument, still ardently defended by Hevelius of Dantzig, was beginning to pass out of use. Within a century Tycho's tables had been thoroughly revised by such astronomers as Halley, the Cassinis, Roemer and Flamsteed. In the interval, however, Kepler's discoveries based on Tycho's work had become recognized. To appreciate the relationship between Kepler and Tycho—the inventive mathematician and the patient observer—it must be realized that the balance of choice between Keplerian and Copernican astronomy is very narrow. Until measurements were available whose accuracy could be relied upon within a range of four minutes, or even less, there was no need to suppose that the planetary orbits were anything other than circles eccentric to the sun. Kepler, in plotting the orbit of Mars, from which he discovered the ellipticity of planetary orbits in general, was able to calculate the elements of a circular orbit which differed by less than ten minutes from the observations. It was only because he knew that Tycho's work was accurate within about half this range that he was dissatisfied and

[1] The point about comets was neglected by Galileo, for he gave them a terrestrial origin. Many of Galileo's other arguments cannot be directed against the "Tychonic" version of the geostatic idea.

impelled to go further. Kepler's famous "First Law" was thus the first instance in the history of science of a discovery being made as the result of a search for a theory, not merely to cover a given set of observations, but to interpret a group of refined measurements whose probable accuracy was a significant factor. Discrimination between measurement in a somewhat casual sense, and scientific measurement, in which the quantitative result is itself criticized and its range of error determined, only developed slowly in other sciences during the course of the scientific revolution.

While Kepler's discoveries would have been impossible without the refinement of observation attained by Tycho Brahe, more than mathematical precision was involved in them. Before the telescope, the only materials available for the construction of a planetary theory were angular measurements—principally determinations of the positions of the planets in the zodiac when sun, earth and planet were in the same straight line. Consequently the most that a planetary theory could achieve was to predict the times at which a planet would return to the same relative situation, and its position at those times. The mathematical analysis of the solar system as a number of bodies moving in three-dimensional space had never been attempted, as such, by the older astronomers, who had been content to assign such problems to philosophers. They had never concerned themselves with the real path of a planet in space, so long as their model predicted with tolerable accuracy the few recurrent situations in which observations could easily be made. The whole tendency of the scientific revolution was to rebel against this view of the astronomer as a mathematician, a deviser of models to save the phenomena, and to see astronomy as a science comprehending the totality of knowledge concerning the heavens and the relations of the earth to the celestial regions. Copernicus had abolished the equant because it was a mathematical fiction, an unphilosophical expedient. Galileo modified Copernicus' universe even further in the direction of physical explicability. Kepler had a true conception of the universe as a system of bodies whose arrangement and motions should reveal common principles of design—or in more modern language, be capable of yielding universal generalizations—which were to be demonstrated from the observations, not from physical or metaphysical axioms. For Kepler the astronomer's task was not to study the universe piecemeal—to construct models for each separate planet—but by studying and interpreting it as a whole to prove that the phenomena of each part were consistent with a single design. His aim was to provide a fitting philosophical pattern for the new discoveries of mathematical astronomy: 'so that I might ascribe the motion of the Sun to the earth itself by physical, or rather metaphysical reasoning, as Copernicus did by mathematical,' he remarked in the preface to the *Cosmographic Mystery*. Exact science might properly make inroads upon the established prerogative of philosophy; it was far from being his purpose to expel natural-philosophical considerations from quantitative science altogether.

Indeed, Kepler's scientific work was critically influenced by his attachment to extra-scientific ideas. He had firm preconceptions, and he was strongly opposed to mere phenomenalism. Even more than Copernicus he was infected with Pythagorean mysticism, and fascinated by the primary, foundational significance of purely numerical relations. He could elaborately interpret his descriptive generalizations in astronomy in the terms of musical harmony— genuine "music of the spheres"; draw the analogy between sun, fixed stars and planets, and God the Father, the Son and the Holy Ghost; or discuss the aspects of the planets at the moment when the cosmos was created. Some of the questions which he sought to answer are, to later minds, absurd or meaningless. Rightly he held that the question, why are the appearances thus and not otherwise, requires a scientific (rather than a purely philosophic) answer, and he was the first astronomer to take a serious grasp of it, but for him also such a question as, why are there no more and no less than five planets, was also urgent. As Kepler handled it, the inquiry involved an unshaken medieval complacency in the certitude of knowledge. His first explanation was given in the *Cosmographic Mystery* (1596), where he adopted the theory that the design of the universe is modelled upon the series of five geometrically regular solid bodies. He had tried in vain to find a rationale for the dimensions of the planetary orbits as calculated by Copernicus by considering them as mathematical series, or as circles described round regular polygons. The number five was certainly accounted for in this theory, and moreover Kepler found that if the regular solids were supposed to be fitted each inside a sphere, and these within each other in a certain order, the dimensions of the six spheres so arranged corresponded approximately to those of the earth and five planets. In the *Cosmographic Mystery* Kepler argued trenchantly in favour of the Copernican system, which he modified in order to make the sun its central point, instead of the centre of the earth's orbit. Believing that the imperfect agreement between his theory and Copernicus' determinations might be due to faulty observation, he had good hope of confirming it with the aid of the more accurate measurements of Tycho Brahe.

As Tycho's assistant at Prague, Kepler was directed to perfect the theory calculated in accordance with the observations on Mars by Tycho's Danish assistant, Longomontanus. The result was published in the *New Astronomy or Celestial Physics*, a book subsidized by the Emperor Rudolph II and published in 1609, eight years after Tycho's death had released Kepler from adherence to Tycho's geostatic system. His first discovery was that the plane of the orbit of Mars passed through the sun (a point in favour of Copernicus) and was invariably inclined to the ecliptic. A major problem was the planet's unequal velocity in its course. Although Kepler restored the equant-point, and so could adjust the varying angular velocity of Mars with respect to the sun in different proportions, he found that no single position of the equant-point would give a rate of variation satisfying all the observations. The same difficulty occurred when the earth's orbit was considered. Kepler found that its motion was certainly faster when near to the sun than when most remote from it, but not in such a way that the angular velocity about any arbitrary fixed point within the circle was uniform.[1] The problem—one after Kepler's own heart—was to find a theorem denoting this variation in velocity, an equation relating the speed of the planet's rotation at any point to its distance from the sun. Here Kepler was assisted by a quasi-philosophical notion that it was a "moving spirit" in the sun itself which caused the planet's circumgyrations. The further the planet receded from this spirit, the more weakly its force would operate, and so the planet's velocity would lessen. This *anima motrix* is referred to in the *Cosmographic Mystery* as hurrying along the stars (i.e. planets) and comets which it reaches, with a swiftness appropriate to the distance of the place from the sun, and the

---

[1] This was the first mathematical proof that the motion of the earth (or, as Ptolemy would have said, of the sun) is strictly identical with that of the planets.

strength of its virtue there.[1] The problem Kepler set himself, of assigning a determinate motion to a body revolving round a fixed point in an eccentric circle so that it moves through equal infinitesimal small arcs in times proportional to its distance from the point, is one that can be solved by integration. He used a method similar to that by which Archimedes had long before evaluated π, and so arrived at his "second" planetary law, that the radius-vector between sun and planet sweeps over equal areas of the orbit in equal times. Though his first proof was open to criticism, Kepler was later able to satisfy himself that the various errors in his method cancelled each other, so that the law was rigorously true.[2]

At this stage in his complex and tedious calculations—involving the geometrical analysis of many theoretical possibilities, as well as the continual checking of the predicted motions against observations selected from Tycho Brahe's great store—Kepler was already convinced that the orbit of the earth or a planet could not be a perfect circle eccentric to the sun. As he said:

> The reflective and intelligent reader will see, that this opinion among astronomers concerning the perfect eccentric circle of the orbit involves a great deal that is incredible in physical speculation. . . . My first error was to take the planet's path as a perfect circle, and this mistake robbed me of the more time, as it was taught on the authority of all philosophers, and consistent in itself with Metaphysics.

In calculations of the earth's angular velocity he could assume the orbit to be circular, because its ellipticity is small (*nam insensile est . . . quantum ei ovalis forma detrahit*), but in the orbits of the other planets the difference would become very sensible.[3] His next problem, obviously, was to define the nature of this non-circular orbit more closely. He therefore returned to the investigations on Mars, in a far more secure position now that he had worked out the movement of the observer's platform—the earth—with greater accuracy than before. Experiment showed that the orbit of Mars could not, indeed, be circular, for this when compared with the observations made the motion of the planet too rapid at aphelion

[1] Kepler: *Gesammelte Werke*, vol. I, p. 77.
[2] *Ibid.*, vol. III, pp. 263–70.
[3] The eccentricity of the earth's orbit is only 0·017, that of Mars is about 5 times as great, and of Mercury 12 times. The error which constituted Kepler's problem increases roughly as the square of the excentricity.

and perihelion, and too slow at the mean distances. After many trials Kepler wrote: 'Thus it is clear, the orbit of the planet is not a circle, but passes within the circle at the sides, and increases its amplitude again to that of the circle at perigee. The shape of a path of this kind is called an oval.' Again, the development of Kepler's thoughts was influenced by his idea of the physical mechanism which could produce such a departure from the perfectly circular form. He supposed that the oval path was traced by the resultant of two distinct motions; the first being that due to the action of the sun's virtue, varying with the distance of the planet, and the second a uniform rotation of the planet in an imaginary epicycle produced by its own *virtus motrix*. The hypothetical orbit would be an oval (or rather an ovoid, since its apses would be asymmetrical) enclosed within the normal eccentric at all points save the apses. Kepler spent much labour in vain attempts to geometrize this hypothesis so that it could be compared with observation; he even used an ellipse as an approximation to the ovoid. But Kepler finally had to confess that the oval orbit and the theory of its physical causation had "gone up in smoke." It was the accidental observation of a numerical congruity that led him to substitute for the oval another ellipse, which fitted the area-law and the observations exactly. Even at this stage he was much disturbed because he could not give a physical meaning to the elliptical orbit, until he satisfied himself that such an ellipse as he required would be traced out by a planet supposed to librate on the diameter of an epicycle.

The third of Kepler's great descriptive theorems, that the squares of the periodic times of the planets are in the same ratios as the cubes of their respective mean distances from the sun, which solved the problem upon which he had originally embarked, was announced in *The Harmonies of the World* (1619). This strange book, resuming the theme of the *Cosmographic Mystery*, has the same concern with esoteric relationships. Kepler compared the instantaneous velocities of the planets at different points in their orbits, and expressed these ratios in terms of musical harmony. He further compared the velocities of the several planets at their nearest approach to the sun; and finally he was induced to compare, not merely the periods, times and distances of the planets, which he had already discovered to be without significance, but the powers of these numbers, and so hit upon the "third law." A century and a half later, in 1722, a purely empirical formula connecting the distances of the planets was used by J. E. Bode to predict the existence of an unknown planet between Mars and Jupiter. He was vindicated by the discovery of the asteroids.

The simplicity and directness which these relations introduced into astronomy need no emphasis. The shapes and dimensions of the planetary orbits, and the velocities of the planet's motions within them, could now be calculated with ease and certitude. Kepler's Laws were the observational axioms upon which Newtonian celestial mechanics was to rest secure. What is not obvious is that Kepler's discoveries were displayed in extremely difficult books, published far from the main foci of scientific activity in France and Italy, and so were passed over by a generation that ignored their true importance. Kepler himself regretted the abstruseness of his subject:

> Most hard today is the condition of those who write mathematical works, especially astronomical treatises. For unless you make use of genuine subtlety in the propositions, instructions, demonstrations and conclusions, the book will not be mathematical; if you do use it, however, reading it will be made very disagreeable, particularly in the Latin language, which lacks articles and the grace of Greek. And also today there are extremely few qualified readers, the rest commonly reject [such books]. How many mathematicians are there, who would toil through the *Conics* of Apollonius of Perga? Yet that material is of a kind that is far more easily expressed in figures and lines, than is Astronomy.[1]

In truth Kepler was the last of the medieval planetary theorists, the last laborious computer and porer over tables. Such methods were too tedious for his own generation, excited by discovery and the new philosophy, and to a careless reader Kepler's discoveries were hidden in the idiosyncrasy of his strange speculation. When they were appreciated, new mathematics and new techniques were framing a new astronomy.

But Kepler was more than a mathematician. Perhaps the importance of his work, apart from the three famous Laws, has not been sufficiently esteemed. The older historians passed politely over Kepler's theorizing on physical mechanisms, his love of analogy, and all that was ancillary to the main mathematical

---

[1] "Astronomia Nova," *Ges. Werke*, vol. III, p. 18.

argument, as so much dross that was best left buried. Now it is not difficult to see that Kepler was as original and stimulating in his sidetracks as when following the plain mathematical road. Certainly his ideas on gravity, on the action of forces at distance, are important factors in the prehistory of the theory of universal gravitation. The Cartesians ridiculed Kepler's mysterious forces seated in the sun, and his appetites of matter, just as they later resisted the notion of gravitational attraction. Kepler does make odd equivalencies between "soul" or "spirit" and force, nevertheless his cosmological theory was mechanistically designed—and designed to give a far more accurate model than those of either Galileo or Descartes. It was Kepler who, in the *Cosmographic Mystery*, followed the example of Tycho Brahe in denouncing the traditional belief in material spheres which had been left unchallenged by Copernicus:

> Neither indeed is to be feared that the lunar orbs may be forced out of position, compressed by the close proportions of [other celestial] bodies, if they are not included and buried in that orb itself. For it is absurd and monstrous to set these bodies in the sky, endowed with certain properties of matter, which do not resist the passage of any other solid body. Certainly many will not fear to doubt that there are in general any of these Adamantine orbs in the sky, that the stars are transported through space and the ætherial air, free from these fetters of the orbs, by a certain divine virtue regulating their courses by the understanding of geometrical proportions.

He went on to ask, by what chains and harness is the moving earth fastened to its orb? and to point out that nowhere on the surface of the globe do men find it embedded in a material medium, but always surrounded by air. Kepler, too, must be credited, at least as much as Descartes, with the perception that there must be some source of force, or tension, within the solar system. It could not be a complex of entirely independent bodies without mutual interaction. It could not be accidental that the planes of all the orbits passed through the sun, nor could the variations of the planet's motion—the differences in its velocity at perihelion and aphelion, for example—be explained without the supposition that some force was acting upon it. For Galileo the universe was simple and dynamically constant: in Kepler's far more realistic picture it was highly complex and its dynamical condition constantly changing. Thus it was the Keplerian picture that enforced the development of celestial mechanics during the late seventeenth century. The descriptive and purely empirical laws of planetary motion presented a problem that natural philosophy could not escape. Kepler had gone far beyond the bounds of the astronomical problem of two generations—does the earth move or not?—to assert principles of celestial motion, set in a pattern of theorizing upon cosmic physics, which displaced traditional doctrines even more thoroughly.

CHAPTER V

EXPERIMENT IN BIOLOGY

WHILE the early stages of the scientific revolution in its physical aspects were strongly positive, the first phase in biology seems by contrast indecisive and inconclusive. The sixteenth century witnessed the promulgation of new ideals and new methods of study in such fields as botany and zoology, or anatomy and physiology, but there was as yet no more than the vague promise of the alternative body of knowledge which the pursuit of the new ideals and the practice of the new methods would construct in due course. Though existing doctrines might be criticized, no others had yet taken shape to displace them. There were criticisms occasionally of exaggeratedly animistic patterns of explanation; but the application of mechanistic philosophy to biological problems was not attempted before Descartes. As the mind abhors a vacuum, the natural result was a lag of theory behind observation. The authority of Galen endured longer than that of Aristotle because it was intrinsically far more difficult to apply the principles of physics and chemistry to the investigation of physiological processes than to apply Galilean mechanics to astronomy; astronomers also had the advantage over physicians that mechanics was the first science to enter a modern stage. The effect of the greater subtlety of the questions handled by the biologist was, so to speak, to distinguish observation from conceptualization as separate branches of scientific activity. Many experiments had been carried out in physical science to confirm or disprove directly the Aristotelean pronouncements, when as yet the theory of humours or the Galenic account of digestion were still untested. Experimentation was not deterred by technical difficulties alone, for imagination was lacking and the whole framework of ideas which would have given meaning to such experiments had still to be created. Those that were planned and successfully carried out—such as the well-known investigation of Sanctorius (1561–1636) into the quantitative gain in weight of the body by ingestion and loss through excretion—though they produced interesting information, carried little or no weight for or against the main strategic ideas of medicine and biology.

In a somewhat similar manner, the advance of the encyclopædic naturalists of the sixteenth century towards a modern scientific method was highly specialized and limited in character. The naturalist's ideas concerning the origin of organic life, the distribution of plants and animals, and the reasons for their wide range of structure and form, were still for the most part non-scientific in origin, or at best derived from very ancient sources. On the other hand, he was progressing towards modern ways of classifying and describing organisms and of defining the subject-matter of natural history. He became less interested in nature-study as an exercise in morality; he made a partial distinction between the *Flora* and the *Pharmacopœia*. This brought the disadvantage, however, that as botanists and zoologists became progressively more efficient in classification and description, they came near to losing interest in all other problems posed by the organic world. The naturalist was limited, in the main, to a particular kind of activity—ultimately inherited from the apothecaries' need to distinguish medicinal herbs—partly, of course, because it was one task worth doing which was within his competence, but partly also because he lacked the imagination which would have freed him from the influence of tradition. A different kind of biology, or such crucial experiments as those of Redi in the seventeenth century on spontaneous generation and those of Mendel in the nineteenth on inheritance, was not technically impossible; it did not necessarily depend altogether upon laboratories, instruments and the discoveries of other sciences. It did require—which is a great deal—intellectual originality. It did require the ability to frame questions about the living state and ways of proceeding to answer such questions. This was very different from the compilation of greater and greater masses of the same type of information.

The vast range of biological and medical science which widened out during the nineteenth century was, therefore, represented in the sixteenth only by medicine and natural history: and even these studies consisted of little more than herbalism and the endeavour to cure disease, these being in turn closely linked. The various branches of medicine, such as physiology (the *function* of organs in contrast to the *arrangement* of organs, anatomy), pathology, or hygiene were hardly distinguished save as subjects of separate treatises by Galen. All parts of medical science, excluding surgery, fell under the general surveillance of the physician. Or he might become herbalist or zoologist—the former in pursuit of the medical virtues of plants, and the latter as a comparative anatomist. Therefore it is not surprising that physicians had a major creative rôle in the biology of the sixteenth and seventeenth centuries, nor that the course of the science was to some extent directed by medical interests. Among the botanists many were medical men—to name only Fuchs, Cordus, Cesalpino, Bauhin, Tournefort, and Linnæus himself. Brunfels, whom Linnæus called the father of botany, and John Ray, Linnæus' greatest predecessor, were exceptions. All the work in human and comparative anatomy, and much microscopy (the famous exception being Leeuwenhoek), was carried out by physicians. Surgeons, being on a lower academic, social and intellectual level, to which they were firmly suppressed by the energetic corporate interest of the physicians, had far less opportunity to add to knowledge. The organization of the scientific movement and the system of the universities protracted the dual relationship of medicine and biology long after it had ceased to be real—when books were already being written purely on botany, or zoology, or physiology. Nowhere was it possible to obtain formal instruction in any of these subjects, save as part of a general course in medicine, until the middle of the eighteenth century. Teachers of botany (or of chemistry) were appointed only to fulfil the needs of the medical faculty.

To the young physician of the later sixteenth or seventeenth centuries, ardent for research, many courses were open. He might venture on original methods in practice, collect case-histories, perhaps contribute to the growing literature of abnormal observations and remarkable cures. Or he might, in the humanistic vein, seek to improve the general understanding of the magisterial texts. Or he might practise anatomy, in which case he would certainly dissect many animals. Or he might embark upon descriptive natural history. But the texture of the scientific work involved in all these courses was far from identical. The humanist-physician was easily assimilated to the type of the scholar, the naturalist-physician to the type of the lexicographer—admittedly with the development of specialized powers of observation. Neither of these courses, at this time, led naturally to the act of experimenting. The problem, however, which comes nearest to the physician's work, the understanding of the functioning of the human (and, by analogy, the animal) body in health and disease, is one that lends itself to observation and experiment. The physician must observe and classify diseases, he must also experiment in his therapy. Admittedly the physician found his normal ratiocinative background in Galen's ideas, admittedly he proceeded in accordance with the accepted theory of the nature of disease and of the measures requisite to effect a remedy; even the ascription of symptoms to the humoral condition, the amount and timing of blood-letting and the preparation of drugs were laid down by rules for his guidance more dogmatically than they are today. Yet, whatever the teaching, in any age a physician must be something of an empiric. He must learn to use his own judgement. He must adapt general principles to particular cases. And in the sixteenth century the art of medicine was far from static. Apart from the great variety of herbal medicaments among which the physician had to make his choice, there were the new inorganic remedies, such as mercury, and new drugs from the East and West Indies. There was a great controversy over the correct procedure in venesection. There were new problems—syphilis, gunshot wounds, scurvy ravaging crews on long ocean voyages, and plagues flourishing with the growth of cities. A physician's practice could be guided by principles—the use of contraries to restore the balance of the humours, or analogy (dead man's skull, powdered, in cases of epilepsy)—but he could be no mere follower of the book, if only because the book was an inconsistent and insufficiently specific guide. The most important part of medicine was learnt through experience, and profitable experience depends on experiment.

Perhaps this lesson was the most enduring contribution made by Paracelsus to true science. Naturally when Paracelsus writes, for example, ' From his own head a man cannot learn the theory of medicine, but only from that which his eyes see and his fingers touch . . . theory and practice should together form one, and should remain undivided. . . . Practice should not be based on speculative theory,' it has to be remembered that "practice" for Paracelsus meant something very different from the rationalist practice of the modern physician. He accepted the doctrine of signatures; he taught the doctrine of the microcosm and the macrocosm that forced the physician to become astrologer; to him medicine was the study of the occult forces that play upon the human body. Within his conception of the physician's practice however, he was empirical. The teachings of Aristotle and Galen provoked his unmitigated scorn, and he castigated the academic physicians who relied upon the theories derived from them. He claimed himself to have learned the art of healing not only from learned men, but from wise women, bathkeepers, barbers and magicians. Ideally, for him, the test of a remedy was its efficacy, though in his devotion to the occult and the esoteric he often fell far short of this ideal. Believing that experience was the best teacher, he did not hesitate to experiment with medicaments of which the academically minded were fearful, and so he became the leader of the chemical school in therapy. The strongest poisons, he held (no doubt arguing from the virtues of mercury or opium), contained hidden arcana, serviceable to the physician initiated into their mysteries. No doubt the latitude which Paracelsus introduced into medical practice was usually profitless and frequently dangerous; but it stimulated a more rationalist empiricism than his own. Not for the first time, the experimentalism of magic was favourable to the growth of natural science.

Of course trial-and-error methods do not constitute a new philosophy of science. Ambroise Paré's use of ligatures and dressings instead of cauterization by fire is not to be put forward as an example of a conscious experimental science—though it was a genuine revolt against authority, and a genuine instance of empiricism. Paré knew no Latin: he was only the royal surgeon. But it is to a certain degree inevitable that the originally minded men who adhered to the more practical aspects of medicine, who were compelled to be empirical (Glauber, after all, must have tested his *sal mirabile*), should have moved more naturally in the direction of experiment than their colleagues whose interests ran otherwise. From dissection for research to experiment of a limited kind is not a great step. Anatomical observations on the veins and arteries suggested simple experiments on the behaviour of the blood in the living body—with which venesection made the surgeon necessarily familiar. Observations involving vivisection had been made long before by Galen and Aristotle, and were repeated in the sixteenth century: wounds occasionally gave opportunities for a glimpse beneath the surface. There was almost a tradition by which poisons and their antidotes were tested upon small animals (and sometimes condemned criminals). Moreover, the Hellenistic tradition in zoology and physiology offered perhaps the best model of experimental science that could be found in the whole corpus of transmitted learning. The Aristotle of the *Generation of Animals* and the *History of Animals* was an experimenter as well as an excellent observer. In embryology especially—again leaving aside all question of theory—the sixteenth-century heritage of experiment is clear. Albert the Great, like Aristotle long before, had opened eggs systematically. The men of the renaissance had only to continue a well-defined course of investigation.

Perhaps it is not stretching imagination to see practical medicine playing somewhat the same rôle in the development of biology as that of technology in the evolution of the physical sciences. The physician, engineer and manufacturer had that practical skill in their encounters with nature which was lacking to the reflective, generalizing philosopher of the study. They wove a strand of empiricism into the web of theory. They were equally (if honest and intelligent men) more interested in the attainment of tangible results than the discussion of means by which such results ought to be attainable. And just as experience with cannon or in industrial chemistry had no simultaneous, directly positive effect upon ideas of motion or the four-element theory of matter, so also empiricism in medical science could not immediately and proportionately modify the broad theory of physiology or pathology. The impact of empiricism was in all cases gradual, subject to variations in emphasis and liable to be different from that which posterity might deduce merely by treating practical experience as the "cause," and change in theory as the "effect."

The history of the discovery of the circulation of the blood is an illuminating instance of the delayed action of observation and experiment upon biological theory. Harvey had little that was new in the way of fact available to him. His great merit was to integrate known but ineffective facts into a new and comprehensive generalization. As he himself constantly reiterates in his treatise *On the Motion of the Heart* (1628)—for he was one of those innovators who had little desire to flout authority unnecessarily—many of the observations on which he relied were already known to Galen. Harvey indeed did not so much contradict Galen as gently convert his doctrines. Other observations must have been made at almost every venesection—if only physicians less intelligent than Harvey had been able to see their meaning. Though their function was imperfectly understood, the valves in the veins had been observed more than sixty years before Harvey's discovery of the circulation was made; the tricuspid and mitral valves in the heart, which were also given a rational function for the first time by Harvey, had been described by Galen himself. In many ways, therefore, Harvey's discovery in biology resembled Galileo's in mechanics in being a new interpretation of familiar data drawn from common experience, and Harvey, like Galileo, had numerous precursors.

Harvey, however, made a far more precise appeal to experimental evidence than did Galileo, and his use of a "critical instance" (though there seems to be nothing to suggest that Harvey was at all influenced by his great patient, Francis Bacon) is not paralleled in mechanics. The greatest physiologist of the sixteenth century, Jean Fernel, had not known how to apply the experimental method. Sir Charles Sherrington has expressly pointed the contrast between him and Harvey:

> Fernel, it would seem, in order to do his work, must find it part of a logically conceived world. His data must be presented to him in a form which, according to his own *a priori* reasoning, hangs together. In that demand of his lies his inveterate distrust of empiricism. "We cannot be said to know a thing of which we do not know the cause." And under "cause" he included not only the "how" but the "why." With Harvey it was not so. When asked "why" the blood circulated, his reply had been that he could not say. Fernel welcomed "facts," but especially as pegs for theory; Harvey, whether they were such facts or not, if they were perfectly attested.[1]

To Harvey, anatomy offered the elementary facts of physiology; Fernel, however, wrote that 'in passing from anatomy to physiology—that is to the actions of the body—we pass from what we can see and feel to what is known only by meditation.' He had liberated himself from the occult influence of the stars, but a mechanistic interpretation of physiological process was still alien to his thought. For him the origins of bodily actions had to be sought in the soul, the non-material entity controlling and

[1] *The Endeavour of Jean Fernel* (Cambridge, 1946), p. 143.

directing the operations of the material parts. In this, of course, he simply followed Galen and the Greek tradition.

The ancients had studied the three most obvious instances of the involuntary physiological process—respiration, the beating of the heart, and the digestion of food—and framed a comprehensive theory linking and correlating the phenomena. This theory contained all that Fernel, or any other sixteenth-century physician, knew of the matter. In the first place they discriminated between three "coctions," the first of which turned the food into chyle, transported through the veins of the intestine from the stomach to the liver. This movement of the chyle puzzled Fernel, since he found the veins full of blood instead of white chyle. In the liver the second coction transformed the chyle into blood, which issued forth to the various parts of the body. In these parts the third coction took place, by which the material absorbed from the veins by the flesh was made flesh itself. The coctions were assisted, if not effected, by the natural heat of the animal body, being thus analogous to ordinary domestic cooking, and each had its specific cause in a faculty of the soul.[1] The nutritive faculty, working through the natural spirits, was the agent of the first coction; an attractive faculty drew the blood from the liver along the veins, and from the veins into the flesh. The liver was the source of the blood, and the centre from which it flowed out to the parts, including the heart. This flow of blood was not constant, but rather an ebb-and-flow alternating motion, by which the humours were uniformly distributed about the body.[2] Thus the Ghost in *Hamlet* speaks of the·

> . . . cursed hebenon
> That swift as quicksilver it courses through
> The natural gates and alleys of the body.

The Galenic theory certainly did not postulate that the blood lay stagnant in the veins, but references to this ebb-and-flow from the liver (to which the Latin *circulatio* was sometimes applied) have been misinterpreted as allusions to the true circulation (L. *circuitio*) of the blood. A principal portion of the output of blood from the liver passed up the great vein of the body, the *vena cava*,

---

[1] Fernel, however, likens the second coction in the liver to fermentation; alchemists similarly spoke of the fermentation of metals.

[2] So the sixteenth century understood Galen, who supposed the blood to move, but not in a tidal manner.

to the heart, into the right side of which the blood was attracted by the heart's active dilatation (diastole) (Fig. 7). At the same time air inspired into the lungs was drawn down the venous artery (pulmonary vein) into the left side of the heart. During the phase of contraction of the heart (systole) blood was squeezed from the right side of the heart into the arterial vein (pulmonary artery) for the nourishment of the lungs, and also through the median septum (the thick wall dividing the heart into two main chambers, or

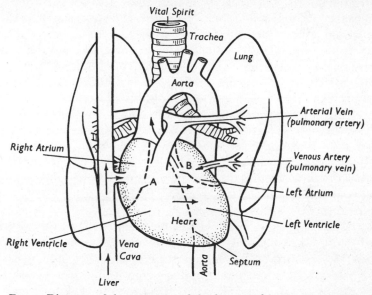

FIG. 7. Diagram of the structure of the heart and lungs, illustrating the Galenic physiology.

ventricles) into the left side of the heart. This was the seat of a most important operation, for there the blood, already containing the "natural spirits" supplied by the liver, was further enriched by taking up "vital spirits" from the air. The blood and vital spirits were conveyed about the body from the left side of the heart by the arterial system. Thus the main function of respiration was to introduce vital spirits into the body *via* the arteries, and of the heart to serve as the organ in which this enrichment of the blood took place—indeed enrichment is an insufficient word, for as venous and arterial blood were distinguished by their colour and viscosity, as well as by their supposed difference in physiological function, it was long denied that they could be the same fluid. The venous, alimentary blood was virtually transmuted by the addition of vital spirit into the spirituous arterial blood.

Another joint task of the heart and lungs was to ventilate the blood and relieve it of its sooty impurities passing along the pulmonary vein and exhaled in the breath. The greater thickness of the aorta and other principal arteries was explained on the grounds that their dense walls had to retain the fugitive vital spirit, but as Harvey emphasized, Galen had not denied that the arteries contain blood as well as spirit. The arterial pulse, the expansion and contraction of the vessels, was not regarded as caused by the similar action of the heart, for Galen judged, as the result of one misleading experiment, that the "pulsive force" was transmitted along the walls of the arteries. Fernel added that if the arteries were swelled by the pulse of blood from the heart they could not pulsate simultaneously along their length, as they do. This belief that the arteries have an active diastole and systole of their own in sympathy with those of the heart was still credited by Descartes, even though he adopted Harvey's circulation. To Fernel this active contraction of the arteries also served to squeeze the vital spirit into the surrounding flesh. He also, like Johann Günther a little earlier, observed that the phases of the heart and arteries are opposite, that when the heart shrinks the arteries swell, and *vice versa*. Some experiments on this seem to have been made by Leonardo. The venous system, arising from the liver, and the arterial system stemming from the heart were thus structurally dissimilar, and since the physiological functions of the blood in the veins and the blood and spirit in the arteries were also different, they were physiologically distinct also. In this theory the active phase of the heart's action was its diastole, by which it drew blood, and vital spirit from the air, to itself.

The Galenical theory was universally adopted by subsequent medical authorities, and became familiar to the Latin West from the writings of Avicenna and Averroes long before the original Greek texts were available or thoroughly understood. Therapeutic directions drawn from the theory varied, but the basic facts were common to all. The two chief physiological statements of the theory: (1) that venous blood nourishes the parts, and (2) that arterial blood supplies the parts with vital spirits, were of course beyond the experimental inquiry of the sixteenth century. The anatomists were, however, able to check upon the agreement between the Galenic conception of the blood's motion and the observed structure of the venous and arterial systems, and of the heart itself. The operation of the valves in the heart offered no problem: their opening and closing was perfectly accounted for. But the density of the septum imposed an act of faith upon Galenical theorists. Berengario da Carpi credulously described the pores or pits in it through which blood passed from right to left ventricle, but Vesalius, probing the pits of the septum, was unable to find a passage, and in the first edition of *De Fabrica* he wrote: 'none of these pits penetrate (at least according to sense) from the right ventricle to the left; therefore indeed I was compelled to marvel at the activity of the Creator of things, in that the blood should sweat from the right ventricle to the left through passages escaping the sight.' In the second edition he expressed his failure to discover Galen's pores even more firmly, and remarked that he doubted somewhat the heart's action in this respect.[1] At least one experiment on the heart is recorded by Vesalius, in which the heart-beat of a dog was restored after opening the thorax by artificially inflating the lungs. Some anatomists, however, still maintained that the passages were easy to find in very young hearts, though concealed in the adult body. Meanwhile, the structures in the veins, later known as valves, had already been observed by Estienne, and from about 1545 were studied by a number of anatomists, such as Amatus Lusitanus (1511–68) who dissected twelve bodies of men and animals at Ferrara in 1547 from which he derived a wholly false theory of their action. As late as 1603 these valves were still misunderstood by Harvey's teacher at Padua, Fabrizio of Aquapendente.

Attention was concentrated, not so much on the motion of the blood, as upon the physiological function of the heart. If the sixteenth-century anatomists had visualized the problem of the ebb and flow of the blood in mechanical terms, as a problem in

---

[1] *De Fabrica* (1543), Bk. VI, Ch. xi, p. 589; (1555), p. 734: 'However conspicuous these pits [in the septum] are, none penetrate (according to sense) from the right ventricle to the left through the intraventricular septum; nor do those passages by which the septum is rendered pervious present themselves to me otherwise than very obscurely, however much they are expatiated upon by the teachers of dissection, because they are persuaded that the blood flows from the right ventricle to the left. Whence also it is (as I shall also advise elsewhere) that I am not a little hesitant concerning the heart's function in this respect.'

hydraulics, the vascular valves would have given them cause to think more profoundly; but this was a post-Harveian conception. The route of the venous blood to the left side of the heart, whence it could issue to the arteries enriched with vital spirits, was however open to discovery. Once the impenetrability of the septum was granted, such blood could only pass *via* the pulmonary artery, the lungs, and the venous artery. This is the so-called "lesser circulation," which is not a circulation at all, for those who discovered this path had no notion that any blood traversed it more than once. It was not, apparently, an original discovery in which Europe has priority. The lesser circulation was accurately stated by an Egyptian or Syrian physician, Ibn al-Nafis al-Qurashi, in the thirteenth century in the course of a commentary on the *Canon* of Avicenna, in which it was deduced specifically from the impermeability of the septum.[1] There is no evidence that his statement was known before very recent years, so that the sixteenth-century discussions appear to be entirely independent. It is significant that in two so different circumstances the same observation elicited the same theory. In Europe the description of the lesser circulation was first printed by the Catalan Miguel Serveto in a theological work, *Christianismi Restitutio* (1553). Serveto was primarily a theologian and though he practised medicine it is not certain that he had a medical degree. In Paris he was associated with the young Vesalius and the veteran Günther, who spoke of him as an anatomist second to none, yet it seems that Serveto's experience in dissection must have been brief. He was greatly interested in medical astrology and his whole knowledge of medicine seems to be somewhat intellectual and literary. There is an element of mystery in the sudden introduction of a physiological heresy into a work that was almost completely obliterated on account of its religious heresy and which was probably written (though not certainly with the passage on the circulation) as much as seven years before its publication. Serveto had some knowledge of Arabic, and it has been conjectured that he may have studied Ibn al-Nafis' text, but it seems unnecessary to postulate that he was less original than the thirteenth-century Syrian. Some scholars judge that Serveto may have been indebted to the Italian anatomists who later described the lesser circulation in print: others think that

[1] On Ibn al-Nafis, cf. Sarton, *op. cit.*, vol. II-2, pp. 1099-1101.

they were indebted to him. This question of priority is not of great importance.[1]

Discussion of the Holy Spirit induced Serveto to write of the three spirits of the blood, the soul of the body (Harvey also wrote "Anima ipsa esse sanguis"). He denied that there was any communication through the septum of the heart: instead, 'the subtle blood, by a great artifice, passes along a duct through the lungs; prepared by the lungs, it is made bright, and transfused from the pulmonary artery to the pulmonary vein. Then in that vein it is mixed with air during inspiration, and purged of impurity on expiration.'

The mixture, he said, takes place in the lungs, where the spiritual blood is given its bright colour, for the ventricle is not large enough for such a copious mixture, nor the elaboration of brightness. He imagined channels connecting the artery and vein in the lung itself, and argued that the artery was far too large for the supply of the lung alone. His physiological conceptions were clearly not very different from those of Galen, save that imbibing of natural spirit by the blood was extended along the pulmonary vein from the heart to the lung. Serveto did not categorically deny that blood sweated through the septum: nor, on one interpretation, had Galen categorically denied that some blood might pass from one side of the heart to the other through the lungs. Such was Harvey's understanding of his words: 'From Galen, that great man, that father of physicians, it appears that the blood passes through the lungs from the pulmonary artery into the minute branches of the pulmonary veins, urged to this both by the pulses of the heart and by the motions of the lung and thorax.'[2] Though it may be doubted that such an interpretation was accepted in the sixteenth century, in Serveto original thinking appears to emerge tentatively from the ideas of the past. His picture of the lesser circulation was very different from that of Harvey.

The same may be said of intermediate presentations of the same idea. In spite of the almost complete destruction of *Christianismi Restitutio*, there is some record of its being read. It has been argued that the treatment of the lesser circulation by another Catalan

[1] For the two arguments, cf. H. P. Bayon: "William Harvey, Physician and Biologist," in *Annals of Science*, vols. III–IV (1938–9); and Josep Trueta: "Michael Servetus and the discovery of the lesser circulation," *Yale Journal of Medicine and Biology*, vol. XXI (1948).
[2] Robert Willis: *Works of William Harvey* (London, 1857), p. 44.

physician, Juan Valverde, in 1554, is imitated directly from that of Serveto, since he stated, like Serveto, that the pulmonary vein contains both blood and air (later, in 1560, he wrote that it contained a copious quantity of blood). Valverde had studied under Realdo Colombo from about 1545 at Pisa and Rome; remarking that he had frequently observed the anatomical appearances with Colombo, he seems to claim no originality for himself. Colombo in turn had been a pupil of Vesalius, succeeding him for a short space in the teaching of anatomy at Padua, and it is possible that the genesis of the idea of the lesser circulation took place there, and so was made known to Valverde. Colombo certainly claimed the new idea as his own, and hitherto unknown, in a treatise published posthumously in 1559, which may well have been written before Valverde's printed in 1556. Certainly Colombo's reasoning on the circulation is superior to any that had preceded it. He made the plain statement that the blood passed from the right ventricle through the pulmonary artery to the lung; was there attenuated; and then together with air was brought through the pulmonary vein to the left ventricle. He relied particularly upon the observation that when the pulmonary vein is opened it is found to be full of bright arterial blood.

From this time the circuit through the lungs from the right side of the heart to the left was described by a number of anatomists, down to the time of William Harvey. It is important to recognize that though these physicians are correctly spoken of as precursors of Harvey (in the sense that this passage of blood through the lungs played a part in the complete theory of the circulation), the lesser circulation as it was understood in the sixteenth century was not a complete fragment of the whole Harveian theory. Harvey understood the lesser circulation in a manner that differed significantly from that of his predecessors. For him it was the path by which all the blood in the body was transferred from the venous to the arterial system: for them it was the path by which a portion of the blood formed in the liver and issuing to the parts became the blood-and-spirits of the arterial system. For him the pulmonary vein contained nothing but arterial blood: for them it contained blood and air. The lesser circulation of the sixteenth century was, as Serveto said, a great artifice for by-passing the impenetrable septum: it led to no other new conception during more than sixty years: it did not suggest the general circulation because it was really not a circulation at all. Galen's physiology was modified, but not entirely displaced. The heart and lungs were still not perceived as the operative organs in the vascular distribution and the identity of the arterial and venous blood was still concealed. The reason for this failure to establish a complete idea of the lesser circulation in the sixteenth century is that the earlier anatomists were attempting to solve a different problem from that of Harvey. They were concerned only to find the route by which blood and vital spirits entered the arteries in view of the impenetrability of the septum. Harvey's problem was two-fold; firstly to account for the function of the valves in the veins (which, as was realized before his time, obstructed the flow of blood outwards along the veins), and secondly to dispose of the large quantity of blood which he knew must enter the heart. The novelty of his approach was that it ignored the question of vital spirits altogether, concentrating upon a wholly mechanical, and partly quantitative, difficulty latent in the accepted doctrine. This difficulty had occurred to no one before, because no one had doubted that the contents of the veins and arteries respectively were absorbed by the parts which attracted them outwards from the central reservoirs, the liver and the heart. The early theory of the lesser circulation was, therefore, useful to Harvey in that, at the proper stage in the development of his own ideas, the transfer of blood from the right to the left side of the heart could be fitted in as a partially complete portion of the puzzle; but that theory in itself was a *cul-de-sac* so long as it was no more than a variation on Galen's.

Harvey began his medical studies at Padua in 1597, the year of his graduation at Cambridge at the age of nineteen. He remained there till 1602. His teacher was Fabrizio of Aquapendente, a late member of the great Italian school of anatomists and embryologists. At this time the lesser circulation was by no means universally accepted, and the valves in the veins were still explained in a variety of mechanically improbable ways. Robert Boyle, in 1688, recorded a conversation with Harvey (d. 1657) in which Harvey had said that he was first induced to think of the circulation by these valves (of whose existence he must have learnt at Padua):

> ...so placed that they gave free passage to the blood towards the heart, but opposed the venal blood the contrary way: he was invited to imagine that so provident a cause as nature had not placed so many

valves without design, and no design seemed more probable than that, since the blood could not well, because of the interposing valves, be sent by the veins to the limbs, it should be sent through the arteries and return through the veins.[1]

This doubtless presents a very foreshortened view of the truth, but it does relate Harvey very definitely to the Italian tradition (as is obvious in many other ways) and it does also indicate that Harvey's view of the problem was from the beginning a mechanical one. The fact that in *De Motu Cordis* the valvular action becomes one argument among many does not impugn the credit of Boyle's statement.

It was natural and fitting that Harvey should have traced the origin of his discovery to the new anatomy of the sixteenth century, for the whole discussion of the vascular system down to his time had been based on advances in observation. While the theory of the lesser circulation was itself framed in accordance with anatomical observation, it had not been examined experimentally, nor did it contain any new physiological interpretation. On both these points Harvey's discovery marks a distinct advance. Firstly, he showed that if the vascular system was analysed hydraulically, considering the heart as a pump, the veins and arteries as pipes, the valves as mechanical valves, the blood itself simply as a fluid, conclusive experiments on the flow of blood could be made. For this purpose he disregarded "spirits" altogether, though he still considered that the heart (not the lungs) restored a spirituous quality to the blood. Secondly, he introduced a new physiological conception in which the arterial blood was revivifying and restorative, while the venous blood was the same fluid returning vitiated and exhausted to the heart where it received its former virtue again. Blood, in fact, was not itself the aliment of the parts, but a vehicle carrying the aliment. Harvey's ideas on this were inevitably vague, and conditioned by the knowledge of his time, but he did conceive of the blood regaining in the heart its 'fluidity, natural heat, and [becoming] powerful, fervid, a kind of treasury of life, and impregnated with spirits, it might be said with balsam.' As cold precedes death, while warmth belongs to life, he saw the heart as the 'cherisher of nature, the original of the native fire' whence new blood, imbued with spirits, was sent

[1] Boyle: *Works*, 1772, vol. V, p. 427.

through the arteries to distribute warmth about the body.[1] On the passage of the blood through the lungs, Harvey's promise to explain his conjectures was not fulfilled: but he indicated that he thought its function was to temper and damp the blood, to prevent it boiling up with its own excessive heat. All Harvey's thought on the physiology of the circulation is obviously pre-chemical, proto-scientific rather than scientific; it does, however, contain the important seminal idea that there is an exchange by which "something" is taken up by the venous blood in the heart (really, of course, the lungs) and given up by the arterial blood to the flesh. Granting the contemporary lack of chemical knowledge, it is only open to the criticism that, in eulogizing the heart, Harvey strangely overlooked the importance of the fact that venous blood becomes arterial in its passage through the lungs from the right ventricle to the left, not in the heart itself. Unable to free himself completely from the error of his predecessors, he could not quite attain the conception of the heart as a pump only, adding neither heat nor spirits nor anything else to the blood passing through it.

Thus Harvey's theory is most perfect in its mechanical aspect, which was fully supported by experiment. His purely anatomical evidence held little that was new, except perhaps his study of the heart as a contractile muscle. He also demonstrated the action of the vascular valves, and the correspondence of the cardiac diastole with the arterial systole, more forcibly than earlier anatomists. Even in anatomy Harvey was successful where his predecessors had failed, in deducing that a mass of discordant observations was made consistent on the single hypothesis of the circulation of the blood; as is most clearly seen in his remarks on the fœtal circulation. The existence of an intercommunication between the pulmonary artery and veins, in the mammalian fœtus, that disappears after birth was familiar to all anatomists, but no one before Harvey had correlated this short-circuiting of the lungs with either the supposed sweating of blood through the septum, or its passage through the lungs. It was left to Harvey to show that the fœtal circulation avoids the lungs because they are collapsed and inactive. He is most original and striking when he uses the comparative method: 'Had anatomists only been as conversant with the dissection of the lower animals as they are with that of the human body, the matters that have hitherto kept them in a

[1] Willis, *op. cit.*, pp. 47, 68.

perplexity of doubt would, in my opinion, have met them freed from every kind of difficulty.'[1]

His admonition was accepted by a host of biologists in the later seventeenth century, including Marcello Malpighi who first observed the blood passing from the arteries to the veins through the capillary vessels in the lungs of a frog—the final link that clinched Harvey's motion in a circle. Harvey found that the action of the heart could be most easily studied through experiments on small animals or fishes, as for instance observing the effect of tying ligatures about the great vessels, in suffusing or draining the chambers of the heart. He correlated the single-chambered heart correctly with the absence of lungs, and the double-chambered heart with the possession of lungs, pointing out that the right ventricle, which only sends the blood through the lungs, is slightly weaker than the left which sends it round the whole body. By experiment Harvey proved that the heart receives and expels during each cycle of expansion and contraction a significant quantity of blood, not a few drops only: by calculation he proved that, on the lowest estimate of the change in volume of the ventricles, all the blood in the body passed through the heart more than once in half an hour. Even this was a generous underestimate. In his second group of experiments, Harvey further demonstrated that the blood moves away from the heart through the arteries, and towards the heart through the veins. These experiments mainly relate to the human subject, and are such as would naturally suggest themselves to a physician practised in phlebotomy. Examining the superficial veins of the arm, he showed that the limb is swollen with blood when the veins are compressed, and emptied of blood when the arterial flow is obstructed. He found that the valves in these veins prevented the flow of blood away from the heart, and that by arterial manipulation it was impossible to force blood through them except in the contrary direction. Blood always filled an emptied vein from the direction of the extremity. Again, he showed that in the jugular vein the valves were so constructed as to permit a unidirectional flow towards the heart only, and that therefore their function was not (as some thought) to prevent the weight of blood falling down to the feet. The experience of wounds and venesection was cited by Harvey to the same general effect, and he further alleged

[1] Willis, *op. cit.*, p. 35.

the experience of physicians as proof that the blood was the mechanical agent by which poisons or the active principals of drugs are rapidly distributed about the whole body.

It is today far more easy, by taking the truth of Harvey's arguments and experiments for granted, to regard his doctrine as an obvious and straightforward deduction from the anatomical history of two earlier generations, than to appreciate the nature of the objections against it. As Galileo remarked in another connection, once this discovery was made its proof was easy: the difficulty was to hit upon it in the first place. Harvey's discovery, like Galileo's, was made at a rudimentary level of science, but the true measure of the intellectual effort involved is the fact that the discovery had escaped all previous anatomists, was greeted with incredulity and scorn, and was not universally accepted even within twenty years. It seems likely that Harvey himself had worked at the problem for at least ten years before he gained the solution. Some, as he said, opposed him because they preferred to endanger truth rather than ancient belief. Others thought that they had discovered technical anatomical arguments against the circulation; or that only a portion of the blood circulated; or that venous and arterial blood could not be the same fluid. Even the basic anatomy of blood-supply to the chief organs of the body (especially the liver) was still doubtful, and its physiological interpretation barely begun; the capillary circulation, and the change in colour of blood, were to remain mysteries long after Harvey's death. His originality was that he preferred to face these new problems, rather than tolerate longer the inconsistencies of the old system, but in this he was followed by few contemporaries. As so often in science, one advance was made not by completely solving an old problem so that no question remained, but by transposing the problem into an answerable form, creating fresh problems by the very act of transposition. Harvey asked a question which, in his precise terms, had perplexed none of his predecessors, and the answer he worked out was important, not only because it was correct, or because it challenged prevailing ideas, or even perhaps because it introduced a new kind of scientific inquiry. Harvey's influence in this last respect was significant (as much in his book on generation as in *De Motu Cordis*) but it was not wholly unheralded, and some of the new methods exploited by later seventeenth-century physiologists, such as bio-chemical research and microscopy, were altogether unknown to him. Perhaps the most important of his achievements was to leave unsolved problems—not blind, impregnable problems, but questions that could be answered in the way he had himself declared. Just as seventeenth-century mechanics was based upon the unsolved (or imperfectly solved) problems left by Galileo, so the experimental problems of biology were inherited from Harvey.

Descartes was not the earliest supporter of the theory of circulation, but he was the first to try to deduce wider implications from it. He himself practised anatomy, and made anatomical experiments. He has been accused of plagiarizing from Harvey in the *Discourse on Method* what he himself did not fully understand. But he assigned to "an English physician" the credit for the discovery of the circulation, and claimed only for himself the elucidation of the mechanism of the heart. In his *Second Disquisition to Riolan* (1649), Harvey had himself commented adversely on the doctrine of spirits: 'Persons of limited information, when they are at a loss to assign a cause for anything, very commonly reply that it is done by the spirits; and so they introduce the spirits upon all occasions. . . .'

Harvey's attitude to authority seems to have sharpened with age, and in this passage it seems clear that he meant to take "spirits" rather as a term in common use, than as having a certain existence. At any rate he declared that the spirits in blood are no more distinct from blood than the spirit of wine from wine itself. Emphatically, in the *Method*, Descartes sought to eliminate the old idea of spirits from physiology altogether.[1] If the human body were purely material, lacking rational or sensitive soul, other than natural heat in the heart (which is compared to the heat of fermentation) it would perform all the functions of the human body except that of thought. Descartes illustrated this deduction by the motion of the heart, which he imagined to work like a crude internal-combustion engine. On contraction, a little blood would be drawn into each ventricle, which being suddenly vaporized in the hot chamber would cause the whole heart to expand and close the inlet valves. This expansion of the blood would also open the outlet valves, so that the blood would pass out into the lungs and arteries, where it would again condense to liquid and the cycle would be repeated. The heat of the heart, about which Harvey

[1] The word was retained, but with a purely chemical meaning.

had written, thus accounted for its purely mechanical cycle of expansion and contraction. In his speculation on the heart, which is neither good engineering nor good physiology, Descartes exceeded Harvey in the functions he assigned to that organ. It supplied heat to the stomach to concoct food; it completed the concoction by distilling the blood in the heart 'one or two hundred times in the day' (according to Descartes, the lungs were the condenser in which the blood was restored to the liquid state); it forced by compression of the blood 'certain of its parts' to pass through pores specially designed like sieves to admit them into the various parts of the body where they formed humours; and it was the hearth where burned a very pure and vivid flame which, ascending to the brain, penetrated through the nerves (imagined as hollow tubes) to activate the muscles. In *On Man* Descartes developed the theory that the flow of the spirits was controlled in the brain by the pineal gland, a sort of valve acting under the direction of conscious volition. According to Descartes' study of the physiology of behaviour, volition played a minor part even in man, who was alone capable of abstract thought and true sensation (that is, sensations capable of objective judgement), and none in the activity of any lesser creature. He devoted much attention to the study of motor mechanisms and reflex actions—as for instance tracing the involuntary mechanisms by which, when the hand is burnt, the muscles of the arm contract to withdraw it from the fire, the facial muscles contract in a grimace of pain, tears flow, and a cry is uttered.[1] He regarded the greater part of bodily activity as due to mechanical processes of this kind, as automatic responses to external stimuli effected by the nervous system; but, though Cartesian physiology was to some extent supported by anatomical investigation of the relations of nerve, brain and muscle, it was in the main a purely conceptual structure. Descartes anticipated some of the conclusions of nineteenth-century physiology without its careful experimental foundation.

Harvey's work was an important step towards a mechanistic approach to biological problems, containing a tentative challenge to the supremacy of spirits founded on a particular experimental investigation. Descartes' more comprehensive and more speculative writings elevated mechanism to a universal truth, in physics and biology alike. Soul and material body could have nothing in

[1] Cf. *De Homine* (Leyden, 1662), pp. 109–10. Sherrington, *op. cit.*, pp. 83–9.

common save a single mysterious point of contact; nothing could be attributed to the soul but thought. The old physiology postulated a variety of non-material souls or spirits each charged with the management of a set of bodily functions; for Descartes those functions were the result of mechanistic processes, as much as the different appearances and movements of an elaborate mechanical clock. This, he said in the *Discourse on Method*, would not appear strange to those acquainted with

> the variety of movements performed by the different automata, or moving machines fabricated by human industry, and that with the help of but few pieces compared with the great variety of bones, muscles, nerves, arteries, veins, and other parts that one finds in the body of each animal. Such persons will look upon this body as a machine made by the hands of God, which is incomparably better arranged, and adequate to movements more admirable than is any machine of human invention.

The body was not maintained alive and active by one or more life-forces, or spirits, or souls, but solely by the interrelations of its mechanical parts, and death was due to a failure of these parts. Therefore, with no non-material factors involved, everything in physiology was potentially within the range of human knowledge, since no more was required than the investigation of mechanistic processes, complex and elaborate indeed. This conception of Descartes' was of course premature, far beyond the scope of the scientific equipment of his age, and it led to no immediate physiological discovery. Except perhaps in his work on the eye, the factual content of his biological theory was wholly misleading. But the influence of his general conception upon the anatomy and physiology of the later seventeenth century was profound. On the whole, those who tried narrowly to demonstrate its truth in particular applications, like Borelli in *On the Motion of Animals* (1680), were least successful, and the attempt to apply mechanical principles to medicine failed. Ultimately the intractability of nature prompted a return to more vitalistic ideas. On the other hand Descartes' justification of experimental inquiry in biology was of permanent value. Terms like "vital force" might conceal deep ignorance without endangering the investigation of those processes through which vital force was supposed to operate. So long as spirits or the Paracelsian *archeus* ruled the body, so long as the functions of its organs had been subject to the influence of the stars and other occult agencies, it had been futile to interpret physiological phenomena in the light of the purely material sciences of physics and chemistry. The barrier between organic and inorganic must have remained for ever absolute. Experiments from which the mystery "life" was excluded would have been useless. To have shown that the transformation of venous into arterial blood can be effected by oxygenation would have been irrelevant to a "spiritual" theory of respiration. The dead liver of a corpse could throw little light on the living liver of a man. Under Descartes' influence, even at the later stage when his mechanism seemed crude to a degree, all this was changed. An organ or a limb could be studied as a part of the whole mechanism, a cog in the works. It could be assumed that what was found to be true of the part in the laboratory must be equally true of the part in the living body; that particular results obtainable from certain experimental processes when observed in the living specimen must be produced by similar processes in its own organization. The basic axiom of experimental science is that, circumstances being unchanged, a like cause will produce a like result because the "cause" releases a chain of events following an unchanging pattern. If this is not so, then the experimental method of inquiry is not one that can usefully be applied to the problem. It was Descartes' discovery (ratiocinative, not empirical) that this was true of physiological phenomena; it could be assumed, *prima facie*, that circumstances were unchanged (e.g. between the living body and the chemist's vessel), and that since functions were automative, like result followed like cause.

The living state was no longer beyond analysis. Descartes' scientific writings, even more than Harvey's, suggested a host of inquiries into the nature of physiological process. Those in which Descartes had been most interested, pertaining to the operation of the nervous system, made little progress before the nineteenth century, though the years immediately after his death saw important work in anatomical neurology. The mechanism of respiration was tackled more successfully, and a pregnant analogy was drawn between combustion and respiration—no doubt owing something to earlier ideas of the heart as the seat of heat. From the members of the Accademia del Cimento through Robert Boyle to the eighteenth century a series of investigators studied the effect of placing small animals *in vacuo*, in confined volumes of air, or of various "elastic fluids" (gases). It was discovered that in the vacuum both combustion and respiration were impossible, and that life ceased. It was discovered that a combustible or an animal consumed air (the carbon dioxide being dissolved in the water which rose up in the vessel), but not all the air, and that the gas remaining after combustion ceased would not support life, as that which was left after respiration ceased would not support combustion. It was further discovered that although vessels could be filled with "fluids" that appeared to be air they would not support life or combustion. Robert Hooke showed (1667) that a dog could be kept alive by blowing into its lungs with a bellows, even with the ribs and diaphragm removed, from which he concluded that the animal 'was ready to die, if either he was left unsupplied, or his lungs only kept full with the same air; and thence conceived, that the true use of respiration was to discharge the fumes of the blood.' Other members of the Royal Society satisfied themselves by experiment that 'the fœtus in the womb has its blood ventilated by the help of the dam'; and that the fœtal circulation depended directly on the maternal.

For a time, as Hooke's words suggest, there was doubt whether the presence of fresh air in the lungs was necessary to remove something from the blood (the "sooty impurities" of Galen's physiology) or to add something to it. On this point the investigations of Richard Lower (1631–91), a physician and an experimental as well as theoretical physiologist, threw new light.[1] In his *Treatise on the Heart* Lower extended Harvey's discovery and defended it against the Cartesian perversions: the heart was not caused to beat by a fermentation of the blood, but by the inflow of spirits from the nerves, and if the nerves were severed the pulsation stopped. The blood, not the heart, was the source of heat, and of the activity and life of bodies—in this Lower, more clearly than either Descartes or Harvey, seems to see the heart as nothing but a mechanical pump. Nor has the heart anything to do with the change in colour of arterial blood, for this can be produced by forcing blood through the insufflated lungs of a dead dog, or even by shaking venous blood in air:

. . . that this red colour is entirely due to the penetration of particles of air into the blood is quite clear from the fact that, while the blood

[1] *Tractatus de Corde* (1669): English translation by K. J. Franklin in *Early Science in Oxford*, vol. IX (Oxford, 1932), especially pp. 164–71.

becomes red throughout its mass in the lungs (because the air diffuses in them through all the particles of blood, and hence becomes more thoroughly mixed with the blood)

venous blood in a vessel only becomes florid on the surface. Lower concluded that the active factor in this transformation of the blood was a certain "nitrous spirit" (elsewhere called a "nitrous food-stuff")[1] which was taken up by the blood in the lungs, and discharged from it 'within the body and the parenchyma of the viscera' to pass out through the pores, leaving the impoverished dark venous blood to return to the heart. Respiration, therefore, was a process whose function was to add something to the blood (Lower remarked that since "bad air" causes disease, there must be a communication between the atmosphere and the blood-stream); but the fuller understanding of the nature of this addition had to await the chemical revolution of the eighteenth century.

The new ideas of blood as a "mechanical" fluid, a vehicle for carrying alimentary substances, constituents of the air, and warmth around the body, suggested the new therapeutic tech-nique of blood transfusion, of which also Lower was a pioneer. The blood had still a semi-magical quality, and as it was thought that "bad" blood could cause debility, frenzy or chronic disease, it seemed logical to suppose that if the blood of a human patient could be replaced by that of a healthy animal, an improvement must result. An Italian who claimed to be the inventor of the method of transfusion (though he admitted he had never tried the experiment) even suggested that it would effect a rejuvenation which should be the prerogative of monarchs alone. Christopher Wren (1632–1723), when an Oxford student, made experiments on the injections of fluids into the veins of animals, by which, according to Sprat, they were 'immediately purg'd, vomited, intoxicated, kill'd, or reviv'd according to the quality of the Liquor injected.'[2] Suggestions for transfusions of blood between animals, and actual attempts to effect it, were made by various Fellows of the Royal Society in 1665, and Lower went into the matter thoroughly, successfully reviving a dog which had been exsanguinated almost to the point of death. Finally, in 1667, Lower performed before the Society the experiment of transfusing the blood of a sheep into a certain 'poor and debauched man . . .

[1] See below, p. 325.
[2] Thomas Sprat: *History of the Royal Society* (3rd Edn., London, 1723), p. 317.

cracked a little in his head,' which the patient luckily survived without any change in his condition. In this Lower had been anticipated by the French physician Jean Denys, whose practice soon after caused the death of a patient, which led to a prohibition of transfusion in France and the abandonment of the English experiments. Several accounts of this time describe the violent reactions produced by the introduction of animal protein into the human blood-stream, which rapidly proves fatal, and doubtless much of the apparent success of these early experiments may be attributed to the clotting of the blood in the tubes used, preventing the passage of more than a small amount. Experiments on transfusion were only resumed in the nineteenth century, when the use of animal blood was abandoned.[1]

While one important aspect of the expanding experimental biology of the seventeenth century was the mechanical and biochemical study of the blood, whose functions figured so largely in the therapeutical theories of the time, in another the essential mystery of "life" was no less involved, and was more directly explored. This was the investigation of generation and the embryonic development of creatures, including man. Just as interest in the motion and functions of the blood may be traced to its prominent place in the Galenic theory of humours, so these embryological researches return in a continuous tradition to the work of Aristotle. Of William Harvey himself, certainly one of the greatest embryologists of the seventeenth century, it has been said that he did not follow the example of some of his predecessors in departing from Aristoteleanism, but on the contrary lent his authority to a somewhat moribund outlook.[2] On one important matter, however, Harvey contradicted Aristotle altogether: he was sceptical of spontaneous generation, and if he did not coin the phrase *omne vivum ex ovo*, it epitomizes his thought. The partial discredit of spontaneous generation (not complete, for the idea was revived in the eighteenth century, when it was refuted experimentally by Spallanzani, and again in the nineteenth in opposition to Pasteur) was one of the most important changes in biological thought of the time; a first step towards modern conceptions of the living state. Formerly organisms had been

[1] Cf. Geoffrey Keynes: *History of Blood Transfusion, 1628–1914* (Penguin Science News 3, 1947).
[2] Joseph Needham: *History of Embryology* (Cambridge, 1934), p. 128.

divided into four distinct groups: (1) those that are generated spontaneously, (2) vegetative, (3) animal, (4) human. The first had only, as it were, a share of the "world-soul"; the others were distinguished according to the "souls" of their class. As long as these distinctions persisted, founded on fundamental unlikenesses attributed to the structure of the world-order reflecting disparities in the original created endowment, it was impossible to approach the modern conception of life-processes depending upon the nature and complexity of the physiological functioning of the organism. To make the generalization that the life-process is transmitted solely and invariably through specific mechanisms of reproduction was a necessary first step towards a rational understanding of the nature of this process, and the differences between matter in the living and the non-living state: indeed, there seems something fundamentally irrational in the supposition that the organization of the living from the non-living might be fortuitous, and commonplace at that. The crude, superficial differences between the life of a plant and that of an animal are at least capable of recognition; but what meaning could be attributed to the difference between the life of a spontaneously generated mistletoe or worm, and that of other analogous forms? The doctrine of spontaneous generation had, moreover, become the refuge for superstitions and fables of the most absurd character, wholly inconsistent with any serious study of natural history.

Harvey, it is true, wrote in *De Motu Cordis* that the heart is not found 'as a distinct and separate part in all animals; some, such as the zoophytes, have no heart,' and he continued, 'I may instance grubs and earthworms, and those that are engendered of putrefaction, and do not preserve their species.' If this was not merely a careless phrase Harvey changed his opinion, for in his later work *On the Generation of Animals* (1651), he declared:

> . . . many animals, especially insects, arise and are propagated from elements and seeds so small as to be invisible (like atoms flying in the air), scattered and dispersed here and there by the winds; and yet these animals are supposed to have arisen spontaneously, or from decomposition, because their ova are nowhere to be seen.[1]

Before such a statement could be given real force and meaning, the arts of natural observation, of comparative anatomy, and of

[1] Willis' *Works of Harvey*, p. 321. Harvey, however, did not give up the use of the term "spontaneous generation."

simple controlled biological experimentation must be developed
to, or beyond, the level which they had reached among the
ancient Greeks. Aristotle's biological knowledge was in many
respects far superior to anything that was available in the sixteenth
century—indeed some of his observations were not to be verified
before the nineteenth. It is astonishing to find, for example, that
Aristotle's sensible and penetrating observation of the process of
reproduction among bees—which itself was not quite correct—
was universally ignored up to modern times, while credence was
given to fabulous tales of their generation in the flesh of a dead
calf or lion which, besides appearing in the works of many Roman
poets and writers on agriculture, were retailed in the sixteenth
century and later by naturalists like Aldrovandi, Moufet and
Johnson, and by the philosophers Cardan and Gassendi. Even the
relatively simple life-cycle of the frog was a mystery, at least to
academic naturalists.

Harvey had conjectured that in some cases the invisible "seed"
of creatures was disseminated by the wind. The man who set
himself to confute the widespread fallacy of spontaneous generation systematically was Francesco Redi (1626–78), an Italian
physician who worked under the patronage of the Dukes of
Florence and was an important member of the Accademia del
Cimento.[1] His observations and experiments were varied and
numerous, but the most telling were the most simple. Thus he
was able to prove, by the simplest means, that decaying flesh only
generated "worms" when flies were allowed to settle on it; that
the larvæ turned into pupæ (which he called eggs) from which
hatched flies of the same kind; and that the adult flies which
infested the putrefying material possessed ovaries or ducts containing hundreds of eggs. Generalizing from such results, Redi
pronounced that all kinds of plants and animals arise solely from
the true seeds of other plants and animals of the same kind, and
thus preserve their species. Putrescent matter served only as a
nest for the eggs, and to nourish the larvæ hatched from them.
However, he had to admit that there were some examples of
generation which he could not explain. Intestinal worms and other
parasites puzzled him, and he failed to discover the cause of the
growth of oak-galls on trees, which was traced later by Malpighi.
This led Redi to speculate somewhat loosely on possible perversions

[1] See below, p. 189.

in the "life-force" of host organisms which might produce parasitic developments. Micro-biology was only coming into existence at the time when he wrote, and its absence set a natural limit to the range of his investigations.

Nevertheless, Redi's demonstrations, combined with the later work of such naturalists as Malpighi and Swammerdam, were generally regarded as sufficiently cogent against the doctrine of spontaneous generation. The second half of the seventeenth century was a period in which, partly through animal and vegetable anatomy, partly through the use of the microscope, and partly also through experiment, many of the mysteries concerning the less obvious processes of reproduction were being cleared up. The sexuality of plants, first asserted by Nehemiah Grew, was established experimentally by Camerarius before 1694.[1] But if the general tendency was for the exclusion of pangenesis, the experimentalists were not inclined to hasten towards a purely mechanistic interpretation. The embryological speculations of Gassendi and Descartes found few followers. Harvey had written, 'he takes the right and pious view of the matter who derives all generation from the same eternal and omnipotent Deity, on whose nod the universe itself depends . . . whether it be God, Nature or the Soul of the universe,' though this did not prevent his studying the phenomena with all attention. Similarly, John Ray in the *Wisdom of God* (1693) related his discussion of the fallacy of spontaneous generation to the fixed, created nature of species. Ray's world was a machine in the sense that he doubted—from the cessation of creation on the sixth day—the divine institution of new species (or the endowment of matter with life *de novo*) but for him life was transmissible only through the recurring generations springing from the original ancestors; since the power of living was confined to the whole group of creatures extant at any moment, it could not be born of any conjunction of purely mechanical circumstances.

Despite the limitations in philosophic outlook, which denied to many experienced naturalists and to Harvey in particular any vision of the ultimate potentialities of the admittedly crude

---

[1] When his *Letter on the Sex of Plants* was published. There were ancient and popular forms of this idea—artificial fertilization of the date-palm had been practised in pre-classical antiquity—but it had not acquired any previous scientific validity.

physico-chemical speculations of the time, the history of embryology offers a useful example of the critical application of observation and experiment to the consideration of scientific concepts
of a complex order. This was possible for a variety of reasons,
which point to some significant analogies between the situation
in this science, and that in the physical sciences where so much
progress was made. It was important in this branch of biology
that there were ideas to be challenged or confirmed, problems
that demanded inquiry, far more obviously than in the purely
descriptive departments. What were the respective contributions
of the male and female parents to their offspring? Were the parts
"formed" or did they merely "grow"? What was the function of
the amniotic fluid, or the fœtal circulation? How was the embryo
nourished, or enabled to breathe? Aristotle's systematic account
had attempted to deal with such questions; his exactness in
biological observation and his acuteness in biological reasoning
were examined no less thoroughly in the sixteenth and seventeenth
centuries than were his doctrines relating to the physical sciences.
As Galileo had wielded the method of Archimedes against
Aristotle, so in effect Harvey and Redi applied the methods of
Aristotle as observer against the conclusions of Aristotle as theorist.
In embryology there was as effective a classical tradition to focus
attention on the critical points as in cosmology or mechanics. Of
course the strategic gains were far less—there was no dramatic
scientific revolution—but the tactical advance in method and
analysis was no less real. Though to later minds some of the
questions asked by the seventeenth-century embryologist are
meaningless, though the teleological cast of his thought has
proved fruitless, the tradition of investigation has continued
unbroken, and some descriptions of observations made at this
time have never been surpassed. The difficulties to be overcome
in any analytical department of biology were far greater than
those which the new mechanics solved, while in addition the
biologist lacked the logical procedures of physical science. Serious
limiting-factors in the development of those subjects were to
disappear only in the nineteenth century: in the late seventeenth,
however, their techniques were greatly enriched by the use of the
microscope, which will be discussed in a later chapter.

CHAPTER VI

THE PRINCIPLES OF SCIENCE IN THE EARLY SEVENTEENTH CENTURY

ONSCIOUS reflection on the relations between man and his
natural environment can only be a product of an advanced
state of civilization in which abstract thought flourishes.
Greek philosophers seem to have been the first to discuss the
problem, how can reason be most successfully applied to understanding the complex phenomena of material things?—and in so
doing they introduced the generalizing of ideas that is essential
to science and distinguishes it from the *ad hoc* solving of practical
problems undertaken in man's struggle with nature. It is generally
agreed that the foundations of scientific knowledge cannot be
settled, or even verified, by the normal processes of science itself.
If such questions are asked as, what is the status of a scientific
theory? or, what is the meaning of the word "explanation" in
science? or, to what extent is science a logical structure? they
cannot be answered without transcending the framework of
science. The scientist must have some idea, which is essentially
philosophical, of how he is going to set about acquiring an
understanding of nature before he can apply himself to this task.
He may in practice be entirely uninterested in philosophy, preferring to regard himself as a compiler of demonstrable facts;
nevertheless, he cannot escape the implications of adopting a
definite scientific method, which teaches him to record particular
kinds of facts, by using certain recognized procedures. Thus the
nature of science in different periods has been determined by the
methods employed in collecting facts and reasoning about them,
and by the prevailing approach to the study of natural phenomena. For example, when the mechanistic philosophy of the
seventeenth century replaced the teleological outlook of earlier
times the change in the character of scientific explanation was
profound: it was no longer sufficient to ascribe the pattern of
events to divine purpose or the necessary conditions for human
existence.

Consequently, when comparing the scientific achievements of one epoch with those of another, it must be recognized that the aims and methods of scientific activity may themselves vary. The fundamental philosophy of science is neither fixed, nor static, nor inevitable. It cannot be claimed that any scientific method is correct, without considering the nature of the objects it seeks to achieve. Both may be subjected to criticism, for it may be asked whether scientists have the proper aims, or whether they are using fit methods; and, indeed, from the thirteenth to the seventeenth centuries there was continuous and effective criticism of science from each of these points of view. During the eighteenth and nineteenth centuries, however, there was a tendency for practising scientists to feel a confident complacency concerning their aims and methods, and to envelop themselves in an impenetrable detachment from any attempt to interpret their activities philosophically. They were scientists, devoted to a peculiarly rigorous pursuit of knowledge, not natural philosophers. They despised metaphysics and logic. Their limited outlook, and their often shallow pragmatism, would have been intolerable to the Greek founders of scientific method.

In essence, the Greek notion of scientific explanation (passing into the European tradition through the medieval dependence on the philosophy of antiquity) did not differ from that of modern science. When a phenomenon had been accurately described so that its characteristics were known, it was explained by relating it to the series of general or universal truths. The most important distinction between Hellenistic science (including that of the middle ages) and modern science is in the constitution of these universals, and the methods of recognizing them with certainty. For Platonists the universal truths were Ideas, the principal task of their philosophy—it cannot properly be called science—being the elucidation of the ideal world of perfect Forms of which the tangible world was a clumsy model framed in imperfect matter. Aristoteleans, on the other hand, denied the separability of Idea (or Form) and Matter (or Substance), but nevertheless were concerned with the processes by which Forms, the generalizations of their science, were detected in the materials provided by sense-perception. Phenomena were then explained by comprehending them within the a priori scheme of Forms. This procedure required, not additions to the already bewildering variety of fact, but the exercise of reason upon the facts requiring organization; thus in the *Physics*, for example, proceeding from the known facts of motion and change, Aristotle discusses the logical meaning to be attached to the idea "motion," and then from the idea of motion deduces the properties of moving things. In this method experience and observation are adduced, less to provide bricks to construct the fabric of the argument, than to give examples of the author's meaning. As Aristotle declares in the opening of the *Physics*: 'Plainly in the science of Nature, as in other branches of study, our first task will be to try to determine what relates to its first principles,' principles however whose validity was tested by the rule of reason not that of experiment. As when (to cite the *Physics* once more) he denies the existence of a vacuum on the ground that bodies would move in it at an infinite speed, he makes use of the logical impossibility of a body being in two places at the same time. Aristotle derived his universal truths *before* the application of intensive inquiry to the phenomena themselves, a procedure which Francis Bacon contrasted with his own inductive method:

> There are two ways, and can only be two, of seeking and finding truth. The one, from sense and reason, takes a flight to the most general axioms, and from these principles and their truth, settled once for all, invents and judges of all intermediate axioms. The other method collects axioms from sense and particulars, ascending continuously and by degrees so that in the end it arrives at the most general axioms. This latter is the only true one, but never hitherto tried.[1]

In building up a body of scientific knowledge other series of steps than those of Aristotle were used by the Greeks. One was a logical process, akin to that of mathematics, applied where mathematical analysis of the phenomena was feasible. This began with a series of axioms or postulates, defining the conditions of equilibrium in statics, or the geometrical character of the propagation of light in optics, then deduced the consequences of these axioms in the same way that the Euclidean geometry of space is deduced from definitions of point and line. The validity of the postulates was purely experiential, not deriving from any more fundamental truths, so that should it be found, for instance, that equilibrium could be produced in other conditions than those

[1] *Novum Organum*, Bk. I, xix.

envisaged the science of statics would be false or incomplete; and in the same way the validity of their consequences might be illustrated from experience. Other Greek writers, notably the successors of Aristotle at the Lycæum, founded their doctrines firmly on a discussion of experimental data—a method which Aristotle had used in his biological works.[1] A large group of later Greek scientists were far from merely speculative in their interests and confined themselves strictly to the discussion of facts, which they sought to extend and enrich by experiment and observation. In this way "the sciences," to contrast them with natural philosophy, came into existence. But their works, though permitting quantitative comparisons with experience (as in Ptolemy's astronomy), were fragmentary in scope and did not combine into a systematic interpretation of the universe as a whole. For this the middle ages turned to the Aristotle of the *Physics* and *On the Heaven*.

As a consequence of his authority the universal truths of Peripatetic science, originally formulated by sorting out and abstracting with the aid of logic the confused information imparted by sense-perception, tended to become the unquestionable arbiters of thought. However, though the attempt to rationalize the phenomena of nature by reference to them might often be productive of nothing more than intricate mental gymnastics, it could occasionally lead to a more factual inquiry. Nor were "the sciences" of the later Greeks wholly neglected by a line of the more critical and heterodox medieval philosophers. Thus arose a problem of method: how were the fruits of the mathematico-logical or experiential procedures in investigating nature to be reconciled with the knowledge gained by deduction from Aristotle's universals? How did knowledge of a phenomenon, in terms of its relationship to Aristotle's theory of nature, differ from knowledge of the same thing in terms of the complete description of the chain of events producing it? The interest of such questions was doubtless heightened by the technological progress of the middle ages, which literary men did not disregard. There was much new knowledge to be systematized (in the field of chemistry, for example), new properties of bodies like magnetism to be explored. And though the main medieval opinion echoed Plato

[1] On Strato, cf. Benjamin Farrington, *Greek Science* (Penguin Edn., vol. II, pp. 27–44).

and Aristotle in its definition of knowledge as "understanding" (passive knowledge) rather than "power to control" (active knowledge), a minority foreshadowed a less contemplative, more practical view of science. Yet the evidence for actual experimentation by medieval philosophers is not massive, even in optics, and some of the warmest advocates of the experimental method, like Roger Bacon, were guilty of misstatements of fact which the most trifling experiment would have corrected. More successfully, they discussed the intellectual problem of the relationship between facts and theories, modifying Aristotle's teaching with regard to the acquisition of knowledge, the nature of causation, and the character of the proof of a proposition. It has been suggested that the middle ages witnessed the philosophical development of the experimental method, through whose systematic application the dramatic changes of the scientific revolution were effected.[1] Attention has been drawn to the empiricism of such philosophers as William of Ockham, who regarded as "real" only that which could be perceived by sensation, and to the notion of explaining an event by giving a history of its antecedents, among other signs of the future onslaught upon the foundations of Aristotelean physics. But traditional ideas continued to satisfy the majority until the seventeenth century, and the philosophic conception of empiricism was a very different thing from its application to scientific problems.

No important re-statement of scientific method was made during the sixteenth century. The medieval tradition continued to run strong, yielding wider divagations from Peripatetic orthodoxy in Leonardo or Copernicus, or the philosophers Cardan and Telesio, but there was no philosopher independent enough to transfer the weight of scientific authority to completely new bases. Instead, it may be noticed that a number of the more interesting developments in the science of the period signify that science was outgrowing its cradle, philosophy. They bear the mark of the practical hand of a Vesalius or an Agricola. Since no true science can remain permanently at the level of simple description, the first half of the seventeenth century witnessed further efforts at

---

[1] e.g., A. C. Crombie: *From Augustine to Galileo* (London, 1952), p. 217, etc.: 'The development of the inductive side of natural science, and in fact the thorough-going conception of natural science as a matter of experiment as well as of mathematics, may well be considered the chief advance made by the Latin Christians over the Greeks and Arabs.'

rationalization. Of these two were properly philosophic and systematic: Francis Bacon aimed at an exclusively experiential method of scientific inquiry, while Descartes fabricated a new logical key to nature. The third was Galileo's conscious reformation of the processes of scientific reasoning in mechanics, which was less a philosophy of science applied to the solving of particular problems than a reconstruction of a department of science which necessitated the introduction of far-reaching philosophic principles.

The types of question which a scientist may ask—which the method he uses enables him to formulate, and possibly to solve— may of course be infinitely varied. But here it is useful to single out two groups among them. The first kind may begin with such words as "How can we demonstrate that . . .", "How may it be proved that . . .". These questions are defined, for the investigator has an idea, however vague, and seeks to test it. The second kind are undefined, taking such a form as "What are the factors involved in . . .", "What is the relationship between . . .", "What are the facts bearing upon . . .". Here the investigator has hardly arrived at the stage of ascertaining and ordering the relevant facts, much less examining a hypothesis. Copernicus' problem falls into the first of these groups, and the problems of seventeenth-century chemistry and physiology into the second. Part of Bacon's significance in the history of science resides in his realization of the insufficiency of the first aspect of scientific method alone, and though he made a notable contribution towards it, the second aspect attracted his main interest. He found the theory of scientific explanation that he encountered insufficient, partly perhaps because his education in renaissance humanism gave him little acquaintance with the natural philosophers of the late middle ages; but still more warmly, he condemned inattention to the methods by which the range of scientific facts could be enlarged, or the facts themselves tested and more closely knit together. It was one of his favourite themes that for many centuries genuine contributions to solid knowledge had been the work of artisans rather than of philosophers, of the weavers of speculation. Bacon's writings have often been described as though his criticisms and proposals were directly and solely the result of his social sense; as though, because he believed that progress in material civilization was a worthy end (a belief shared by both Galileo and Descartes), he reasoned that the single function of science was to enhance man's command over natural forces. He has thus been depicted as the first philosopher to appreciate the potentialities of science as the servant of industrial progress. The truth seems to be rather more complex. It is not even true that Bacon was the first to see science as a powerful agent in improving material welfare, for this point had been made by the empiricists of the middle ages and was part of the common descent of magic and science. Nor was Bacon merely a philosophical technologist; if he wrote

> the true and lawful goal of the sciences is none other than this: that human life be endowed with new discoveries and power,

he also declared, more emphatically, that as

> the beholding of the light is itself a more excellent and a fairer thing than all the uses of it—so assuredly the very contemplation of things as they are, without superstition or imposture, error or confusion, is in itself more worthy than all the fruit of inventions . . . we must from experience of every kind first endeavour to discover true causes and axioms, and seek for experiments of Light, not for experiments of Fruit.[1]

Many passages in Bacon's writings indicate that he had a philosophic appreciation of the value of knowledge for its own sake, not merely for its utilitarian applications. The test by works, in Bacon's thought, assumed a particular importance not because works were the main end of science, but rather because they guaranteed the rectitude of the method used. A discovery or explanation which was barren of works could hold no positive merit not because it was useless to man, but because it lacked contact with reality and possibility of demonstration. Since Bacon's science was to deal with real things, its fruits must be real and perceptible.

Measured by these standards, Aristotelean science was a hollow structure, dealing with abstractions rather than real things, justified by no fertility in works. Bacon did not deny that there was truth in the content of orthodox science—he was quite as certain as Aristotle of the stability of the earth—but these truths were buried in a misleading and sterile philosophy. His remedy was to return to a consideration of the bare facts, and above all to increase vastly the range of facts available. Only when *all* the

---

[1] *Novum Organum*, Bk. I, lxxxi, cxxix, lxx.

material upon a particular phenomenon, or natural process, had been collected, classified and tabulated could any general conclusions be drawn from it and generalizations be framed. The facts might be collected from experience, from reliable reports, from the lore of craftsmen, but above all from designed experiment. For Bacon clearly conceived of experiment not merely as a trial "to see what happens," but as a way of answering specific questions. The task of an investigator was to propose questions capable of an experimental answer, which could then be recorded as a new fact appertaining to the phenomenon under study. In this way the lists of "instances" were to be built up, as Bacon attempted himself to construct tables of instances of heat and motion. Other aids were then required in the intellectual process of finding order in the mass of fact compiled, for which also Bacon made suggestions. In the *New Atlantis* the work of the fact-gatherers is separated altogether from the work of the fact-interpreters, and this has been criticized as a defect in Bacon's system. Yet in practice in science it has often happened that a new generalization in theory has interpreted a mass of evidence assembled by a line of earlier experimental investigators.

Some comments on science have denied categorically that there is any such thing as a specific "scientific method," by saying, for example, that science is organized common sense. Bacon's method at least seems to suffer from excessive formalization and a top-heavy logical apparatus. Even in his own ventures into scientific research Bacon did not observe his complex rules very strictly, and hence the once popular notion that he invented and described the method of experimental science is no longer acceptable. Modern science was not consciously modelled upon Bacon's system. Mathematical reasoning especially, so freely and successfully exploited from the earliest stages of the scientific revolution, he never understood so that its essential rôle was hidden from him. It has also been said, with less justice, that the integration of theory and experiment typical of modern research was not allowed for in his system and that he did not foresee the importance of hypothesis in the conduct of an investigation. In fact Bacon did envisage the situation where reflection on the facts suggests several possible theories, and discussed the procedure to be adopted for the isolation of the correct one by falsification of the others. And certainly he understood the decisive nature of a "crucial experiment" in judging the merit of an idea taking shape in the investigator's mind. It should not be forgotten, too, that pure fact-collection (the first stage in Bacon's system) has been a most important fraction of all scientific work up to the present time. Even the routine verification of measurements, or the establishment of precise constants, has been productive of original discoveries. It is true that the main course of physical science in the seventeenth century ran in a very different direction, that in the new mechanics of Galileo the plodding fact-gathering imagined by Bacon had little significance; elsewhere in science, however, where the organization of ideas was less advanced and the material far more complex and subtle, the straightforward acquisition of accurate information was a more fruitful endeavour than premature efforts at conceptualization. This is most clearly true of the biological sciences; no Galileo could have defined the strategic ideas of geology or physiology which only emerged from the wider and deeper knowledge of facts obtained in the nineteenth century. Bacon's advice that solid facts, certified by experiment, should be collected and recorded was sound and practical; this task occupied chemistry and biology till towards the end of the next century. But in the long run the great generalizations in these fields did not follow from the kind of digestion and sublimation of fact that Bacon had described.

While Bacon's works gave a useful impetus to the growing interest in science, especially in England, his attempt to define the intellectual processes involved in the understanding of nature was limited and only partially helpful. Empiricism alone is an insufficient instrument in science. The history of the scientific revolution shows the fertility of the critical examination of concepts and theories, even when the modification of the simple account of the facts is insignificant (as with Galileo's new concepts of inertia and acceleration). Bacon's views were characterized by his approach to science, which was that of a philosopher rather than that of an experienced investigator. His own ventures in research are notoriously uninteresting and unproductive, for except in his leaning towards an atomistic materialism he was out of sympathy with the progressive ideas of the time, and remarkably indifferent to those developments which posterity has found most significant. His logical system was for the most part ignored by those who were finding how to make discoveries regardless of logical systems, and consequently modern science did not so much grow up through Bacon as around him.

Galileo offers a very different picture. On the one hand he was mainly occupied with purely scientific matters and the discussion of specific problems. He did not construct a methodical philosophy of science, though the elements of such a philosophy may be extracted from his works. On the other hand he may properly be described as a philosopher, for his conscious reflection on the obstructions to be overcome in arriving at a clear and confident understanding of nature is explicit in a number of passages and implicitly conditions the revolution in ideas that he effected. Like other major critics of Aristotle, Galileo was faced with two inescapable problems: on what foundations was the intellectual structure of science to be built, and what criteria of a satisfactory explanation were to replace those of Aristotle? With Galileo these questions were not answered in prolonged metaphysical or logical analyses—though it seems clear that his ideas were shaped by just such analyses carried out by his predecessors—but the answers were given as they became necessary in the progress of his attack on the prevailing ideas of nature. As scientist Galileo's aim might be to detect Aristotle's errors in fact or reason, while as philosopher he demonstrated more fundamentally how these errors had arisen from weaknesses in method that were to be avoided by taking a different course. The negative exposure of an isolated mistake by means of experiment or measurement was not, in Galileo's view, the sole advance of which the new philosophy was capable.

Galileo's two greatest treatises are polemics. They do not relate how certain conclusions were reached, instead they seek to prove that these conclusions are certainly true. Their arguments are therefore synthetic, and the texture of reasoning and experience is so woven that experience appears less as a peg upon which a deduction depends, than as an ocular witness to its validity. It is universally the case that the methods by which a discovery is made and expounded differ, in varying degrees, and Galileo rarely used the direct technique of reporting and inference, so much favoured later by the English empiricists. In the *Dialogue* and *Discourses* the foundations of scientific knowledge are shown to reside in phenomena and axioms conjointly. By its attention to actual phenomena Galilean science was made real and experiential; by its use of the capacity of the mind to apprehend axiomatic truths its logic was made analogous to that of mathematics. The latter were indeed generalized from the former, but the process might involve historical as well as philosophical elements. Thus a fundamental axiom of the *Dialogue* is that heavenly bodies participate in uniform circular motion, while in the *Discourses* successive propositions in dynamics are deduced from the axiomatic definition of uniform acceleration. Such axioms, illustrated and confirmed by experiment, become the starting-point for arguments through which their implications are unfolded (in the manner of Euclidean geometry or Archimedean statics) and again in turn verified by experience, or applied to specific problems, such as the isochronism of the pendulum.

Galileo's remarks on the procedure to be adopted in arriving at these principal generalizations are therefore of special interest. The most important step is that of abstraction. The essential generalizations are not to be taken as the end-product of the logical examination of an idea, in the manner of Aristotle, but are obtained by abstracting everything but the universal element in a particular phenomenon, or class. So far Galileo agrees with Bacon, though he offered no comparable set of logical rules for effecting this operation. He went on, however, to insist emphatically that by abstraction it is learnt that the real properties of bodies are purely physical, that is, size, shape, motion, propinquity, etc., not colour, taste or smell so that as he stated in the *Saggiatore*, the "accidents, affections and qualities" attributed to them are not inherent in the bodies at all, but are names given to sensations stimulated in the observer by the physical constitution of that which he perceives. Galileo noted that this failure to abstract from sensations to the underlying physical reality had given rise to much confusion in the study of heat; physically considered (he says) there is no mystery in heat, which is merely a name applied to a sensation produced by the motion of a multitude of small corpuscles, having a certain shape and velocity, whose penetrations into the substance of the human body arouse such sensation.[1] In these opinions the influence of Epicurean atomism is evident, one might say that this whole approach to the question

---

[1] *Il Saggiatore* (Bologna, 1655), pp. 150–3. Bacon also agreed that 'heat is an expansive motion restrained, and striving to exert itself in the smaller particles.' (*Novum Organum*, Bk. II, xx.)

of primary and secondary qualities is determined by a mechanistic notion of the composition of matter. The explanation of a scientific problem is truly begun when it is reduced to its basic terms of matter and motion—the transformation which remained the ideal of classical physics. The name *heat* could not be a cause, since as Galileo pointed out there is nothing between the physical properties of bodies with the varying motions and sizes of their component particles and the subjective perceptions of the observer. He found other instances in conventional science of this tendency to believe that matters could be explained by juggling with abstract names, as when in the *Dialogue* gravity is defined as only the *name* of that which causes heavy bodies to fall; naming does not contribute to understanding. Of course Galileo did not mean that there is no purpose in classifying phenomena; his argument is that classifications based on superficial characteristics and naïve analyses are misleading because they conceal physical realities. They had concealed the universal generalizations on motion, whether caused by gravity or any other force, which it was Galileo's main achievement to reveal.[1]

In the process of abstraction an important aid was mathematics. If the elements of a problem were capable of statement in numerical terms then the most exact definition had been framed and the most general case considered. Moreover, by transposition into mathematical language the conditions of the problem could be exactly prescribed, in order to remove the imperfections and minor variations that always occur in actual experience. As in geometry the areas of triangles can be calculated more precisely than they can be measured, so in mechanics the properties of the lever, the inclined plane, the rolling sphere could be calculated by reducing these physical bodies to geometrical forms whose behaviour could be established by abstraction from that of their physical counterparts. Galileo knew that this was a function of the imagination; a calculus may solve a problem, but the due conditions must be postulated before the calculation can begin. The motion of a perfect sphere on a perfect plane must be inferred imaginatively from the motion of physical spheres on physical planes, or the motion of an ideal pendulum from that of actual

---

[1] Similarly, a chief problem of the early botanical taxonomists was to penetrate below the superficial differences and similarities in plants to more "real" morphological distinctions.

oscillating bodies, but this, for Galileo, was strictly analogous to Euclid's abstraction in the definition of space, or Archimedes' assuming that the cords hanging from the ends of a balance are geometrically parallel. Mathematics, serving as a guide to the imagination as well as handling the abstracted properties, could yield further statements which, through reference to experience, might confirm the process of abstraction and the generalizations derived from it. In this ideal world of abstraction, without resistance or friction, in which bodies were perfectly smooth and planes infinite, where gravity was always a strictly perpendicular force and projectiles described the most exquisitely exact parabolas, the principles of Euclidean geometry held absolutely. The world of Galileo's imagination in mechanics was in fact Euclid's geometrical space with the addition of mass (later defined precisely by Newton), motion and gravity. The secret of science, in Galileo's outlook, was to transfer a problem, properly defined, to this abstracted physical universe of science which, as ever greater complexities are added to it, approximates more and more closely to the actual universe.

This procedure was not quite new. Optical writers had always treated light-rays and reflecting and refracting surfaces purely geometrically; Archimedes, Galileo's model, had submitted statics to geometry. But no one before had extended the mathematical method of reasoning to the motions of real bodies, nor been so bold as to declare that this method was valid through the whole range of physics: that indeed it was the *only* valid method. For (to Galileo) the mathematical method alone offered certainty of proof. It should be used in preference to all others, even to experiment, since experiment and observation often deceived those who sought to interpret them. When a problem had been set out in mathematical form and a given conclusion reached without error, that conclusion must be correct if the first assumptions were correct. Nor did this method desert the reality of the physical world, since to Galileo the book of nature was 'written in mathematical language . . . the letters being triangles, circles and other figures without which it is humanly impossible to comprehend a single word.' The architecture of the real world was no less geometrical than that of abstract Euclidean space. Nor was there any distinction between "real truth" and mathematical truth. If efforts to mathematize nature fail, it is merely because the task has been undertaken improperly. A physical plane is not a geometrical plane, indeed, but its departures from geometrical planeness are in turn expressible in mathematics. It is a question simply of having skill to unfold the successive layers of mathematical complexity in nature. Reasoning thus, Galileo has been called a Platonist because he sought for the mathematical ideal in nature; but Galileo also perceived that while mathematical logic is infallible, it may rest on false assumptions, like those of the Ptolemaic system, which 'although it satisfied an *Astronomer meerly Arithmetical*, yet it did not afford satisfaction to the *Astronomer Phylosophical*.'

By the method of abstraction, moreover, the scientific concept of "laws of nature" was simply and neatly accommodated. This concept, unknown both to the ancient world and to the Far Eastern peoples, seems to have arisen from a peculiar interaction between the religious, philosophic and legalistic ideas of the medieval European world. It is apparently related to the concept of natural law in the social and moral senses familiar to medieval jurists, and signifies a notable departure from the Greek attitude to nature. The use of the word "law" in such contexts would have been unintelligible in antiquity, whereas the Hebraic and Christian belief in a deity who was at once Creator and Law-giver rendered it valid. The existence of laws of nature was a necessary consequence of design in nature, for how otherwise could the integrity of the design be perpetuated? Man alone had been given free-will, the power to transgress the laws he was required to observe; the planets had not been granted power to deviate from their orbits. Hence the regularity of the planetary motions, for example, ascribed by Aristotle to the surveillance of intelligences, could be accounted for as obedience to the divine decrees. The Creator had endowed matter, plants and animals with certain unchangeable properties and characteristics, of which the most universal constituted the laws of nature, discernible by human reason. This conception is clearly capable of association with a mechanistic philosophy, and irreconcilable with animism; as Boyle put it:

> God established those rules of motion, and that order amongst things corporeal, which we call the laws of nature. Thus, the universe being once framed by God, and the laws of motion settled, the [mechanical]

philosophy teaches that the phenomena of the world are physically produced by the mechanical properties of the parts of matter.[1]

If this transcendental status be granted to the laws of nature— so that one may inquire what they are, but not why they hold— the question may still be asked, "How may a given proposition be recognized as a law of nature?" In other words, how does a law of nature differ from any other generalization which happens to be true because no instance to the contrary has yet been discovered? Modern philosophers of science, having deprived the laws of nature of their transcendental status, present their own answers to this problem. Galileo, and after him Newton, obtained an answer to it by application of the method of abstraction. When Galileo created by abstraction the essential model of the phenomena of motion which he studied, he transformed the pragmatic validity of a generalization appropriate to the world of experience into the absolute validity of the law of nature in the intellectual model. Thus Newton, following Galileo, formulated his laws of motion as laws of nature having complete applicability within the fabric of mathematical physics, whose conclusions as a whole can be confirmed by direct observation. In this way the difficulty that the perfect universality of the laws of nature cannot be established by experience of countless instances was overcome. For Galileo and the later scientists who adopted his method such laws had a greater force than descriptive generalizations could attain, because they had acquired a fundamental systematic status in the scientific picture of the universe.

Hence laws of nature could be considered in theory as being rigorously exact, although the ascertainable correspondence between laws and experience in the physical world is limited by probability-factors in experiments and by the intervention of a multitude of complications. Such limitations were discovered, for example, by the early experimenters in mechanics who discovered that Galileo's theorems on motion could not be rigorously confirmed when applied to the movements of physical bodies. Galileo had clearly appreciated the exceptional usefulness of the concept of laws of nature in which they were taken to represent that which

---

[1] *The Excellence and Grounds of the Mechanical Philosophy*, in *Philosophical Works*, abridged by Peter Shaw (London, 1725), vol. I, p. 187 (condensed).

the intellect, aided by scientific abstraction, conceives as the essence of certain phenomena, but it fell to others to demonstrate more satisfactorily the precautions which the scientist must take in relating the laws to the crude evidence of the senses.

How can the investigator in mathematical physics be sure that his theories are applicable to the real world of experience? Galileo's answer, prepared by earlier logicians, was the appeal to observation and experiment. If theory leads to a certain result, it is only necessary to ascertain that it occurs in nature. So, in the *Discourses*, Galileo verified the law of acceleration by experiments on the inclined plane. But such a clear instance is rare. With Bacon the appeal to experiment is a remedy for ignorance:

> Let further inquiry be made as to the comparative heat in different parts and limbs of the same animal; for milk, blood, seed and eggs are moderately warm, and less hot than the outward flesh of the animal when in motion or agitated. The degree of heat of the brain, stomach, heart, and the rest has not yet been equally well investigated.[1]

With Galileo experiment is also used as a test of theory, or as a device for convincing doubters. He may even declare: "If you perform such an experiment, then you will obtain such a result," although he has never made that experiment himself. In some passages he refers his readers to their experience of reflection from mirrors, of motion in a ship under way, of the flow of fluids; there are ideas for experiments, but none are made. Galileo derived his new ideas of motion not from experiments but from his analysis of the nature of movement according to common experience, and so using the same process he teaches his readers how to make true sense of their own experience by means of these thought-experiments. They are more common in his books than real ones, for (Galileo thought) we know enough to grasp the truth, if only we learn to reason rightly. Being above all things a theoretical scientist he was no crude empiricist and (unlike Bacon) he sought not for more facts, but for deeper understanding. He was well aware that experimentation is a double-edged weapon, deceiving those who use it crudely, as when he writes of the "sublime wit" of Copernicus, who

> did constantly continue to affirm (being perswaded thereto by reason) that which sensible experiments seemed to contradict; for I cannot

[1] *Novum Organum*, Bk. II, xiii.

cease to wonder that he should constantly persist in saying, that *Venus* revolveth about the Sun, and is more than six times further from us at one time, than at another; and also seemeth to be always of equal bigness, although it ought to shew forty times bigger when nearest to us, than when farthest off.[1]

Sheer empiricism, therefore, could not uncover physical reality, which could only be glimpsed through the alliance of analytical reasoning (especially of the mathematical kind), scientific imagination, and cautious experiment always safeguarded by reason.

From the critique of empiricism it emerges that in Galilean science experiment is incompetent to confirm the *whole* intellectual structure, whose conceptual elements transcend experiment. For example, the concept of acceleration which science owes to Galileo cannot be proved in the laboratory, though its applicability in representing phenomena can be illustrated. For the definition of acceleration involves the further concepts of time and velocity, the latter a function of time and the concept distance. There is periodicity in nature, and there are intervals in nature, but nature offers no ready-made dimension-theory embracing the concepts of time and distance. These can have no other status than that of ideas or mental constructs which help to form the world-picture, having the advantage that unlike concepts of beauty and justice they can be understood in the same sense by all men. But their definition is of mind, not innate in the fabric of the universe. The concept time gives order to certain kinds of experience, the concept distance to others, and from these arise velocity and acceleration rationalizing others still, so that the first test for a definition of acceleration must be its assimilability in logic to existing dimension-theory; moreover, in a second test, by experiment, the usefulness of the new construct cannot be distinguished from the usefulness of the existing constructs, time and distance, so that effectively the whole system of constructs must be tested together, if at all. Though the Galilean scientist seeks to penetrate ever deeper into physical reality, the nodes of his exposition of nature can never be more than mental constructs, time, acceleration, the chemical element, or the electron, which give order and significance to the experimental data. This incidentally provides the justification for Galileo's thought-experiments; the constructs are equally valid when they give order to the facts of experience, duly analysed, as

[1] *Dialogue* (ed. G. de Santillana), p. 347.

when they apply to the most delicate determinations of the laboratory.

If the theoretical part of science, leaving aside its practical success in operating with materials and instruments, is a framework of mental constructs giving order to experience, then it follows that the only kind of explanation that is possible is one that arranges the constructs in a logical pattern; as when the properties of the molecule are traced to its atomic structure, and the properties of the atom to its electrons. The sole method of assigning a *cause* to any particular phenomena is to invoke the constructs applicable to it, that is, the generalizations derived from the study of less complex phenomena. In this way Galileo used his concepts of motion to describe and account for the trajectory of a projectile, as a modern physiologist more elaborately uses the concepts of cytology, biochemistry and even the physical notions of matter and energy to describe and account for the functioning of a part of the body. That it is never possible to touch any cause more fundamental than a construct or generalization derived from the description of some definite phenomenon, and that therefore explanation and description have no really distinct significance in science, was the great methodological discovery upon which the scientific revolution flourished. In Galileo's works its full powers appear for the first time, but it was only gradually extended from mechanics to the non-physical sciences. In the *Discourses* (Third Day) Galileo disposed of the causes of the acceleration of freely falling bodies imagined by philosophers as fantasies unworthy of examination: 'At present it is the purpose of our Author merely to investigate and to demonstrate some of the properties of accelerated motion (whatever the cause of this acceleration may be).' This does not mean that in replacing the question *why* by the question *how* Galileo has excluded the study of phenomena in terms of cause and effect—it was his pupil Torricelli who proved that the cause of the *horror vacui* (so called) was the pressure of the air. Mere simple description, like that of anatomy, was not the sole end of the new sciences Galileo created, for the formulation of constructs and generalizations is a necessary feature of the full description of a class of events, e.g. accelerated motion. Galileo seems rather to be making the minor point that if the cause of B is called A, the first subject of study must be B itself (since it is from B that the very existence of A is wholly or in part inferred), and the more serious point that to describe and account for the A-B relationship, the investigator must be able to make a number of statements concerning A independently of B in order to establish its character. In other words, since causation and full description are synonymous, the "cause" of B (acceleration) is a property of A (gravity), not of B, and can be sought only in the description of A. This is Galileo's attitude throughout the *Discourses*, as it is Newton's throughout the *Principia*. The explanation of phenomena at one level is the description of phenomena at a more fundamental level, that is, one nearer to the primary realities of classical physics, matter and motion.

Following the example of Galileo, the scientist may as it were work either upwards or downwards; he may seek for a more fundamental construct (like the law of inertia, or the laws of thermodynamics) or he may examine the applications of the construct to the details of a complex phenomenon (like the isochronism of the pendulum). In either case he may have to handle constructs which are not reducible to the ultimate physical realities, as was the case for instance with Newtonian mechanics where the law of gravitation had to be taken as descriptively correct, though gravity was not explicable in terms of matter and motion. For Galileo there was no anomaly in recognizing that certain constituents of the physical world had to be accepted as axiomatic; descriptive analysis can only advance gradually from the coarse to the refined, from the lower to the upper levels, each with its appropriate generalizations. In the period between Galileo and Newton, however, the validity of the purely descriptive generalization (which rests upon scepticism concerning the possibility of arriving at a final indubitable truth serving as the single origin of scientific thought) was challenged in the philosophy of Descartes, and rejected by the systematists who expounded Cartesian ideas. The chief difference between Galileo and Descartes lay in this, that while the former believed that a body of knowledge successful in organizing sense-perceptions (duly refined and analysed) and in framing generalizations based on them gave an adequate understanding of nature, the latter believed that there was no reliable test of the significance of sense-perceptions other than that which issues from a deeper metaphysical certainty. The mind, being extra-nature, was capable of doubting anything external to itself in nature.

As Descartes relates in the *Discourse on Method* (1637), after completing a thorough education, in which, 'not contented with the sciences actually taught us, I had read all the books that had fallen into my hands, treating of such branches as are esteemed the most curious and rare,' he found himself involved in many doubts and errors, persuading him that all his attempts at learning had taught him no more than the discovery of his own ignorance. In philosophy, despite all the efforts of the most distinguished intellects, everything was in dispute and therefore not beyond doubt, and as for the other sciences 'inasmuch as these borrow their principles from philosophy,' he reasoned that nothing solid could be built upon such insecure foundations.

In this perplexity, Descartes proposed to himself four "laws of reasoning" which he applied in the first place to the study of mathematics:

> In this way I believed that I could borrow all that was best both in geometrical analysis and in algebra, and correct all the defects of the one by the help of the other. And, in point of fact, the accurate observance of these few precepts gave me such ease in unravelling all the questions embraced in these two sciences, that in the two or three months I devoted to their examination, not only did I reach solutions of questions I had formerly deemed exceedingly difficult, but even as regards questions of the solution of which I remained ignorant, I was enabled as it appeared to me, to determine the means whereby, and the extent to which, a solution was possible.

Mathematical ideas, then, could be understood with perfect clarity and mathematical demonstrations accepted with absolute confidence. These principles, to which Descartes held firm in all his scientific activities, ally him with Galileo in the attainment of the ideal of mathematization throughout science. Indeed, Descartes' most valuable contribution to the scientific revolution was the co-ordinate geometry described for the first time in the same volume as the *Method*. But his grasp of a starting point for the comprehension of *fact*, rather than the abstractions of mathematics, depended upon a form of psychical crisis from which he emerged possessed with the metaphysical force of the statement, *I think, therefore I am*:

> I thence concluded that I was a substance whose whole essence or nature consists in thinking, and which, that it may exist, has no need of place, nor is dependent on any material thing; so that "I," that is

to say the mind by which I am what I am, is wholly distinct from the body, and is even more easily known than the latter, and is such, that although the latter were not, it would still continue to be all that is.

This led Descartes to inquire why he had found *Cogito, ergo sum* an infallible proposition, whence he convinced himself that all things clearly and distinctly perceived as true, are true, 'only observing that there is some difficulty in rightly determining the objects which we distinctly perceive.' Further, he declared that since the mind is aware of its own imperfection, there must be a being, God, which is perfect and that since perfection cannot deceive, those ideas which are clearly and distinctly perceived as true are so because they proceed from perfect and infinite Being. So much more certain are the fruits of reason, says Descartes, that we may be less assured of the existence of the physical universe itself, than of that of God, 'neither our imagination nor our senses can give us assurance of anything unless our understanding intervene . . . whether awake or asleep, we ought never to allow ourselves to be persuaded of the truth of anything unless on the evidence of our reason.'

After this denunciation of empiricism, this declaration that all knowledge of truth is implanted by God, this assertion that the task of the scientist is to frame propositions as clearly and distinctly true as those of geometry, what suggestions can be made for the deciphering of the enigma of nature? According to Descartes, it is necessary to follow exactly that procedure which Bacon had condemned in Aristotle, that is, to establish the prime generalizations that are 'clearly and distinctly true.'

I have ever remained firm in my original resolution . . . to accept as true nothing that did not appear to me more clear and certain than the demonstrations of the geometers had formerly appeared; and yet I venture to state that not only have I found means to satisfy myself in a short time in all the principal difficulties which are usually treated of in philosophy, but I have also observed certain laws established in nature by God in such a manner, and of which he has impressed on our minds such notions, that after we have reflected sufficiently on these, we cannot doubt that they are accurately observed in all that exists or takes place in the world.

Thus the science of Descartes is a centrifugal system, working outwards from the certainty of the existence of mind and God to embrace the universal truths or laws of nature detected by reason, and then from the "concatenation of these truths" revealing the mechanisms involved in particular phenomena. It is systematic, unlike the "new philosophy" of Bacon or Galileo, because its aim is not to enunciate a correct statement here and there as it becomes accessible to intellect, but to provide an unchanging fabric whose relevance to particulars is the sole remaining subject of inquiry. In this respect, despite his contempt for scholasticism, Descartes sought for himself the commanding authority of a new Aristotle. Indeed, among Cartesian scientists, and still more among Cartesian philosophers of later generations, a new scholasticism flourished through the dissection, embroidering and expansion of Descartes' doctrines, until they, like Aristoteleanism in the sixteenth and seventeenth centuries, were in turn regarded as a bulwark against dangerous innovations and as the philosophic justification of religious orthodoxy.[1]

Apart from his researches in optics and mathematics—by far the portion of the whole which proved of greatest value to science —Descartes preferred to express his ideas in the form of a model, whether of a man or of the universe. Superficially this procedure seems to resemble the Galilean process of abstraction, but in reality it is very different. Having settled his principles, Descartes believed that to philosophize about mathematical abstractions was to promote delusions; his object was the real world, but as he did not know from experience what the mechanisms of the real world, or the actual human body, are, he was forced to imagine what they must be to accord with the principles and such knowledge as he had. He did not claim that in describing the model he was describing the real world, only that from the identity of their properties the real world could be understood in terms of the model. Its mechanism was absolute, since it followed from the dualism of mind and matter that all the phenomena of nature resulted from the properties of matter, especially its motion, which in turn were fixed by natural law. The laws of nature, including the definition of matter by extension and the impossibility of a vacuum, the law of inertia, and the laws of impact between particles, were derived by Descartes as ideas 'clearly and distinctly perceived to be true.' Thus Descartes and Galileo agreed that the only physical reality which science can study is that of matter in motion; unlike Galileo, however, Descartes did not

[1] Cf. A. G. A. Balz: *Cartesian Studies* (New York, 1951).

hesitate to extrapolate far beyond the limitations of mathematical analysis or experimental inquiry. In the end, elaboration of the principles, guided only by the criteria of "clear and distinct," yielded in Cartesian cosmology, chemistry and physiology nothing other than subtle scientific fantasy.

Again, with regard to the functions of experiment Descartes and Galileo adopted antithetical positions. The pillars of Cartesian science were "clear and distinct" ideas formulated as laws of nature; it was to fit these, and not experimental evidence, that its subsidiary theories were shaped. It was essentially deductive from these natural laws, and if knowledge did not supply the requisite materials, then they had to be invented with the aid of reasoned deduction, as the celestial vortices carrying the planets about the sun, the three kinds of matter and the variously contrived pores of substances were invented in accordance with the exigencies of experience and reason. Of course, experience was respected in the sense that Descartes sought to explain in his model the sum of the phenomena of nature as he knew them, for it is obvious that he could not have deduced magnetism within his system had he not known of its manifestations. But Descartes made no attempt to confirm his mechanisms in detail by experiment. The foundations of knowledge, he thought, were best settled without it:

> for, at the commencement, it is better to make use only of what is spontaneously represented to our senses, and of which we cannot remain ignorant, provided we bestow on it any reflection, however slight, than to concern ourselves about more uncommon or recondite phenomena; the reason for which is, that the more uncommon often mislead us so long as the causes of the more ordinary remain unknown. . . .

Experiments indicating some conclusion detached from a deductive system Descartes distrusted; hence nothing that Galileo did had value for him, because Galileo did not know the cause of gravity. This must not be taken to mean that either Descartes or his successors were totally blind to the merits of experimentation, though it could only be an adjunct when the application of clear and distinct ideas failed, or in an obscure inquiry. In his own experimental researches Descartes revealed great talent, and those who were influenced by him, like the Dutch physicist Christiaan Huygens, included some of the great exponents of experimental science of the later seventeenth century—though indeed they owed much less in this respect to Descartes, than to the empirical temper of the age, and the emulation of Galileo's example in mechanics.

It was Huygens who later described Descartes as the author of 'un beau roman de physique.' Though its doctrines were recited into the mid-eighteenth century the Cartesian system of science proved sterile, and it may be doubted whether the Cartesian philosophy of science ever produced a single useful thought, save in the mind of its originator.[1] The optimistic metaphysical belief that what is clear and distinct must be true proved unfounded. The deductive method, subjected to the destructive criticism of the neo-Baconians of the Royal Society, was again convicted of fostering works that were shallow, speculative and remote from real things. Clearly when Descartes devoted himself to systematics his status as a scientist diminished. But the importance of his systematic works—especially the *Principles of Philosophy* (1644), the text-book of the Cartesian school for nearly a century—must not be underestimated, for in the mid-seventeenth century the intellectual appeal of Descartes throughout France, Britain and northwestern Europe was immensely greater than that of Galileo, while Bacon was almost unknown to continental scientists. The very fact that Descartes wrote as a philosopher gave his scientific ideas greater currency, and to many places where Aristotelean and humanistic conventionality lingered untroubled, Cartesian notions brought the first breath of a new outlook, a fresh vitality in natural philosophy. Thus Oxford, because Descartes was read there, was regarded about 1650 as being much in advance of Cambridge. His was undoubtedly the pre-eminent intellect in the swelling movement, ebullient with ideas and discoveries, that made Paris the scientific focus of Europe from about 1630 to 1670. Even among those who cannot be enrolled with the expositors of Descartes' science, there were many who, like Boyle or Newton, though they learnt through the use of an empirical or Galilean method to criticize his theories, had found in those same theories their point of departure; indeed, the main activities in physical

---

[1] It may be remarked that Descartes' metaphysic of scientific discovery, as described in the *Method*, is essentially individualistic, i.e. it shows how each man may learn to frame his own idea of nature. But, just as those sects which claimed to be founded on the free interpretation of the Bible imposed the sternest discipline in order to safeguard the interpretations of their founders, so the later Cartesians instead of going through this process of discovery adhered rigidly to the idea of nature developed by Descartes himself.

science for more than a generation after Descartes' death can be interpreted, without gross distortion, as a commentary upon Descartes' works. And if the *Principles of Philosophy* proved ephemeral compared with Newton's *Mathematical Principles of Natural Philosophy*—even the title suggests a reaction—its influence in leading later seventeenth-century science to entertain ideas of mechanism, of the corpuscular structure of matter, of the importance of "natural laws," was creative of further progress.

Perhaps it may justly be said that Descartes' successes in science were due less to any peculiar merits in his method, than to his native genius for investigation. There is one point, however, both in the method and the texture of his thinking on scientific subjects that deserves to be singled out. Descartes well understood the importance, in any work of research, of scientific imagination, a faculty with which he himself was so well endowed that he hardly perceived its limitations when controlled by reason alone without cautious experimentation. Bacon had recognized that imagination or intuition might surmount an inconvenient obstruction; Galileo also admitted that in demonstrative sciences a conclusion might be known before it could be proved:

> Nor need you question but that *Pythagoras* a long time before he found the demonstration for which he offered the Hecatomb, had been certain, that the square of the side subtending the right angle in a rectangle triangle, was equal to the square of the other two sides: and the certainty of the conclusion conduced not a little to the investigating of the demonstration. . . .[1]

In Descartes there is a more overt appreciation of the function of directed imagination, playing on the problem in hand, in formulating hypotheses to be tested by experiment or other means:

> . . . the power of nature is so ample and vast . . . that I have hardly observed a single particular effect *which I cannot at once recognize as capable of being deduced* in many different ways from the principles, and that my greatest difficulty usually is to discover in which of these ways the effect is dependent upon them; for out of this difficulty I cannot otherwise extricate myself than by again seeking certain experiments, which may be such that their result is not the same If It is in one of these ways that we must explain it, as it would be if it were to be explained in another.[2]

[1] *Dialogue* (ed. G. de Santillana), p. 60.
[2] *Discourse on Method*, Part VI; my italics.

Here experiment is put forward, not as by Bacon to uncover the unknown, or as by Galileo to confirm the known, but as a means of
eliminating all but one of the mechanisms suggested by imagination as the explanation of a particular phenomenon. And as
Descartes correctly stated, the imagination is directed because it
is referred to certain known principles (or constructs), and further
because the mechanisms suggested must be susceptible in the
first place of deductive check, since science does not admit of
idle guessing. If Descartes had realized that even when only a
single hypothetical mechanism seems deductively feasible it
remains a hypothesis until confirmed by experiment, and if he
had applied this test more meticulously, his thought would have
been less liable to run into speculation. In any case the liberty to
frame hypotheses (in spite of Newton's famous dictum), with the
rigorous attention to the findings of experiment and observation
which Descartes himself neglected in his encyclopædic survey of
nature, was to prove a creative factor in the accelerating progress
of science.

The scientific method of the seventeenth century cannot be
traced to a single origin. It was not worked out logically by any
one philosopher, nor was it exemplified completely in any one
investigation. It may even be doubted whether there was any
procedure so conscious and definite that it can be described in
isolation from the context of ideas to which it was related. The
attitude to nature of the seventeenth-century scientists—especially
their almost uniform tendency towards a mechanistic philosophy
—was not strictly part of their scientific method; but can this be
discussed except in connection with the idea of nature? In large
part the character of the method was determined by the mental
range of the men who applied it; hence Bacon's method bore
fewer fruits in his own hands because his conception of the facts
of nature was still Aristotelean. The influence of Descartes, too,
was so great because he produced a mechanistic world-system of
infinite scope (enriched with some genuine discoveries) which was
welcomed by his age, not because he outlined a remarkably clear
or satisfactory way of proceeding in scientific research. Even
Galileo's observations on method were probably less important
than direct imitation of the kind of mathematical analysis he
initiated in mechanics. Over the broad area of scientific activity
the influence of content on form was more significant than the

reverse effect. Methods changed, because different questions were asked, and a new view of what constitutes the most useful kind of scientific knowledge began to prevail. Perhaps this is most effectively revealed in the biological sciences, where the century witnessed a progressive change in the content of investigations unaccompanied by conscious discussions of the methods to be employed. Here there was no parallel to the criticism of the methods of Aristotle and the scholastics in physics, though of course the medieval neglect of the descriptive sciences was often commented on adversely. The far-reaching inter-action between the content and the techniques of science was also uncontrolled by any very explicit conceptions of method. This interaction had a profound effect on the quality and extent of the information available; but Bacon alone explicitly recognized the importance of accurate fact-gathering in science. It seems most natural to believe that in any effective step the method, the philosophy and the discovery itself were carried along together in the subsequent impact, for though there is nothing that can reasonably be called a specific method of science to be found in the works of Harvey, or Kepler, or Gilbert, these men changed the character and form of future studies. Who, for instance, could ignore the challenge of the phrase with which Gilbert opens his Preface to *De Magnete*: 'Clearer proofs, in the discovery of secrets, and in the investigation of the hidden causes of things, being afforded by trustworthy experiments and by demonstrated arguments, than by the probable guesses and opinions of the ordinary professors of philosophy. . . .' Yet the meaning and weight of experimental testimony was still open to discussion a century later. A scientific approach to problems must be the sum of its many aspects—experimentation, mathematical analysis, quantitative accuracy, and so on—varying according to the nature of the problem; and this the seventeenth century drew from many varied sources. Its implied implementation in practice was more important than its explicit formulation, with the somewhat curious result that scientific method, shaping itself to the needs of practising scientists and vindicated rather by results than by preconceived logical rigour, has remained something of an enigma to philosophers from Berkeley onwards. In the long run the obstinate empiricism of a Gilbert or the unpredictable intuition of a Faraday have successfully broken the rules of both inductive and mathematical logic.

CHAPTER VII

THE ORGANIZATION OF SCIENTIFIC INQUIRY

EACH phase of civilization tends to produce its own institutions of learning. In the ancient empires science was attached to the temples of religion; Greece saw Plato's Academy, the Lycæum founded by Aristotle, and the vast library of Alexandria; the middle ages created the common school and the university in a structure of education which has not wholly vanished. Lastly, in the modern era, the learned society, with its international affiliations and specialized journals, has profoundly influenced the stratigraphy of research. Neither the learned society nor the learned journal (the terms are convenient, if pompous) was altogether the creation of the scientific revolution, but in both cases the course of events was very much determined by the necessities of the scientific movement, and scientific organization was taken as a model by those who worked in other fields of knowledge. From the end of the seventeenth century the majority of active men of science were members of some active scientific group; publication in one of the ever more numerous journals gradually became the recognized manner of announcing the results of investigation; and the national scientific society was accepted as the vehicle for the state's concern in scientific matters. Although the Fellowship of the Royal Society, for example, was much less indicative of intellectual distinction in the eighteenth century than it has since become, as institutions the Royal Society or the Académie Royale des Sciences enjoyed a prestige even greater than that of the universities in humane studies. Indeed, the evolution of modern science outside academic walls was the main cause of the lack of cohesion, and of the difficulty in the communication of ideas, whose correction was one of the principal objects of the founders of the scientific societies. In the unity of medieval learning the scholar enjoyed communion with others of similar interest in the university of which he was almost invariably a member, and wherever his studies might lead him. The sixteenth century saw the new phenomenon of scholars, literati and scientists whose interests were no longer embraced in the work of the university, and who were also more uniformly distributed throughout a highly cultivated society. The landed gentleman, the country physician or clergyman, the apothecary, the soldier and the lawyer, played a new and important part in the advancement of knowledge or the patronage of literature. Expecially in northern Europe, where the universities were less numerous and more conservative than those of Italy and France, intellectual leadership passed to a class which was not merely outside the orbit of the university, but was apt to regard academic learning as old-fashioned and sterile.

> Oxford and Cambridge are our laughter,
> Their learning is but pedantry:
> These Collegiates do assure us,
> Aristotle's an ass to Epicurus.

wrote the "wit" who composed the *Ballad of Gresham College* about 1667.[1] From the first, the connection of the new scientific movement with practical arts rendered it in some degree independent of the universities. As the friends of the "new philosophy" became increasingly critical of Aristotle and conventional education, they found their opponents the more firmly entrenched behind academic walls; hence the innovators tended to seek a more congenial intellectual environment elsewhere.[2] Scientific knowledge was no longer in the mid-seventeenth century limited to the religious and medical classes, but was widely diffused through a diversified and exuberant society. Many biographies relate the feeling of confidence, the depth of intellectual satisfaction, the release of a creative drive that was experienced by members of the new class of laymen, educated and leisured, when they passed from the confines of academic disputation to the methods of experimental science.

Scientific discovery is (or was, until recent times) an act of the individual, with of course a greater or less indebtedness to his intellectual inheritance. So, equally, in the seventeenth century, was the adoption of the novel scientific outlook, critical of orthodoxy,

---

[1] Gresham College, founded in 1598 by the merchant financier Sir Thomas Gresham, was the first meeting-place of the Royal Society, whence the Fellows were known as the "Gresham philosophers."

[2] In Europe, this was partly due to the control of education exercised by certain religious orders, but everywhere the university was deeply committed to Aristotelean thought.

which can be seen almost as a conversion in many instances, as when Galileo became an adherent of the Copernican system. But since men naturally assemble to indulge a common taste, and since wits are sharpened by contact, the groups of intellectuals who collected in a tavern, a lecture-room, or about an enterprising patron tended to assume a more formal character, to look for recognition and privilege. The first of such groups to acquire an organization and a history were products of Italian humanism. Their interests were literary rather than scientific. About a century later the first national academy, the Académie Française founded by Richelieu in 1635, was also a literary institution: its main task was the conservation of the purity of the French language. The early literary societies met for discussion and criticism; their object was rather the extension of knowledge and the refinement of taste than anything resembling research or analysis, and there was nothing foreshadowing the modern presentation of papers. The first scientific societies, which also originated in Italy, followed the same pattern. Occasional meetings of groups of experimenters, like that which is supposed to have collected about William Gilbert at his London house, or that centred about Giovanbattista Porta in Naples (the so-called Accademia Secretorum Naturæ) were hardly societies at all.[1] The first assembly emphatically of this character was the Accademia dei Lincei in Rome, of which Galileo was a member, which lasted with one break through the first thirty years of the seventeenth century. The *Lincei*[2] rose to a membership of thirty-two, and planned to set up branches everywhere, equipped with printing-presses, botanic gardens and laboratories. The patron of the society was Duke Federigo Cesi, a naturalist, and much of its activity was diverted to natural history. One member, Francesco Stelluti, published the first zoological studies made with the aid of the microscope. Two of Galileo's early books were published by the Lincei, but the society did not approve his later cosmological ideas.

The Accademia dei Lincei, like earlier literary societies, and some later scientific groups, did not engage in any form of corporate activity. The members followed their own investigations, whose

---

[1] Baptista Porta (d. 1615) was the author of *Magiæ Naturalis Libri IV* (1558, enlarged edn. 1589: Englished as *Natural Magick*, 1658), and a great exponent of esoteric experimentation.

[2] So called, because the Lynx symbolized the clear-sightedness of science.

results they discussed at the meetings. However, the great Florentine society of the mid-seventeenth century, the Accademia del
Cimento, followed the alternative plan, which had already been
described by Francis Bacon in the *New Atlantis*. The object of
Bacon's model organization was not merely to bring men together,
but to set them to work in common on the tasks most important
for science, so that it resembled a scientific institute more than a
modern scientific society. The vast realm of natural knowledge,
he felt, was too vast for one man to tackle single-handed, while
concentration on a single problem or set of problems was likely
to produce a myopic picture of single trees, not a survey of the
forest. To the efforts of individual pioneers, as Sprat put it later
in speaking of the Royal Society, 'we prefer the joint Force of
many Men.' Other advantages of the Baconian plan were that,
as it ensured that due attention was always paid to each division
of science, so it made certain that none could be carried on in
complete isolation from the rest; and it also provided a means by
which, it was hoped, a quantity of necessary apparatus beyond
the means of a private purse could be gathered together.

In Bacon's view, the assembling of pure information, the
preliminary to the elucidation of natural truths, was such a
formidable task that it could only be tackled by a co-operative
endeavour. Otherwise science was likely to be for ever deluded
by theories enunciated on the basis of an insufficient mass of
digested fact.[1] Later the Royal Society was for a short time to
embark on such a project of fact-collection. The Accademia del
Cimento, on the other hand, was not committed to this Baconian
conception of procedure in science. It concerned itself with the
experimental development of the scientific ideas of Galileo, and
of his two most successful pupils, Torricelli and Viviani, while the
nine members also pursued their own problems independently.
One large fraction of their total activity was directed to the proof
of the theorems on motion that Galileo had demonstrated mathematically, and another to the study of the barometric vacuum

---

[1] Thomas Sprat, in his *History of the Royal Society* (1667), echoed the then
typical opinion that in Bacon's writings were 'everywhere scattered the best
Arguments, that can be produced for the Defence of experimental Philosophy,
and the best Directions, that are needful to promote it,' but he did not hesitate
to confess that Bacon's natural histories were far from accurate, because Bacon
seemed 'rather to take all that comes, than to choose, and to heap, rather than
to register ' (1722 edn., pp. 35–6).

discovered by Torricelli. As the business of the society was experiment, its activities had little effect on the development of theoretical science though its work on mechanics and on the vacuum confirmed the new views on these controversial questions. Unfortunately the Florentine experiments were not well known outside Italy until later. Part of the achievement of the Accademia del Cimento—in spite of its short duration of ten years from 1657 to 1667—was due to the richness of the apparatus at its command, for it made use of what was really the first physical laboratory in Europe.[1] The academy was founded by the Grand Dukes Ferdinand II and Leopold of Florence, who used the remaining wealth of the Medicis to buy the services of the finest instrument-makers, to procure the most perfect lenses, and equip their colleagues with a most elaborate series of barometers, thermometers, time-measuring devices and whatever else was required for their work.

The book in which this was described—the *Saggi di Naturali Esperienze* (1667)[2]—was an early piece of experimental reporting. Interpretation of the results gained was strictly limited and (as with the Royal Society) a good deal of the experimenting was haphazard. The experiments recorded are various as well as numerous. Attempts were made to verify Galileo's theory of projectiles, and the time-keeping properties of the pendulum were studied—without, however, leading to its application to a mechanical clock, which was made by the Dutch physicist Christiaan Huygens. Various forms of the thermometer, hygrometer and barometer were tested, and the design of optical instruments improved. Experiments were made on the "radiation" of cold from a lump of ice; on the thermal expansion of many substances; on the incompressibility of water; on the force of gunpowder; and on capillary attraction. The researches of Torricelli and Pascal, showing that the observations formerly explained by the statement that nature abhors a vacuum should be attributed to atmospheric pressure, were confirmed. It was proved that neither combustion nor respiration was possible in a space exhausted of air, that magnetic attraction was transmitted through it but not sound, and a rather unsuccessful attempt to construct an air-pump was made. In this field of activity the Accademia was largely repeating work that had already

[1] Many of the instruments are still preserved in Florence.
[2] Translated by Richard Waller as *Essays of Natural Experiments* (1684).

been done by Robert Boyle at Oxford, and described in his *New Experiments Physico-Mechanical on the Spring of the Air and its Effects* (1660).

Naturally the benefits to be derived from closer co-operation among those interested in the new scientific movement were not perceived in Italy alone. In both England and France informal groups had existed for many years before the Accademia del Cimento was founded. The generation of Frenchmen which included Descartes, Gassendi, Fermat, Desargues, Roberval, Pascal and Mersenne (*c.* 1630–60) was particularly active, fertile both in new ideas and new experiments. Paris was their centre, and there was a continuous tradition of scientific gatherings which was ultimately formalized in the Académie Royale des Sciences. One of the earliest groups was held together by the personality of Marin Mersenne, a Minim friar of the convent in the Place Royale, which was not only a meeting place at which important discussions took place, but also the centre from which Mersenne conducted his vast correspondence, maintaining communication between his colleagues even more effectively than the actual gatherings. It is a significant historical fact that, since the late sixteenth century, transport facilities and postal organizations had much improved, rendering possible regular and frequent exchanges of letters.[1] By this means news of the latest developments could be spread more rapidly: problems could be exposed for general consideration: and criticism could be provoked and collated. In the mid-seventeenth century a number of men occupied a prominent position, less on account of their own intellectual capacities, than because of their indefatigability as correspondents. Their function was to be acquainted with everyone of importance in science, to gather information and to re-distribute it to those of their friends who were likely to be interested. Of these Fabri de Peiresc at Montpellier was one, and Mersenne in Paris another. Later Henry Oldenburg, first Secretary of the Royal Society, carried on the same rôle, his talent as a "philosophical merchant" (as Robert Boyle called him) greatly strengthening the bonds between English and continental scientists. Even in the eighteenth century the private correspondence of a great public figure in science (like Sir Joseph Banks) was still of international importance.

---

[1] Tycho Brahe was one of the first scientists to leave an important mass of material of this kind, which has been edited by J. L. E. Dreyer.

In London, as in Paris, informal groups preceded a formal scientific society. These seem to have had a stronger common interest in the mathematical sciences than in moral and natural philosophy, partly because geometry and astronomy were well represented at Gresham College (the natural focus for scientific pursuits in London), partly in continuation of the Elizabethan tradition of developing the practical sciences (navigation, surveying, cartography, etc.). While there was little evidence of English concern for the ideas of Bacon, Gilbert and Harvey until near the close of the first half of the seventeenth century, there was considerable activity in the field where science and technology overlap. Cornelius Drebbel, an ingenious engineer as well as an experimenter of some repute, was acclaimed at the court of James I,[1] whose successor, besides patronizing Harvey's researches in embryology, established an experimental workshop at Vauxhall. Even the mathematician Napier of Merchistoun, inventor of logarithms, in a fit of Protestant fervour invented a series of terrible war-like devices, the plans for which he destroyed on his death-bed. On the whole, the more important exponents of pure science in England at this early period, like Thomas Harriot (d. 1621) and Jeremiah Horrocks (d. 1641), though by no means isolated, had slender contact with the great scientific movement on the Continent.[2] Dilettanti, philosophers and literary men (such as Thomas Hobbes, Sir Charles Cavendish and Sir William Boswell) did more to make its literature known in England. Indeed, the "grand tour" was a serious and necessary education, as may be seen in the life of Robert Boyle.

The history of the emergence of the Royal Society from these groups has been told many times. It now seems probable that there were two at least of these, whose members became the fathers of real science in England; some men entered more than one circle, but there was far from complete unity of ideas. In the London of the Commonwealth and Protectorate men of antagonistic religious and political persuasions could have very similar

[1] Who paid a state visit to Tycho Brahe's observatory at Hveen, and received the dedication of Kepler's *Harmonices Mundi*.
[2] Harriot was an inventive algebraist, and perhaps an independent discoverer of the usefulness of the telescope in astronomy. He was closely connected with the Elizabethan explorers. Horrocks, in a very short life, proved himself a theoretical and practical astronomer of genius. He was an early student of Kepler.

scientific aspirations.[1] Something, at least, of Bacon's influence may be detected in all; many wished to see some sort of specifically scientific institution established, not a few were firmly convinced that civilization could be powerfully advanced through scientific knowledge. These last were particularly evident in the group that Boyle called the "Invisible College" in 1646, which apparently devoted especial attention to agriculture, as well as to natural philosophy and mechanics. The leading figure in this group was Samuel Hartlib, a Polish refugee, a man of great learning and wide connections, an enthusiast for the union and defence of the Protestant churches. Hartlib was a great advocate of the application of science to technology, but no scientist himself. His hopes broken by the restoration of the monarchy, he appears to have played no part in the foundation of the Royal Society that soon followed. Another group included men of greater weight. As John Wallis, the mathematician, recollected the events of 1645:

> We did, by agreement, divers of us, meet weekly in London on a certain day and hour, under a certain penalty, and a weekly contribution for the charge of experiments, . . . of which number were Dr. John Wilkins . . . Dr. Jonathan Goddard, Dr. George Ent, Dr. Glisson, Dr. Merrett (Drs. in Physick), Mr. Samuel Foster . . . Mr. Theodore Haak . . . (who I think first suggested these meetings) and many others.

Haak was another German refugee, probably the most important of Mersenne's English correspondents. All those named by Wallis (including himself), except Foster who died in 1652, were Original Fellows of the Royal Society. The group met in term-time at Gresham College, and Wallis's list of topics in the "New or Experimental Philosophy" which came up for discussion recalls the subjects treated by the Accademia del Cimento:

> Some were then but New Discoveries, and others not so generally known and embraced as now they are, with others appertaining to

---

[1] Thus, a circle close to the government included the poet Milton, Oldenburg, probably John Pell (mathematician), Lady Ranelagh (Boyle's sister). In touch with this was another group (Boyle's "Invisible College"?), which included Hartlib, Boyle, Dury, Oldenburg, Plattes, Dymock, Petty, and evidently others. Then Wallis's group at Gresham College, also deeply committed to the republican régime, was linked with the universities and the "Invisible College." Finally, the Royalists (Evelyn, Brouncker, Moray) seem to have maintained amicable relations with individuals (at least) in these groups.

what hath been called the New Philosophy which from the times of Galileo at Florence, and Sir Francis Bacon in England, hath been much cultivated in Italy, France, Germany and other parts abroad, as well as with us in England.

It is interesting to note that the Copernican hypothesis was still debated by these philosophers. About the year 1649 they were divided, some moving to Oxford, where they formed the Oxford Philosophical Society, which was joined by Boyle on his removal there in 1654. In Oxford they were reinforced by some brilliant students, among them Christopher Wren and Robert Hooke. Thus there is considerable evidence that before the restoration there existed an extensive ramification of personal connection among at least thirty men, many of whom were prominent in academic and public life, and that their common interest was in mathematics and science, not in the promotion of a religio-political movement. The arch-royalist Evelyn could even visit Wilkins, Cromwell's brother-in-law. Few were too deeply committed to the republic to adjust themselves to the restoration of the monarchy, which provided an opportunity for effecting the formal organization long discussed among these amateurs.[1] Charles II's dilettante interest in science was well known, and his scientifically minded courtiers, Sir Robert Moray[2] and Viscount Brouncker (later first President of the Royal Society), were able to win his patronage.

The *Royal Society of London for the promotion of Natural Knowledge*, which received its first charter in 1662, was a wholly private creation, very different from other major societies of the century. Royalty gave patronage, but nothing more. The Fellows enjoyed neither privilege nor pension. They were granted no buildings or funds. Therefore the Society remained, to the nineteenth century, impoverished and inadequately housed. It has never possessed

---

[1] In addition to the proposals of Bacon, Comenius and Hartlib, plans were made by Evelyn, Petty and Cowley. In Cowley's plan sixteen resident professors were to teach 'all sorts of Natural, Experimental Philosophy, to consist of the mathematics, mechanics, medicine, anatomy, chemistry, history of animals, plants, minerals, elements, etc.; Agriculture, Art Military, Navigation, Gardening. The mysteries of all trades, and improvement of them; the Facture of all merchandises, all natural magic, or Divination; and briefly all things contained in the catalogues of natural histories annexed to my Lord Bacon's *Organon*.' (*A proposition for the Advancement of Experimental Philosophy*, in Cowley's *Works*, 1680, pp. 43–51).

[2] Moray was a close friend of Huygens, and like Oldenburg he had attended sessions of the Montmor academy in Paris.

laboratories, nor other than honorary means of promoting research, and was never able to implement the Baconian conception to which many of the Founders were attached. For over a century and a half the qualifications for a Fellowship included wealth and influence as well as scientific merit, because without such support the Society would have collapsed. On the other hand, it had a corresponding sovereignty over its actions. It was independent of government, and though the specialist knowledge of the Fellows was often placed at the service of the state (especially in the eighteenth century), state officials guided neither its elections nor its business. By contrast, in France the Académie Royale des Sciences was the creation of the first minister, Colbert, who arranged the appointments and suggested problems in accordance with political interests. The members were pensionaries—when occasion demanded, they became civil servants. That the Dutch physicist Huygens, who retained his Fellowship of the Royal Society throughout the Anglo-Dutch wars, found his position in the Académie des Sciences inconsistent with Louis XIV's anti-Dutch policy, and resigned from it, sufficiently indicates the difference in character between the two institutions. And these in turn correspond to the different social and constitutional structures of the two states.

The middle-class intellectuals, who in England combined to form their own clubs in which they met as equals, were in France, as in Italy, more dependent on the good offices of a patron. Thus, at the beginning of the seventeenth century, one of the groups most notable in Paris for literature and learning met regularly at the residence of the historian de Thou. His patronage, which included the use of his valuable library, was continued by his relatives the brothers Dupuy to about 1662. Less exalted gatherings met at the *Bureau d'Adresse* managed by the journalist Renaudot. At these, as at the *Cabinet* of the Dupuys, literary and political news was more eagerly awaited than discussion of scientific topics. Mersenne's circle, however, confined its attention almost entirely to mathematical and scientific affairs: it was he, for example, who made the discoveries of Galileo and his pupils known in France, who gave currency to the Cartesian system, and publicized Pascal's problem on the cycloid.[1] After Richelieu's

---

[1] i.e. the calculation of the area bounded by the curve and a straight line, which proved to be three times the area of the generating circle.

foundation of the Académie Française there was some feeling among those who cultivated the sciences that encouragement ought to be given to a similar non-literary institution. Among their number was Habert de Montmor, a man of great wealth who had offered his patronage to both Descartes (who declined it) and Gassendi. Not long after the death of Mersenne in 1648 weekly meetings were taking place in his house, presided over by Gassendi. Their discussions were not limited to science, and it was required only that those who took part should be 'curious about natural things, medicine, mathematics, the liberal arts, and mechanics.' The Montmor Academy, which gave itself a formal constitution in 1657, soon became a fashionable resort; at the meeting in 1658 when Huygens' paper announcing his discovery of Saturn's ring was read, there were present 'two *Cordon Bleus* . . . both Secretaries of State, several Abbés of the nobility, several Maîtres des Requêtes, Conseilleurs du Parlement, Officers of the Chambre des Comptes, Doctors of the Sorbonne,' after which the amateurs, mathematicians and men of letters seem rather insignificant.[1] Science, even the most abstruse mathematics, had become respectable, and apparently interesting, even in the upper levels of Parisian society. The new philosophies of Descartes and Gassendi were victoriously allied against Aristoteleanism. But the course of the Academy was not altogether smooth; the amateurs were more ready to discuss the latest marvels in science than to work for its advancement, and there were sharp clashes of personality.

Mazarin, who had shown far less concern for the intellectual eminence of France than his master Richelieu, died in 1661 and the supreme power was then committed to the young King, Louis XIV. There was thus the possibility of acquiring for science the greatest of all patrons. Since the Royal Society had begun to take shape in 1660 the Montmor Academy had followed its fortunes with some envy, and had even to some extent modelled its own proceedings upon the Royal Society's example. The links between the two bodies were close, for Oldenburg corresponded with several members of the Montmor Academy. Huygens and Sorbière (the Secretary of the Academy) were members of both societies, and a number of the Parisians visited London.

The works of Boyle and other Englishmen were carefully

---

[1] Saturn's ring had been seen in very distorted form by other astronomers, but Huygens first interpreted its nature correctly.

studied in Paris, where the usefulness of the empirical attitude was gradually more highly esteemed. While the Royal Society had grown from its own independent and varied origins, the Académie des Sciences was certainly inspired by its success. In 1663 Sorbière sent to Colbert, who was virtually Louis' Minister for Internal Affairs, a copy of his memoir on a proposed reform of the Montmor Academy. He regarded an experimental organization without royal support as hopeless; soon afterwards the Academy did indeed cease to exist, partly as a result of tension between the experimentalists and the philosophers. Meetings continued, however, at the house of Melchisédec Thevenot,[1] who also found the expense of providing for experiments too great, and appealed to Colbert. In a situation where co-operation between the Cartesians, the Gassendists, the amateurs, the mathematicians and the experimentalists in a joint undertaking seemed increasingly impossible, the last-named group turned to the monarchy. Their first scheme planned in an ample Baconian fashion, was severely pruned by the minister. Ultimately the Académie des Sciences consisted of two classes only, mathematicians (including astronomers and physicists) and natural philosophers (including chemists, physicians, anatomists, etc.), meeting jointly on Wednesdays and Saturdays in two rooms assigned to their use in the Royal Library. Appointments of the academicians were made during 1666, and sessions began at the end of that year.

Most of the active members in the English groups preceding the Royal Society became Fellows,[2] but few of those associated with earlier Parisian assemblies were received into the new Académie. The systematic Cartesians were carefully excluded; on the other hand, three foreigners, Huygens, Cassini and Roemer, were among the most distinguished of Colbert's appointments. Thus the Académie des Sciences was not strictly a continuation of any previous body, nor did it include the amateurs and dilettanti admitted by the Royal Society. No rules or constitution exist earlier than the reorganization effected by the Crown in 1699, but it is evident that as in England the precepts of Bacon were not without weight. Huygens, especially, advocated the preparation

---

[1] (1620–92), traveller, linguist, author, student of many sciences, and inventor of the familiar spirit-level.

[2] The few who did not seem to have retired into obscurity for political or religious reasons, as Milton did; the election of John Ray (1667) shows that the Royal Society exercised considerable latitude in these respects.

of a complete Natural History, and the examination of new inventions was an important aspect of the Académie's work. Having sketched a common programme, the pensionaries proceeded to their experiments and discussions in concert. These proceedings were strictly secret. As in London, the private researches of the academicians gradually assumed a greater significance than their co-operative undertakings. Yet the Académie, thanks to its greater wealth, was able to attempt ventures beyond the resources of the Royal Society. Two of them, the measure of a degree of a great circle about the earth, giving an accurate estimate of its diameter for the first time, and the expedition to Tycho Brahe's observatory at Uraniborg, were conducted by the astronomer Picard. A third, the expedition to Cayenne (in which it was discovered that the length of the pendulum beating seconds was less in southern than in northern latitudes) had as its object the measurement of the earth's distance from the sun by means of simultaneous observations on Mars.[1]

Indeed, the study of astronomy by the members of the Académie was particularly successful. They developed the telescope of very long focal length to its useful limit, the application of the telescope to measuring instruments, and the use of the telescope micrometer. Cassini's observations on Saturn, and Roemer's on Jupiter (from which he correctly deduced the finite velocity of light) won great fame. Some of these observations made at the Paris Observatory provided important evidence for the system of the universe which Newton was to substitute for that of Descartes. They were made possible by royal generosity in equipping the Observatory, built in 1666, which became an experimental institution as well as an astronomical observatory in the modern sense, fitted with the finest instruments. For astronomy was the first of the sciences to reach the stage where the results obtained bear a definite ratio to the expenditure permitted. The Royal Society, by contrast, was far less well provided for. It was not placed in control of the Royal Observatory at Greenwich (founded in 1675), though it was granted some vague surveillance over it.[2] The equipment at Greenwich, provided in the first instance by John Flamsteed, the

[1] The result obtained was about 6 per cent. too small.
[2] This somewhat anomalous situation provoked a serious conflict between Flamsteed and the Society early in the eighteenth century.

Astronomer Royal, and his friends, was limited though good of its kind. He himself was forced to take a country living for support, and could never afford proper assistance. His main work in determining the positions of the stars was not available for more than thirty years, but he supplied observations on the moon used by Newton in his gravitational theory. The credit for Flamsteed's great achievement in reducing the errors of astronomical measurement to a new low order, despite all the obstacles he had to overcome, clearly attaches to himself alone. The Royal Society had other observers of note, such as Robert Hooke and Edmond Halley (Flamsteed's successor at Greenwich), but these had to do what they could with their own resources. In fact the Society could offer, at this period, no fit facilities for scientific work of any kind, apart from its library and museum.

Everywhere in Europe the formation of scientific societies illustrates a dual tendency, on the one hand towards the crystallization of a specifically scientific organization out of informal groups having broader and more superficial intellectual interests, and on the other towards the preponderance of the experimentalists within the organization. In Italy, France and England there was a transition from the discussion of natural-philosophic systems or hypotheses to the verification and accumulation of fact; as the course of the scientific revolution laid more emphasis on deeds than on words, on the laboratory rather than the study, and as the preparation of commentaries and criticisms of ancient texts gave place to the writing of memoirs describing the results of systematic investigation, so the characteristics of scientific organization changed accordingly. In the first half of the seventeenth century the function of a scientific assembly had been to promote discussion and dissemination of the new idea of science, and to provide a forum in which, not merely before an audience of enthusiasts, but before the broadest cross-section of educated and literate society, the original thought of a Galileo or a Descartes could challenge conventional opinions in science. Patrons like the Medici brothers, or a newsgatherer like Mersenne, amassed the accumulative weight of innovation against the science of colleges and text-books. They presented the total case for the "new philosophy," in an intellectual environment whose dogmatic traditions were already disintegrating, to a new learned class freed from the sterner discipline of the old professional scholar, ready to admire acuity of wit, subtlety of reasoning and fertility of imagination more than allegiance to orthodox "sound" views. If the "new philosophy" was obstructed in the university, there was appeal through the scientific assembly to the more tolerant, eager and wealthy intellectual circles of court and capital. But the alliance between modern science in its early stages and the whole turbulent current of cultural development in the seventeenth century was inevitably incomplete and of short duration. Important, creative scientific work rapidly outdistanced the dilettante and virtuoso: rarely, for example, does the name of John Evelyn occur in the proceedings of the Royal Society printed by Birch in his *History*. A man of general culture could not see the point of detailed scientific labours, for whereas he might enjoy a debate on Descartes' concept of the animal as machine, he tended to find the naturalist grubbing in ditches for insects' eggs merely comic. The exploitation of the shift of intellectual perspectives, so fascinating in general outline, inevitably sank to tedium and pedantry in the eyes of those who sought entertainment and striking novelty. As a result, the Montmor Academy broke up, and the Royal Society failed in wide appeal after its first fifteen years.

Consequently, in the second half of the seventeenth century the rôle of the scientific society changed considerably. Having become a thoroughly professional body, it served as a focus for the discussion of works rather than ideas. Its aim was to develop the sciences, rather than promote a "new philosophy." Opposition from Aristotelean university teachers, or the medical profession, was hardly serious any longer. The scientific movement required countenance less than means—buildings, apparatus, money for the maintenance of research, and methods for exchanging its results. It was found, for instance, that as scientific books became more truly technical, more fully devoted to describing research (rather than useful textbooks or practical manuals), the publishing trade refused to handle them unless large sums were laid down. Or again, while there was an expanding commercial market for ordinary watches and clocks, navigational instruments, and even telescopes and microscopes, financial encouragement alone would induce craftsmen to hazard their profits in efforts to improve instruments for the advancement of science. In short, the task before a scientific society was less to secure the scientific revolution, than to maintain its momentum and to reap its harvest. The founders of the Académie des

Sciences were perhaps the first to appeal to national interest in this connection.[1] Earlier exponents of the utility of scientific discovery had rather looked to the improvement of the condition of mankind—to a shift in the precarious balance between human powers and the forces of nature. As Bacon had put it, the ambition to exalt one's state was a degree less noble than the ambition of the natural philosopher to elevate mankind. But in the proposals put forward by Leibniz, for example, to establish a scientific academy in Germany, there seems to be a clearer statement of the case for investment in science as a national benefit. There the first scientific academy, founded at Berlin in 1700, did not develop from the efforts of earlier groups of amateurs.[2] The capital of Brandenburg-Prussia was indeed far removed from the main centres of culture in Germany, its university not being founded until more than a century later, but Leibniz had in the Elector (later King) Frederick I a patron willing to realize his long-matured plans. These had always aimed at furthering the interests of the German nation and raising its technological standards through the encouragement of the vernacular language and the reform of education in a practical direction. To attain these objects a national academy which should concern itself with practical applications as well as with the pure sciences was the first necessity. In Leibniz' view Germany had once enjoyed pre-eminence in useful arts, especially in mining and chemistry,[3] but also in horology, hydraulic engineering, goldsmith's work, turnery, forging etc. Astronomy was restored by the Germans, and the "Nieder-Deutschen" (Netherlanders) had invented the telescope and mastered navigation. The only remedy for the subsequent deterioration was the generous encouragement of science, which Leibniz coupled with the enforcement of a strictly mercantilist economic policy, by which the

[1] 'Sans exclure de son programme d'études les sciences pures et spéculatives, Colbert essaie de l'orienter vers les sciences appliquées à l'industrie et aux arts." P. Boissonade: *Colbert* (Paris, 1932), p. 28.

[2] There were already active scientific groups in Germany in the late seventeenth century such as the *Collegium Naturæ Curiosorum*, founded by Lorentz Bausch in 1652 and dealing with the medical sciences, and the *Collegium Curiosum sive Experimentale*, founded by Christopher Sturm of the University of Altdorf in 1672 and dealing with physical sciences. Neither was a national academy of science in any sense.

[3] 'Denn weil keine Nation der Teutschen in Bergwergssachen gleichen Konnen, is auch Kein Wunder, dass Teutschland die Mutter der Chimie gewesen.' L. A. Foucher de Careil: *Œuvres de Leibnitz* (Paris, 1859–75), vol. VII, pp. 64–74.

state should become self-sufficient.[1] As he put it in a letter to Prince Eugene, discussing the proposed scientific academy in Vienna:

Pour perfectionner les arts, les manufactures, l'agriculture, les deux espèces d'architecture [i.e. civil and military], les descriptions chorographiques des pays, le travail de minières, item pour employer les pauvres au travail, pour encourager les inventeurs et les entrepreneurs, *enfin pour tout ce qui entre dans l'œconomique ou mécanique de l'etat civil et militaire*, il faudrait des observatoires, laboratoires, jardins de simples, menageries d'animaux, cabinets de raretez naturelles et artificielles, une histoire physico-médicinale de toutes les années sur des relations et observations que tous medecins salariés seraient obligez de fournir.[2]

Leibniz, historian, mathematician, philosopher, diplomatist and confidential adviser of princes, in his dual devotion to science and Germany saw the scientific academy as a necessary instrument of the modern state, through which science could be made to play its due part in social and economic policy. He had little patience with those who 'considèrent les sciences non pas comme une chose très importante pour le bien des hommes, mais comme un amusement ou jeu,' and criticized the Académie des Sciences for this reason.[3] Science as a factor in creating national prestige, its rôle in war and in the commercial rivalry of states, were appreciated in England and France as well as in Germany; but no one who could claim high rank as a philosopher and scientist announced the importance of scientific organization to jealous statesmen more clearly than Leibniz.

Thus, one alleviation of the obstructions to scientific progress was sought through the conversion of Bacon's appeal to the interests of humanity into an appeal to the interests of the state.[4] One

[1] On mercantilism, cf. E. Hecksher: *Mercantilism* (London, 1935). The Berlin Academy was intimately linked with Leibniz' typically mercantilist project for the establishment of a silk industry in Brandenburg. (Foucher de Careil, *loc. cit.*, pp. 280 *et seq.*)

[2] *Ibid.*, p. 317, italics inserted.

[3] Letter to Tschirnhaus, January 1694 (C. I. Gerhardt, *Mathematische Schriften*, in *Gesammelte Werke* hrsg. von G. H. Pertz, vol. IV, p. 519).

[4] In the later seventeenth century gunpowder was still quoted as one of the great discoveries of the modern age. War was accepted by scientists, as by all men, as an inevitable human evil, and the development of the war-like potential of the state did not provoke moral condemnation, perhaps because this generation experienced warfare which was more technically efficient than that of previous generations, but which the lessening of religious fanaticism had rendered less horribly destructive. Many scientists, however, commented that it was more noble to advance the arts of life than those of death.

other must be briefly treated. The Royal Society approved a step which was calculated to increase its support in a rather different way, by the publication of its *Philosophical Transactions*, begun in 1665. Partly a profit-seeking venture on the part of the Secretary-editor, Henry Oldenburg, the *Transactions* were also intended to attract the "curiosi" and "virtuosi" to the work of the Royal Society, and to stimulate the submission of original reports. The publication consisted of discourses read by Fellows at meetings of the Society, of letters on scientific subjects (translated, if necessary, and printed more or less verbatim), both from Britain and abroad, most of which also had been read at meetings, and of book-reviews. The modern scientific "paper," of which the first examples appear in the *Philosophical Transactions*, has in fact a double origin. One of its ancestors exists in the interchange of scientific correspondence discussing original work, which may be traced back to the late sixteenth century. Very many of Oldenburg's articles were letters with hardly more than the "Dear Sir" and "your obedient servant" deleted: but they were letters intended for publication. The other ancestor was the formal essay or discourse read to scientific groups from the early seventeenth century onwards. At the time, this form of presentation required the development of a new art of composition, for no suitable literary or academic models existed and its perfection, leading to the complex machinery for scientific reporting existing today, is an event of historical importance. At the end of the century, however, the learned journal was still far from being the accepted means for the announcing of discoveries. Huygens, for instance, though he stated his results in bald (or even enciphered) language to the societies of London and Paris, published them completely only in full treatises which sometimes appeared many years later.

Nevertheless, the *Philosophical Transactions* was immensely successful. Latin translations were produced at Amsterdam, and the Académie des Sciences prepared its own French version. Though Antoni van Leeuwenhoek, the great microscopist, never visited London and knew no language other than his native Dutch, he sent the accounts of his astonishing discoveries for publication in English in the *Transactions*. Oldenburg had created a new field of literature, which was rapidly extended; for if the *Transactions* was not the first journal to appear in regular numbers, it was the first to print original communications. Its predecessor by two months, the *Journal des Sçavans*, surveyed the whole field of learning, devoted much space to summaries of books, and merely reported the business of the scientific societies. The series of original *Mémoires* of the Académie des Sciences was begun only after the reorganization of 1699. Many other reviews, of which the most famous was Bayle's *Nouvelles de la république des lettres* (1684) followed the broad pattern of the *Journal*; a few others (the *Miscellanea* published by the *Collegium naturæ curiosorum* from 1670, and the *Acta Eruditorum* founded by Leibniz in 1682, the former restricted to the medical sciences and the latter embracing many aspects of learning) printed communications, but none was so purely descriptive in its content as the *Philosophical Transactions*. While the reviews enjoyed a steady repute among intelligent readers, and gave a superficial picture of the total scientific activity in Europe, private communication in correspondence, and the frequent exchange of their respective publications, was for the leading figures far more important.

From the preceding account, it will be clear that during the mid-seventeenth century there was a tendency for all effective scientific activity to be focused upon some society or group. In the small area of England a single organization embraced everyone, but in France and Germany, both before and after the foundation of national academies, less magnificent societies flourished, as they did also in the Italian and other small states. Each of these had its own character and major preoccupations. Naturally also each proclaimed, with varying degrees of emphasis, the principal tenets of the scientific revolution. Strategic concepts, like that of natural law, gradually gained a universal validity; the practical benefits to be expected from the cultivation of natural science were canvassed in all parts of Europe; Aristotelean philosophy was everywhere condemned, and the virtues of mathematical analysis everywhere exalted; nearly all the societies embarked on elaborate programmes of experiments. The societies differed in character, and individual members of the same society differed in their opinions; in France Descartes, and in England Bacon, were regarded as peculiarly magisterial figures, but it is commonplace to find members of different societies working on the same or allied problems in similar ways, or to discover the reaction of some discussion in Gresham College upon the work of the Parisian academy, and *vice versa*. Only to a strictly limited degree did local or personal traditions bar the complete amalgamation of the new scientific spirit.

Perhaps this may best be illustrated in the development of a mechanistic view of nature during the seventeenth century. This proceeded at three different levels. In the first place, a mechanistic theory of the universe was fully described by Descartes in 1644 (in the *Principles of Philosophy*), to be supplanted by the far more perfect theory of Newton's *Principia* (1687).[1] Spheres and intelligences were finally banished, and the secrets of the heavenly motions were traced to the properties of matter and the rule of laws of nature. Secondly, in biology, the theory of the organism as a machine was taken up by the Cartesian school, and exercised a wide influence; this theory, of course, was the product of a shift in philosophical outlook. Newtonian mechanism could be shown to satisfy all the minutiæ of evidence; biological mechanism was a profound hypothesis, but no demonstration of it in physiological terms as yet existed, or was possible. In the third place, mechanistic views in relation to physics and chemistry fall into an intermediate category. They could not be demonstrated completely, but they could be applied successfully in a number of particular instances. They were certainly philosophical in origin, and not deductions from definable experimental investigations, but they were applicable in a wide area of experimental research.

At each of these levels, the attitude of the mechanistic scientist to the complexity of nature enabled him to cope with different fundamental problems. Pure and celestial mechanics, the most highly mathematical branches of science, treated of the nature of forces and motion. Mechanistic biology gave a new interpretation of the nature of life, of growth and of sensation. In relation to physics and chemistry, the problem to which the scientist sought an answer was the nature and constitution of matter, in so far as these sciences dealt with the properties, changes and transformations of inorganic substance. The distinctions between the three states of matter; differences in density and mass, hardness and

---

[1] The *Principia* exposes, of course, no mechanistic theory of the nature or cause of gravitational forces. Newton's conjectures, on various occasions, merely indicate the possibility of some form of mechanistic interpretation. See below, pp. 273-5.

brittleness; the nature of magnetism, electricity, gravity and heat; hypotheses of light and colour, all seemed to require interpretation in the light of some general theory of what matter is, which would account for the variations in observed properties between different kinds of material substance. Similarly the philosophical chemist, studying solution, volatilization, fusion, and the analysis and synthesis of substances effecting striking qualitative changes in macroscopic properties, necessarily tried to form some picture of what was happening to the very nature of the materials in his vessels. Starting from the axiom that matter is neither created nor destroyed, what internal modification was implied by a change in observable qualities?

During the sixteenth and seventeenth centuries there was, as might be expected, a sharp reaction against the answers which Aristotle had furnished to this type of question. In the period of the foundation of the Royal Society and the Académie des Sciences various versions of a mechanistic, particulate theory of matter were widely entertained. As in other matters, there was a tendency for the scientific revolution to revert to a Greek view of nature older than Aristotle's. Medieval philosophers had, very largely, respected Aristotle's condemnation of atomistic doctrines, but the scholars of the renaissance and the scientists of the seventeenth century, with the full text of Lucretius' exposition of Greek atomism in their hands,[1] preferred its use of the concepts of structure and physical texture in matter to Aristotle's theory of forms and qualities. The Greek atomists had explained the complexities of substances without making any assumptions of a non-material nature; indeed, if Lucretius was right in his declaration that the only realities are atoms and the vacuum in which they move, qualities were mere illusion, the perceptual registration of physical reality.[2] This was a virtue of the "mechanical" or "corpuscular" philosophy particularly attractive to Galileo, for example, who sought to penetrate by means of the process of

[1] There were about thirty editions of De Natura Rerum (editio princeps, Brescia, 1473) before 1600. There was a French translation in 1677, and an English in 1683.

[2] By contrast, in the older philosophy, forms and qualities were real, existing entities. If it was said that snow had the "form of whiteness," this did not describe the snow, but explained its optical properties. Similarly for the alchemists the "form of gold" was something that could be separated from the "substance" of gold and transferred to the "substance" of another metal, e.g. lead deprived of its own distinctive "form."

abstraction from the evidence of sensation to the basic reality of nature.

Before Descartes' influence became significant a number of writers had expounded or commented on ancient particulate theories, drawing on the texts of Democritus, Epicurus, Lucretius and Hero of Alexandria. The earlier ones treated the atomistic doctrine purely as a theory of matter, which they freely combined with Aristotelean physics. Pierre Gassendi, from about 1625, was the first writer to attempt to develop a completely mechanistic physics founded on Epicurus and rejecting Aristotle, but his success was hardly greater than that of Lucretius, whom, except in matters touching on religion, Gassendi very closely followed. Physical properties were simply traced to the imagined size and shape of the component particles. In Galileo and Bacon, on the other hand (as already mentioned), are found the beginnings of a true kinetic theory, especially in relation to heat. As Isaac Beeckman stated it, 'all properties arise from [the] motion, shape, and size [of the fundamental particles], so that each of these three things must be considered.'[1] For both Bacon and Galileo an important aspect of the "new philosophy" was its endeavour to establish, through experimental study, a theory of particulate mechanisms to replace the doctrines of forms and qualities, yet neither of them was an atomist in a narrow sense. Bacon, indeed, wrote that the proper method for the discovery of 'the form or true difference of a given nature, or the nature to which nature is owing, or source from which it emanates' would not lead to 'atoms, which takes for granted the vacuum, and immutability of matter (neither of which hypotheses is correct), but to the real particles such as we discover them to be.'[2] In short, there is ample evidence that in the early seventeenth century the conception of matter as consisting of particles, whose aggregate might be a solid, a liquid, an air or a vapour, was commonplace and generally acceptable and that the properties of the aggregate were attributed to the nature and motions of the particles.

In considering the development, extension and diversification of this generalized concept in the second half of the century, three traditions may be distinguished. The first is that of the strict

[1] *Journal tenu par Isaac Beeckman de 1604 à 1634*, ed. Cornelis de Waard (La Haye, 1939–45), vol. I, p. 216 (1618).
[2] *Novum Organum*, Bk. II, Aphorisms i, viii.

atomists, adhering firmly to the indivisibility of the ultimate particle and the absolute reality of the vacuum between atoms. They were the followers of Gassendi. Next, Cartesian scientists continued the peculiar corpuscularian doctrine of Descartes, which admitted the infinite divisibility of matter, and denied the vacuum. Finally, the experimentalists refused to commit themselves to any precise body of philosophic preconceptions. They learnt from both Gassendi and Descartes: the form of their particulate theory was closer perhaps to that of Gassendi, but in its application to physics it borrowed much from Descartes' concrete imagination. This experimentalist tradition in corpuscularian physics grew most rapidly in England, and was especially fostered by the works of Robert Boyle. Never a dogmatic system, it was a late development which played an important part in the relations between the French and English scientific societies. It was, in fact, a major product of the impact of French scientific ideas upon Englishmen about the middle of the century—an impact which conditioned the nature of the scientific achievement of the Hooke-Boyle-Newton generation.

Starting from the assumptions that in nature there are no occult forces, like gravity or magnetism, and that the universe is continuously and completely filled with matter, Descartes developed in the *Principles of Philosophy* (1644) an extraordinarily elaborate mechanistic and corpuscularian "model" of the physical universe. His particles, of which there were three species, were not atoms, because he imagined them as divisible, though in nature not normally divided. The *first element* was a fine dust, of irregular particles so that it could fill completely the interstices between the larger particles. The *second element* (*matière subtile* or æther) consisted of rather coarser spherical particles, apt for motion, and the *third element* of still more coarse, irregular and sluggish particles. These three elements corresponded roughly to the Fire, Air and Earth of Aristotle, and as they were composed of the same matter, the elements could be transformed one into another. This was regarded by the Cartesians as no arbitrary hypothesis, but as a truth (according to Rohault) 'necessarily follow[ing] from the Motion and Division of the Parts of Matter which Experience obliges us to acknowledge in the Universe. So that the *Three Elements* which I have established, ought not to be looked upon as imaginary Things, but on the contrary, as they are very easy to conceive, and we see a necessity of their Existence, we cannot reasonably lay aside the Use of them, in explaining Effects purely Material.'[1] The nature of a substance was mainly determined by its content of third element, its properties by the second. Since Descartes denied that the third-element particles had intrinsic weight or attraction, hardness (the cohesion of these particles) was attributed to their remaining at rest together, fluidity to their relative motion; but this motion also was not intrinsic but imparted by the first and second elements. Thus in solution the grosser particles of the solvent by their agitation dislodged those of the dissolved material; if however the particles of the solvent were too light, or the pores of the solid material too small to admit them, the latter would not dissolve. When the pores between the third-element particles were large enough to admit a large quantity of the second element, the substance was an "elastic fluid" (gas), whose tendency to expand was caused by the very free and rapid motion of the second-element particles. Flame itself consisted of matter in its most subtle form and under most violent agitation, and was therefore the most effective dissolvent of other bodies, while the sensation of heat increased with the degree of motion in the particles of the heated body. Rohault notes that, when filed, copper grows less hot than iron because, copper being the softer metal, its particles do not require such violent agitation to separate them as do those of iron.[2] The greater motion associated with heat was also the cause of thermal expansion. Light was thought to be '*a certain Motion of the Parts of luminous Bodies* whereby they are capable of pushing every Way the subtil Matter [second element] which fills the Pores of transparent Bodies,' and secondary illumination was attributed to the tendency of this matter to recede from the luminous body in a straight line. Transparent bodies had straight pores through which the *matière subtile* could pass, opaque bodies blocked or twisted pores. If this exertion of pressure by the luminous body were confined or resisted, it would grow hot. Refraction and reflection of this pressure (or

---

[1] John Clarke (trans.): *Rohault's System of Natural Philosophy Illustrated with Dr. Samuel Clarke's Notes Taken mostly out of Sir Isaac Newton's Philosophy* (London, 1723), vol. I, pp. 115–17. Jacques Rohault (1620–72) was the greatest of all expositors of Cartesian physics; his *Traité de Physique* appeared first in 1671. Clarke's *Notes* strongly oppose Newtonian corpuscularian ideas to those of Descartes.

[2] *Ibid.*, p. 156.

rather pulse) which is Light were explained by analogy with the bouncing of elastic balls.[1] In dealing with magnetism Cartesians were careful to emphasize that 'though we may imagine that there are some Particular Sorts of Motion which may very well be explained by *Attraction*; yet this is only because we carelessly ascribe that to *Attraction*, which is really done by *Impulse*'; as when it is said a horse draws a cart, whereas it really *pushes* it by pressing on the collar.[2] Magnetic effects were actually caused by streams of screw-like particles, entering each Pole of the earth and passing from Pole to Pole over its surface, which passed through nut-like pores in lode-stone, iron and steel, and thus were capable of exerting pressures on these magnetic materials.

By theorizing in this way on the different motions of the three species of matter the Cartesian physicists tried to account for all the phenomena of physics as known in the second half of the seventeenth century. They had some striking contemporary discoveries on their side: for instance, the discovery that the rising of water in pumps and analogous effects were not due to *horror vacui* or to attraction, but simply to the mechanical pressure of the atmosphere. They also explained gravitation mechanically as a result of pressure, and extended their corpuscular ideas to chemical reactions. The idea of particulate matter-in-motion was therefore the very foundation of Cartesian science, the basis of a homogeneous system of explanation. The fact that (in Rohault's words) 'the few Suppositions which I have made . . . are nothing compared with the great Number of Properties, which I am going to deduce from them, and which are exactly confirmed by Experience' was a strong reason for believing 'that *That* which at first looks like a Conjecture will be received for a very certain and manifest Truth.'[3] As expounded by Descartes and his successors this "mechanical philosophy" was illustrated by many qualitative experiments; but these could hardly be said to *prove* the Cartesian system, which always remained, in addition, entirely non-mathematical.

Nevertheless, Cartesian science had a great influence upon the Fellows of the Royal Society. Just as, in the past, Aristotle's teaching was the inevitable starting-point for scientific thought, so they

[1] Clarke, *op. cit.*, vol. I, pp. 201 *et seq.*
[2] *Ibid.*, vol. II, p. 166.
[3] *Ibid.*, vol. I, p. 203; vol. II, p. 169.

often found their point of departure in Descartes. For Boyle, indeed, 'the Atomical and Cartesian hypotheses, though they differed in some material points from one another, yet in opposition to the Peripatetic and other vulgar doctrines they might be looked upon as one philosophy.' While older philosophers had given but superficial accounts of natural phenomena, relying on an incomprehensible theory of forms and qualities, the moderns agreed in explaining 'the same phenomena by little bodies variously figured and moved.' The differences between the modern schools were rather metaphysical than physical, and did not greatly affect the study of the world as it actually is.[1] Apparently the English experimentalists who formed the Royal Society already held eclectic opinions: 'they found some reason to suspect,' wrote Hooke in his Preface to *Micrographia* (1665), 'that those effects of Bodies which have been commonly attributed to *Qualities*, and those confess'd to be *occult*, are perform'd by the small machines of Nature.' Over this wide and in many ways strategic area of scientific thinking the English and French societies spoke the same language and shared a common inheritance of ideas. The differences between Cartesians and Gassendists were found in both alike, and the appeal to a particulate theory of matter was as common in England as in France.

Though the experimenters of Gresham College were far from being pure empiricists, in giving high praise to Descartes' system they did not forget their other allegiances to Bacon, Galileo and Gilbert. Few were dogmatic or literal followers of Descartes. Accepting the general form of the "mechanical philosophy," they measured Cartesian science rigorously by the experimental test. Hooke challenged its theory of light and colours, Boyle its theory of elastic fluids, and Newton its cosmology. Yet even the last of these suggested a space-filling æther, and questioned whether the most subtle effects of nature were not obtained by purely mechanical means. The very titles of Boyle's works indicate the tendency of his thought: *The Excellence and Grounds of the Mechanical Philosophy*; *The Origin of Forms and Qualities*, an introduction to the same; *The Mechanical Origin of Volatility and Fixedness*; *The Mechanical Production of Electricity*; *The Mechanical Origin of Heat and Cold*; with many more in a similar vein. And during the greater part of his

---

[1] *Certain Physiological Essays* (publ. 1661 but written some years earlier). *Works*, ed. T. Birch (London, 1772), vol. I, p. 355.

scientific career, from about 1655 at least, he devoted himself to trying 'whether I could, by the help of the corpuscular philosophy . . . associated with chymical experiments, explicate some particular subjects more intelligibly, than they are wont to be accounted for, either by the schools or [by] the chymists.'[1] In his *Excellence and Grounds of the Mechanical Philosophy* Boyle defined his position exactly:

God, indeed, gave motion to matter; . . . he so guided the motions of the various parts of it, as to contrive them into the world he designed to compose; and established those rules of motion, and that order amongst things corporeal, which we call the laws of nature. Thus, the universe being once fram'd by God, and the laws of motion settled, and all upheld by his perpetual concourse, and general providence; the [mechanical] philosophy teaches, that the phenomena of the world, are physically produced by the mechanical properties of the parts of matter, and that they operate upon one another according to mechanical laws.[2]

Here Boyle appeals for justification of his natural philosophy to the divine plan of creation (thus withdrawing himself from the atheistic connotations of Epicureanism, and from the evolutionary suggestions of Descartes); seeing the world as a machine indeed, but a machine whose complex processes are continually supervised by a divine providence. Though his world-picture transcended the evidence of experimental science, Boyle was convinced that the details of the mechanism of nature could only be revealed through experimental study. It might even be more useful not to attempt at this stage to relate observed physical phenomena to the 'primitive and catholick affections of matter, namely bulk, shape, and motion,' but rather to intermediate physical properties such as hardness, temperature and so forth. Therefore he did not hesitate to doubt the necessity of Descartes' rectitude. He doubted whether the air in an exhausted vessel was really replaced by a *matière subtile*, of whose existence he was generally sceptical.[3] He attributed the elasticity of air to the springiness of its particles. He was much more vague in his pronouncements on the basic

[1] *Some Specimens of an Attempt to make Chymical Experiments useful to illustrate the Notions of the Corpuscular Philosophy. Works*, 1772, vol. I, p. 356.

[2] Peter Shaw: *Works of Boyle abridged* (London, 1725), vol. I, p. 187.

[3] The *matière subtile* (or æther) in an exhausted space was used by Cartesians generally to explain the transmission of light, and by Huygens to account for effects of surface tension *in vacuo*.

structure of matter than Descartes, indicating only that he thought it composed of fundamental particles, and larger aggregates of these, the real corpuscles. The particulate theory was related by Descartes to the Aristotelean four-element doctrine, but not by Boyle. He was sceptical also of the Cartesian interpretation of magnetism and electricity. He carried through a far more thoroughgoing mechanistic attack on the doctrine of "forms" than the Cartesians. Moreover, while the Cartesians sought to illustrate a theory of nature by experiments, Boyle sought to interpret his experimental researches in the light of the corpuscular philosophy. The distinction is perhaps subtle, but it is real. For besides his sense of the ultimate truth of a mechanistic view of nature, Boyle was also imbued with Bacon's conception of the scientist's compiling histories of nature, and promoting the progress of material civilization. In the last resort his devotion to the corpuscular philosophy seems to have been grounded less on an opinion that it was a key to the final understanding of nature, than on the conviction that it provided a broad framework of ideas within which scientific research could most rapidly progress towards this final understanding.

If Boyle presented the corpuscular philosophy more elaborately than any other Englishman, the influence of Newton on non-Cartesian particulate theories of matter was perhaps of even longer duration. His successors did not hesitate to read the *Queries* appended to Newton's *Opticks* (1704) as though they were statements of his considered opinions. Newton's views on the structure of matter are indeed shadowy. The famous *Hypotheses non fingo* is not to be taken too literally, but he certainly would not have accepted Rohault's easy dictum that a conjecture agreeing with the properties of things may be taken as very probable. The questions which Newton felt impelled to ask, and the answers to them at which he hinted, were indeed only made public because Newton knew that his scientific career was over. The experiments needed to gain further insight into these problems he would never perform. Yet Newton had confidence enough in his opinions to declare:

> It seems probable to me, that God in the Beginning form'd matter in solid, massy, hard, inpenetrable, movable Particles, of such Sizes and Figures, and with such other Properties, and in such proportion to Space, as most conduced to the End for which he form'd them. . . .

These, then, were true atoms; endowed with 'a *Vis inertiæ*, accompanied with such passive Laws of Motion as naturally result from that Force,' and with 'certain active Principles, such as is that of Gravity, and that which causes Fermentation, and the Cohesion of Bodies.' Such principles were not occult qualities, like those of the Aristoteleans, because they were made precisely known by experiment; only the causes of them were hidden.[1] Of the particles Newton asked, have they not also 'certain Powers, Virtues, or Forces, by which they act at a distance, not only upon the Rays of Light for reflecting, refracting, and inflecting them, but also upon one another for producing a great Part of the Phænomena of Nature?'

He confessed that these "Virtues" might be veritably performed by impulse, but their cause was only to be unfolded through study of the "Laws and Properties of the Attraction."[2] To attraction between particles he attributed cohesion and the strength of macroscopic bodies, the force being very strong in immediate contact, but reaching 'not far from the Particles with any sensible Effect.' Thus the particles were compounded (as Boyle had suggested) into corpuscles of weaker attractive force, and so successively into the largest aggregates 'on which the Operations in Chymistry, and the Colours of Natural Bodies depend, and which by cohering compose Bodies of a sensible Magnitude.'[3] Newton was the first to point out that in any apparently solid body the volume of matter is small compared with the volume of space between the particles, and to think of each particle as surrounded by a field of force. Again, he asked:

> Is not . . . Heat . . . conveyed through the *vacuum* by the vibrations of a much subtiler Medium than Air, which after the Air was drawn out remained in the *Vacuum*? And is not this Medium the same with that Medium by which Light is refracted and reflected. . . . And is not this Medium exceedingly more rare and subtile than the Air, and exceedingly more elastick and active? And doth it not readily pervade all Bodies? And is it not (by its elastic force) expanded through all the Heavens?

But this Newtonian æther was very different from the Cartesian. As the cause of gravity Newton imagined it as being more rare

---

[1] Sir Isaac Newton: *Opticks* (with Introduction by Sir E. T. Whittaker, London, 1931), pp. 400–1.
[2] *Ibid.*, pp. 375–6.   [3] *Ibid.*, pp. 389, 394.

in solid bodies than in free space; where it must be of the order one million times less dense than air, but more elastic in the same proportion.[1] With regard to the theory of light, it seems futile to try to harmonize the different notions referred to in the *Queries*. Newton's views involved a compromise between the "undulatory" and "corpuscular" theories, and in different passages he seems, as it were, to be operating at different depths of thought. But whether he spoke of light as a vibration in the æther, or as a stream of particles issuing from the luminous body, he was at once mechanistic and anti-Cartesian. Similarly he gave a purely mechanical account of the physiological nature of vision.

The *Queries* are highly suggestive. The reader half-glimpses entrancing vistas of the territory to be conquered by the "mechanical philosophy." In the early eighteenth century a number of rather unfruitful attempts to take Newton's ideas further were made, especially in relation to chemistry with the notion of corpuscular attraction as a precursor of affinity. He has been regarded as one of the founders of nineteenth-century atomic theory, and perhaps the very fact that the mighty Newton *had* speculated in this way made atomism more respectable at a time when the Cartesian system of science had passed into oblivion, and many matter-of-fact chemists distrusted John Dalton as a weaver of idle fancies. Certainly the less specialized corpuscularian and mechanistic concepts had the best chance of survival; Cartesian mechanism, the dinosaur of seventeenth-century scientific thought, could not adapt itself to a new intellectual environment. It was committed dogmatically in too many points of detail where it proved to be false.

Only with the passage of time, and usually in relation to points of precise detail, did the development of scientific activity in Britain after the foundation of the Royal Society begin to follow this definitely less specialized course, yielding a mechanistic philosophy that was ultimately anti-Cartesian. More than two decades passed, after the death of Descartes in 1650, before his works acquired their greatest fame and authority, and before the character of his system became rigid. Meanwhile, the prestige of Gassendi was falling in France, and that of Descartes in England —especially through the success of new ideas concerning light and gravitation. The Académie des Sciences became, before the

[1] *Ibid.*, pp. 349–52.

end of the century, more deeply committed to Cartesian thought; the Royal Society more ready to criticize it. It is important to realize that the somewhat singular character of the Royal Society during its first half-century was due, not solely to the Fellows' greater assiduity or success in experimentation—emphasis on the purity of their empiricism is certainly to be suspected—but to experimentation combined with a definite eclecticism of outlook, within the broad framework of a mechanistic, corpuscularian science which was becoming almost a commonplace. The intellectual tension between the English scientific groups, on the one hand, and the French and German on the other, certainly increased between about 1665 and 1720, the substantial difference between Cartesians and non-Cartesians being exacerbated by the adventitious dispute between Leibniz and Newton over the invention of the calculus, but this should not be allowed to conceal the fundamental similarity of their attitudes, on which Boyle had insisted, and of their approach to scientific research. Although the national organization of science facilitated a deplorable kind of national partisanship most vicious in the early years of the eighteenth century, beneath this there existed a fundamental community of thought and activity. Broadly, the tendencies of science were everywhere in the same direction, and in the later eighteenth century, in a different situation, the friendly cooperation between scientific societies was once more of the greatest importance.

CHAPTER VIII

TECHNICAL FACTORS IN THE SCIENTIFIC REVOLUTION

T HE renaissance of science in the sixteenth century, and the strategic ideas of the first phase of the scientific revolution, owed little to improvements in the actual technique of investigation. Before the beginning of the seventeenth century there is little evidence, except perhaps in anatomy and astronomy, of any endeavour to control more narrowly the accuracy of scientific statements by the use of new procedures, still less to extend their range with the aid of techniques unknown to the existing tradition of science. Even the refinement of observation, begun in anatomy by Vesalius and his contemporaries and in astronomy by Tycho Brahe, hardly involved more than the natural extension and scrupulous application of familiar methods. Since the apparatus and instruments available were crude and limited the means were not at hand for gaining knowledge of new classes of phenomena, or eliciting facts more recondite than those already studied. Though greater reliance was placed on observation and experiment, the change in the content of science could not be dramatic and other sources of information were, at least till the latter part of the sixteenth century, largely traditional. Aristotle, Pliny, Dioscorides, Theophrastus and Galen were still very highly respected. Gradually, however, the tendency to supplement this book-learning, checked by personal examination where possible, by the experience of various groups of practical men gained ground. The wealth of fact was augmented by admitting the observations of craftsmen, navigators, travellers, physicians, surgeons and apothecaries as worthy of serious consideration, and thus the status of purely empirical truths, hardly inferior to that of the systematic truths of physics or medicine, was in time enhanced.

In this respect, as in others, the work of Galileo gives a useful indication of a turning point, displaying in various ways the operation of new technical, as well as conceptual, factors in the development of science. Galileo's conceptual achievements were of the greater importance, and imply a new metaphysics rather than the total absence of metaphysics, but he also admired the technological achievements of his time and appreciated the scientific problems suggested by them. By revealing the value of mathematics as a logical instrument in scientific reasoning, he transformed, if he did not actually create, an important method of inquiry. His exploration of the potentialities of the telescope and other instruments shows his concern for the enlargement of the scope of observation and experiment through newly invented techniques. It is typical of the evolution of the apparatus of science during the seventeenth century that Galileo's results were more notable for their qualitative originality than for quantitative accuracy, since the necessity for precision in measurement was less apparent than the strange novelties which the new techniques unfolded. Though the perspective in which science regards nature changed markedly in the sixteenth century, it was only in the seventeenth that a significant qualitative change occurred in the image itself, to which the technical resources used by Galileo contributed profoundly.

It has already been pointed out that the ideal of social progress was also a commonplace among seventeenth-century scientists, and that with varying degrees of assurance the attainment of this ideal was linked with the application of scientific knowledge to technology. Conversely, it is clear that scientific research is itself dependent upon the level of technical skill, especially when the endowment or organization of science compels the experimenter to rely upon the skills acquired by the craftsman in the normal course of his trade, as was the case before the nineteenth century. Perhaps, in the early stages of a science, it is even more important that the investigator should be amply provided with both problems and the materials for solving them by the technological experience to which he has access. This is partly a question of attitudes—the ability to receive the stimulus from a merely practical quarter—partly of the richness of the techniques. Galileo makes Sagredo remark, on the first page of the *Discourses*:

> I myself, being curious by nature, frequently visit [the Arsenal at Venice] for the mere pleasure of observing the work of those who, on account of their superiority over other artisans, we call "first rank men." Conference with them has often helped me in the investigation of certain effects including not only those which are striking, but also those which are recondite and almost incredible. At times also

I have been put to confusion and driven to despair of ever explaining something for which I could not account, but which my senses told me to be true.

It can hardly be doubted that the dialogue of the First Day in this work was influenced by such practical observation, and it was from a workman that Galileo learnt of the break-down of the *horror vacui* theory when the attempt was made to lift water through more than thirty feet by means of a suction pump. Bacon also wrote of the knowledge concealed in skilled craftsmanship. In the next generation Boyle thought that only an unworthy student of nature would scorn to learn from artisans, from whom knowledge could best be obtained; for

> many phenomena in trades are, also, some of the more noble and useful parts of natural history; for they show us nature in motion, and that too when turn'd out of her course by human power; which is the most instructive state wherein we can behold her. And, as the observations hereof tend, directly, to practice, so may they also afford much light to several theories.[1]

Such opinions did not spring from theoretical reasoning alone. They express the new philosophy's concern for *realia*, but they also recognize a genuine historical fact, that many of the ordinary operations of household and workshop were quite beyond the reach of scientific explanation. To remedy this, Galileo began the theory of structures and Boyle the study of fermentation in food-stuffs. Many of the problems suggested by the "naturalist's insight into trades" could not, of course, be very profitably handled in the seventeenth century and some of the most intractable—like fermentation—were in any case very old. On the other hand, the inquiry into geomagnetism begun in the late sixteenth century is an example of a branch of science originating in the recent observations of practical men and followed up with profit to both theory and practice. Time-measurement also was both a scientific and a commercial problem, especially in relation to navigation. More obviously, skill in glass- and metal-working, especially grinding, turning and screw-cutting, could be readily applied to scientific purposes. Improvements in such arts were sought by scientists and craftsmen together, as when Robert Hooke collaborated with the famous clock-maker, Thomas Tompion.

[1] *Considerations touching the Usefulness of Natural Philosophy;* Shaw's *Abridgement,* vol. I, pp. 129–30.

In three related sciences, chemistry, mineralogy and metallurgy the pre-eminence of art over science was very marked at the opening of the sixteenth century. In natural philosophy there was a rudimentary knowledge of the classification of gems, earths and ores together with a wholly useless theory of the generation and transformation of substances. The pseudo-science, alchemy, had its own theory of the nature of metals and their ores, and contained some sound information on chemical processes and the preparation of simple inorganic compounds. But during the previous three centuries its originally useful content had become garbled and obscured through the growth of esoteric mysticism and the propagation of absurdities in its name. By contrast, great progress in chemical industry, at a time when this represented almost the only rational body of chemical knowledge, was scarcely reflected at all in scientific writings before the mid-sixteenth century. There were changes, permitting the use of new materials, economy of manufacture, or the improvement of the product, in a long list of trades, all of which depended on chemical operations, such as the extraction of metals and the refining of precious metals, glass- and pottery-making, the manufacture of soda and soap, the re- fining of salt and saltpetre and the manufacture of gunpowder, the preparation of mineral acids, and distillation. Other chemical arts, like dyeing and tanning, were probably less improved; some later innovations, like sugar refining, immediately aroused scien- tific interest. The knowledge of the craftsmen concerned was, of course, wholly empirical; they were uninterested in theory, and given to superstition and prejudice. Part of their skill may have derived from the Greek scientific tradition through Islamic sources—the art of distillation was clearly derived in this way, but it was perfected by artisans, not by philosophers or alchemists. Much of their skill was the tardy fruit of long experience. Taken altogether, craft knowledge in chemistry and related sciences im- plied a far greater acquaintance with materials and command over operations than were available to the philosopher or the adept.

By the end of the sixteenth century something like a rational chemistry was coming into existence, though sixty years later Boyle could still write:

There are many learned men, who being acquainted with chymistry but by report, have from the illiterateness, the arrogance and the impostures of too many of those, that pretend skill in it, taken

occasion to entertain so ill an opinion as well of the art as of those
that profess it, that they are apt to repine when they see any person,
capable of succeeding in the study of solid philosophy, addict himself
to an art they judge so much below a philosopher . . . when they
see a man, acquainted with other learning, countenance by his example sooty empirics and a study which they scarce think fit for any
but such as are unfit for the rational and useful parts of physiology
[science].[1]

In the course of that century a number of books had appeared
which, although primarily concerned with technological processes,
had a significant influence on the chemical group of sciences.
Avoiding theory, they threw off the air of mystery. They described
in a matter-of-fact way how mineral substances were found in
nature, extracted, and prepared, and how further commercial
products were obtained from them by the operations of art. The
processes described required mineralogical and chemical knowledge, manipulative skill, and often a complex economic organization. Some of the German mines already absorbed heavy
capital expenditure, and some processes, like the manufacture of
nitric acid needed for the separation of gold from silver, were
conducted on a considerable scale.

The first of these treatises was a small German work known as
the *Bergbüchlein*, printed at Augsburg in 1505.[2] Before this, in the
fifteenth century, there had been in circulation manuscripts
written in German dealing with pyrotechnics, the preparation of
saltpetre and the manufacture of gunpowder, but these were never
printed and seem to have been unimportant in science. Possibly
there were similar "handbooks," earlier than the invention of
printing, dealing with mining and metallurgy. The *Bergbüchlein*
describes briefly the location and working of veins of ore, and is
followed in the *Probierbüchlein* (first printed about 1510) by an
account of the extraction, refining and testing of gold and silver.
Their usefulness is proved by the many editions published. The
same subjects were treated by Biringuccio in 1540, by Agricola in
1556, and by other German authors later in the century. The best
informed of these was Lazarus Ercker, superintendent of the mines
in the Holy Roman Empire, whose *Treatise on Ores and Assaying*

[1] *Works* (ed. Birch, 1772), vol. I, p. 354.
[2] A. Sisco and C. S. Smith: *Bergwerk- und Probierbüchlein* (New York, 1949).
Cf. also Anneliese Sisco in *Isis*, vol. 43, 1952.

(Prague, 1574), was translated into English as late as 1683.[1] Ercker's thoroughly practical book is chiefly concerned with the precious metals, but has chapters on working with copper and lead, on quicksilver, and on saltpetre. The *Pirotechnia* of Vanoccio Biringuccio and the *De re Metallica* of Agricola both cover a wider range of topics.[2] Biringuccio, for instance—the only Italian author of an important work of this type—describes the blast furnace, bronze- and iron-founding, and glass manufacture, but the technical information is somewhat unspecific. Agricola's book, massively detailed in its account of geological formations, mining machinery and chemical processes is justly regarded as the masterpiece of early technological writing. Agricola [*germanice* Georg Bauer] was a scholar, corresponding with Erasmus and Melanchthon, writing good Latin, enriching his observations with appropriate quotations from classical authors. He wrote also *On the Nature of Fossils* and on other scientific subjects. His knowledge of mining and industrial chemistry was gained through long residence, as a physician, in the mining towns of Joachimsthal in Bohemia and Chemnitz in Saxony. About the first third of *De re Metallica* is given to a discussion of mining methods. Then Agricola passes on to describe the assaying of ores to determine their quality, and the operations of preparing and smelting them. Iron, copper, tin, lead, bismuth, antimony and quicksilver are considered as well as the precious metals. The testing of the base metals for gold and silver content is the next topic, followed by an account of the separation of precious and base metals, and of gold from silver. Here the various processes of cupellation, cementation with saltpetre, liquation with the use of lead, amalgamation with mercury, refining with stibnite, and extraction with what Agricola calls *aqua valens* are described at length. This last was, apparently, a mixture of mineral acids prepared by distillation of different mixtures of vitriols, salt, saltpetre, alum and urine. The last section of the work treats of the preparation of "solidified juices"—salt, potash and soda, alum, saltpetre, vitriols, sulphur, bitumen and glass. Here Agricola was on less firm ground and was guilty of some confusion and error.

[1] Modern translation by A. Sisco and C. S. Smith (Chicago, 1951).
[2] The former translated into English by M. Gnudi and C. S. Smith (New York, 1943); the latter by H. C. and L. H. Hoover (London, 1912; New York, 1950).

This series of technical books reflects a tradition in applied science that had grown slowly in the later centuries of the middle ages, that was still gradually increasing in skill, and was capable of producing new techniques for handling the unprecedented richness of the South American mines. The authors, like the contemporary anatomists and herbalists, took full advantage of the art of wood-cut illustration. They did their work so well that it lasted into the early eighteenth century, when a new era of technology was beginning; it was over Agricola's great folio that Newton pored when he was investigating the chemistry of metals. Chemical industry did not merely furnish the chemists of the late sixteenth and seventeenth centuries with the materials in their laboratories; it supplied them with a factual account of the occurrence of minerals in the natural state and the methods of their preparation. More than this, the technical treatises provided, in contrast with the fanciful symbolic language of the alchemists, a precise account of basic chemical operations and reactions. Besides the works already mentioned, the philosophical chemist and virtuoso could turn to the *Distillation-book* of Hieronymus Brunschwig (1512) and its successors for instruction in this most necessary, and most difficult, of chemical arts. The alchemists, even when honest, wrote on the principle that if the reader had not been admitted to the secrets he would fail to understand, and if he had he would scarcely need further guidance. These writers, however, set forth the best of their knowledge as plainly as possible; and it was likely to be sound, for as Boyle remarked, 'tradesmen are commonly more diligent, in their particular way, than any other experimenter would be whose livelihood does not depend on it.' Only in its practical applications, stripped to its bare essentials of preparing this from that, was chemistry on a really solid foundation, independent of the misleading implications of false, and often fantastic, theories. But the chemical operations of industry were not merely qualitatively reliable and instructive. The application of *quantitative* methods to a chemical reaction was the essence of assaying, for example in calculating the quantity of gold in an alloy by carefully drying and weighing a precipitate.

The assayer deserves as much credit as the observational astronomer for providing numerical data and establishing the tradition of accurate measurement without which modern science could not have arisen. Though more of a craftsman than a scientist and more concerned with utility than with intellectual beauty, the assayer nevertheless collected a large part of the data on which chemical science was founded.[1]

When, in the eighteenth century, the balance was recognized as an invaluable tool in research the chemist was only extending a technique whose specialized usefulness in assaying was long familiar. Even the law of the conservation of mass was no more than a theoretical statement of a truth on which the operations of this craft were founded.

Boyle once spoke of German as the "Hermetical language," because so many alchemists had used it. It is perhaps a more useful observation that rational chemistry began with accounts of the elaborate chemical industry in Germany, and was continued by German experimenters, some of them inspired by Paracelsus, himself a German-Swiss. Here there seems to be a clear case for believing that the development of a technical art to the necessary point in complexity and achievement provided much of the basis of fact and method from which experimental sciences could arise. Of course, the roots of modern chemistry, mineralogy and metallurgy are also to be found in alchemy, in pharmacy, and in philosophy. To the formation of chemical theories in the seventeenth and eighteenth centuries the description of practical operations contributed very little. Ideas were derived from different sources; and there was even a muddled, wrong-headed tradition of laboratory work in alchemy parallel to operations on the industrial scale. Yet in many ways the outlook of Black or Lavoisier resembles that of a practical assayer more than it does the esoteric perspective of Raymund Lull, Paracelsus or "Basil Valentine." The influence of the artisan, conceived as closer to the realities of nature than the abstracted philosopher, was an important element in many of the nascent sciences, but nowhere more than in chemistry, which most of all required an alliance of sane thought and reasoned activity.

In the second aspect of the rôle of technical factors in the scientific revolution, the technique of mathematical analysis offers a good example of a factor internal to science itself influencing the progress of certain branches, in this instance mechanics and

[1] Smith: *Lazarus Ercker's Treatise on Ores and Assaying* (Chicago, 1951), p. xv.

astronomy. In a similar manner chemistry was another internal factor affecting the progress of physiology, but this was scarcely realized as yet. The ambition to formulate theoretical propositions and experimental results in the form of mathematical functions, nourished in some men by reflections of Platonic or Pythagorean philosophy, was commonly entertained in the seventeenth century, even by those little learned in mathematics. As Boyle foresaw, 'A competent knowledge in mathematics is so necessary to a philosopher that I scruple not to assert, greater things are still to be expected from physics, because those who pass for naturalists have been generally ignorant in that study.'[1] The singularly happy synthesis of mathematical reasoning and experiment in geometrical optics and mechanics was regarded as a model to be imitated in other parts of science, and it was recognized that a mathematical demonstration has a rigour, and sometimes a generality, not easily obtained in other arguments. The study of mathematics, for the sake of its beneficent effect upon the intellectual powers, as an introduction to natural and moral philosophy, and as a necessary preliminary to certain professions, was by degrees granted a more prominent place in education. It was readily seen that the practical sciences—navigation, cartography, surveying, gunnery—owed their origin to the combination of mathematical method with more exact instrumental measurement. They had thus gained a certainty impossible with rule-of-thumb procedures. About the middle of the seventeenth century it was found that the vagaries of chance, in the throw of a die or the odds in betting, were not immune to the laws of mathematics. The calculus of probabilities was begun. Closely connected with this were the first essays in statistical analysis, by John Graunt and Sir William Petty in England, which proved that even the hazards of human life were not beyond computation. From such crude beginnings developed an impressive mathematical apparatus of immense importance to theoretical physics in the nineteenth century.

These analogies, of very different kinds, suggested that where quantitative observation and experiment were possible, a mathematical formulation ought to be adopted. Galileo had shown the splendour of the prize that might be won. If the workings of nature

[1] *Usefulness of Natural Philosophy; Philosophical Works, abridged by Peter Shaw*, vol. I, p. 118.

were regular and uniform, should not they ultimately reveal a mathematical harmony, as Kepler thought? In the study of acoustics, during the seventeenth century, it was indeed found that harmonies pleasing to the ear were produced by the vibrations of different strings when their lengths, thicknesses and tensions agreed with simple mathematical ratios. More generally, Cardano in the sixteenth century had reviewed the physical applications of the rules of proportion, while Petty examined, in a paper in the *Philosophical Transactions*, the special significance of the quadratic ratio in nature $\left(\text{that is, functions of the type } a = kb^2, a = \dfrac{k}{b^2}\right)$. Again, if the mechanical philosophy was justified in its belief that the physical properties of macroscopic bodies result from the shape, size and motion of the particles composing them, ought not all these properties of the particles to be susceptible of mathematical discussion, which was already making great progress with the study of motion? And, indeed, by making certain measurements in connection with the phenomenon of optical interference, and examining them mathematically, Newton was able to make quantitative statements about the nature of light. Furthermore, mathematical analyses enabled him to prove that some consequences of the Cartesian theory of matter were quite incompatible with observation.

The limitations to the extension of the mathematical method, flourishing in dynamics, to the whole of physics and *a fortiori* to still less "exact" parts of science were obviously of two kinds. The first lay in the ability of physicists and others to design experiments and produce results of a form suitable for mathematical analysis; the experimental physicist had to discover in what ways mathematics could help him before he could become a mathematical physicist. This kind of limitation had vitiated all attempts towards a mathematical theory of projectiles before the time of Galileo.[1] The second limitation is in the nature of mathematics itself, in its ability to perform the operations required. For example: Boyle's experiment on the compression of a volume of air in one limb of a U-tube by a column of mercury poured into the

---

[1] The parabolic trajectory of a projectile in a vacuum was first established by Bonaventura Cavalieri in *Lo Specchio Ustorio* (1632), six years before Galileo's much fuller treatise appeared. The problem was easily solved mathematically with the aid of Galileo's theorem on acceleration.

other (described in *A Defence of the Doctrine touching the Spring and Weight of the Air*, 1661) gave quantitative results which could be simply interpreted to yield "Boyle's Law" ($pv = k$). It was also found that the height of the mercury column in a barometer falls as the instrument is carried progressively higher above sea-level, because the weight of the atmosphere above decreases. The celebrated experiment was carried out by Périer, on the suggestion of Pascal, in 1648.[1] From a combination of these facts, it was realized that it should be possible to frame a mathematical theory which would enable the vertical distance between two stations to be calculated from the difference in atmospheric pressure found from simultaneous readings of two barometers. To make a rough approximation is sufficiently easy, but an accurate calculation involved purely mathematical difficulties which were not fully overcome in the seventeenth century.

Discussion of the first group of these factors limiting the application of the mathematical method to science must, naturally, belong to the history of the separate branches of science. It would be necessary to discuss, in each case, the steps by which experiments were designed and instruments invented to obtain quantitative results of various types, and the methods followed in the formulation of a variety of theoretical postulates, before materials suitable for mathematical study were assembled. Till the nineteenth century no parts of science, other than mechanics and astronomy, were so highly organized and coherent that the application of mathematics was possible, otherwise than in elementary computations. But with regard to mechanics and astronomy the second limitation—latent in the resources of mathematics itself— becomes significant; the great achievements which signalized the triumph of the scientific revolution would have been impossible in the absence of the enormous elaboration of pure mathematics which took place, and was to some extent inspired by realization of the fact that it was essential. Nearly all the great mathematicians of the sixteenth and seventeenth centuries, from Tartaglia and Stevin to Cavalieri, Descartes, Newton and Leibniz, were at least partly interested in the physical sciences. One of the unexpected

---

[1] Pascal *deduced* that the atmospheric pressure would fall progressively above sea-level from his own repetition of Torricelli's experiments with the barometer. The deduction was confirmed by Périer's ascent, with the instrument, of the Puy-de-Dôme, in Auvergne.

discoveries of the time was that a number of regular mathematical curves (some long familiar), such as the ellipse, the parabola and the cycloid, or the algebraic functions which Descartes associated with such curves, appeared in the investigations of the astronomer and the physicist, so that their study proved to have a double interest. The calculations performed by the physical scientist frequently required the calculation of an area bounded by a curve of some type, and so in turn stimulated investigation of the operation of integration in which perhaps the success of seventeenth-century mathematicians was most striking. A number of their advances in method were first denoted by the solution of some problem in mechanics, which offered from about 1650 onwards a most rewarding opportunity for the display of mathematical inventiveness, formerly more commonly devoted to the improvement of the mathematical procedures in astronomy.[1]

The progress made in mathematics in the seventeenth century can be very easily illustrated from the fact that, about 1600, it had hardly as yet reached a form intelligible to modern eyes. The writing of Arabic numerals was indeed nearly stabilized in the modern style, but the Roman were still commonly employed, especially in accounting. The use of the modern symbols for the common operations of multiplication, division, addition and so forth was standardized only in the second half of the seventeenth century. Previously mathematical arguments were set out in a diffuse rhetorical form. Algebraic notation was settled at about the same time, the practice of employing letters to denote unknown or indeterminate quantities having been introduced by the French mathematician Viète shortly before 1600. Arithmetical operations, particularly those involving long division or the handling of fractions, were still performed by cumbersome methods, and "reckoning with the pen" (rather than with the abacus or other aid) was still regarded as a somewhat advanced art. One of the earliest calculating-devices, the so-called "Napier's bones," was designed to obviate the memorization of multiplication tables and the labour of handling long rows of figures. Again, at a higher level, tables of functions were very deficient. In trigonometry, the

---

[1] It has been pointed out that almost all the progress made in the theory of structures, for instance, before about 1770, was made as a by-product of sheer mathematical ingenuity. The same might be said of the applied science of ballistics, which also had a negligible experimental foundation before 1740.

Greeks had known tables of chords alone; during the late middle ages tables of sines and tangents became available, and during the sixteenth century other trigonometrical functions were tabulated. But the methods of computing and of using the tables were both very tedious. From an attempt to facilitate calculations involving these functions developed the invention of logarithms, perhaps the most universally useful mathematical discovery of the seventeenth century, as it was certainly the least expected. Napier's tables, published in 1614 (*Mirifici logarithmorum canonis descriptio*), gave logarithms of sines which are effectively powers of a base e^{-1}, the reciprocal of the base of modern Napierian logarithms.[1] A table of logarithms to the base 10 for the first thousand numbers was published by Henry Briggs in 1617. Logarithms offered a compelling instance of the utility of the decimal system of fractions, of which Stevin had been a most forceful advocate some thirty years earlier.

Since the Greeks had excelled in geometry and trigonometry, little more than the assimilation, with some slight extension, of their methods in these branches of mathematics had been accomplished by 1600. Renaissance scholarship had devoted itself to the recovery of the pure classical tradition in this as in other departments of learning, with the result that the texts, particularly those describing the more advanced Greek studies of geometrical analysis and the conic sections, were far more completely available by the middle of the sixteenth century than before. Here, at least, pure scholarship caused an immediate rise in the level of competence. Even in the mid-seventeenth century the practice of "restoring" a lost or fragmentary work by means of the methods presumably used by the ancient author did not seem absurd, and when Newton wrote the *Principia* the synthetic geometry of the Greeks was still held to be a more reliable form of mathematical demonstration than the recently developed analytical method.

Algebra, on the other hand, represents a native European development from Hindu and Islamic sources made known to the Latins by medieval translators. Considerable progress in the sixteenth century, for example in the solution of equations of higher powers than the quadratic, was unaffected by humanistic influences; indeed, Greek geometric procedures for solving equations were supplanted by algebraic methods. Operations with

[1] Hence $\log_N a = 2.30259 \log_{10} \frac{1}{a}$.

proportions and series, also known to the Greeks solely in the geometric form, were similarly transformed into the more convenient algebraic symbolism. Islamic mathematicians (notably Omar Khayyám, c. 1100) had recognized a partial equivalence between algebra and geometry, that is, that certain geometrical problems could be represented by algebraic functions; and this equivalence was more thoroughly exploited by the European mathematicians of the sixteenth and early seventeenth centuries. The next step, upon which analytical geometry is founded, was the representation of any function graphically by the use of rectangular co-ordinates. Graphs with such co-ordinates had been used in the middle ages for a few special purposes (e.g. to represent the motions of the planets), and Oresme's calculus of varying qualities was itself an elementary form of co-ordinate geometry. Far more general and precise methods had been sketched in private by the English mathematician Harriot, and by Pierre Fermat, before the publication of the first treatise on the subject by Descartes—the *Geométrie* annexed to his *Discourse on Method* (1637). In this he showed that a constant relationship exists between the co-ordinates x and y of any point on a regular mathematical curve which can be expressed as the algebraic function $y = f(x)$, and that certain patterns of function corresponded uniformly with certain classes of curves.

This discovery was capable of immediate and fruitful application to mechanics, in which many problems could be most easily postulated in an initial geometric form, and then analysed with the aid of the appropriate algebraic functions. Thus, to cite a simple instance, in ballistics it was no longer necessary by the end of the seventeenth century to work out a range by supplying the necessary constants in the geometric construction used by Galileo, since it could be obtained directly from the function

$$y = x \tan e - \frac{gx^2}{2V^2 \cos^2 e},$$

appropriate to the parabolic trajectory, from which all the theorems established by the elaborate geometry of Galileo and Torricelli are readily deduced. Such functions can, of course, be derived far more easily by use of the differential and integral calculus than by the application of algebraic notation to conventional geometrical reasoning, and the usefulness of the "calculus" to the engineer and technician as well as to the mathematical scientist has therefore been clearly recognized since about the middle of the eighteenth century. Both Kepler (1616) and Cavalieri (1635) made use of infinitesimals in integration, the former in connection with practical problems such as the calculation of the volumes of casks. Thus Cavalieri imagined that the area of a surface was made up of an infinite number of lines, and the volume of a solid of an infinite number of superposed surfaces, calculating these quantities by the summation of an infinite number of their geometric elements. Such methods had a certain practical value, but the concepts were badly defined and of doubtful validity. In co-ordinate geometry the problems for which they were designed were first set out in the modern systematic form. Perhaps even more important were the methods for finding the maximum and minimum values of curves (or the equivalent functions), and for drawing tangents to them, discovered by Fermat (1638) and others about the middle of the century, for they used operations identical in principle with what was later known as differentiation. Barrow, for example, in his method of tangents, obtained by geometrical means the derivative $\left(\dfrac{dy}{dx}\right)$ of the function ($y = f(x)$) represented by the curve, and equated this derivative to zero.

Before 1670, therefore, the crude components of new mathematical processes for dealing with quantities having a non-linear variation were already in existence. The priority in time in perfecting them undoubtedly belongs to Newton (whose *Method of Fluxions* was written in 1671), but a more useful notation and a somewhat clearer presentation of the matter were later developed independently by Leibniz, whose first essay on the calculus was published in 1684. The charge maintained by a group of English mathematicians, with considerable covert assistance from Newton himself, that Leibniz had plagiarized Newton's ideas from unprinted manuscripts which had circulated privately provoked in the early years of the next century a sordid and bitter dispute between them and most continental mathematicians. Its only significance is that the transmission to England of continental discoveries in pure and applied mathematics was inhibited for over a century, until soon after 1830 the Newtonian method of fluxions was replaced by the virtually equivalent differential and integral calculus which had advanced very far from Leibniz' original invention.[1] In principle, both Leibniz and Newton were successful in the same task: they generalized the methods of differentiation and integration which were already in existence, recognizing that the latter operation was the inverse of the former; they developed the new notation required for these new operations; and they greatly clarified mathematical thinking about their nature. Both succeeding in solving by the new methods problems which had been nearly intractable to the old.

The first coherent treatise explaining the calculus was only written at the close of the seventeenth century, and its general extension to mechanics and physics was the work of more than one succeeding generation. Consequently its full impact on science was not immediate, though Newton referred in passing to his method of fluxions in the *Principia* and the mathematical demonstrations he offered might have been impossible without it. Nevertheless, mathematical methods tending towards those of the generalized calculus, or even anticipating it, were widely applied in mechanics from about 1650 onwards, along with other new developments in advanced algebra (for example, the study of series). Huygens offers perhaps the neatest instance of a physicist making by means of co-ordinate geometry many complex calculations that would now be dealt with by means of the calculus, through the development of his own methods which were virtually equivalent to differentiation and integration. In this way Huygens was able to investigate some of the properties of motion in a resisting medium, which were also examined (rather more thoroughly) by Newton in Book II of the *Principia* with the aid of fluxions. It seems that, with the exception of celestial mechanics, the development of no branch of science was seriously obstructed by the absence of a suitable mathematical technique, even before the formulation of the true calculus. Some known problems remained unsolved, naturally, partly owing to imperfect conceptualization, but partly also owing to the fact that their solution

[1] The truth seems to be that Leibniz was in close contact with men who understood something of Newton's new ideas; that he saw some of Newton's manuscripts; but that he could have learned little in this way that he could not have gathered from other sources, and that his ideas were already at the significant time shaping towards the form in which they were later formulated. Leibniz has long been cleared of the charge of plagiarism; any historical obscurities that remain are of little significance. (For bibliography cf. D. E. Smith: *History of Mathematics*, vol. II, p. 691.)

required difficult approximations involving constants which had not yet been determined experimentally. On the contrary, it seems rather that the nature of the possible mathematical processes exercised a powerful guiding influence at least over the science of mechanics. Even had ideas been more precise, and the experimental material more complete, it would have been impossible for the seventeenth-century physicist to work out a mathematical corpuscular theory of matter because he lacked the necessary technique of statistical analysis; but mechanics acquired an increasingly theoretical character because its problems could be stated mathematically, and solved by known procedures. The propositions of the *Principia* are not less certain, although they may not be the result of mathematical analysis applied to a mass of experimental data; but it must be recognized that Newton's method was to deduce, by mathematics, the consequences of selected axioms and then to show by reference to observation or experiment that these consequences are actually confirmed in nature. Already in Galileo's writings it is apparent that in science the mathematical and experimental approaches are far from identical, though they may be closely correlated. He himself attached greater importance to his mathematical theories than to his experimentation (which is never precisely described, and often appears casual). The early researches of the scientific societies laid a greater emphasis on exactitude and consistency in experiment; but mechanics was again sublimated into the regions of higher mathematics in the last years of the century. The consequence of this may be illustrated by Huygens' ingenious proof that a pendulum is perfectly isochronous when describing a cycloidal arc—a discovery, made with the object of perfecting the mechanical clock, which proved completely irrelevant to the experimental search for a scientifically reliable time-keeper; or by the history of the industrial revolution, which was scarcely at all promoted by the elaboration of theoretical mechanics during the eighteenth century. There was, in fact, a strong tendency for the first truly mathematical branch of science to lose touch with its roots in experiment by becoming no more than a specialized department of mathematics.

It may perhaps seem surprising that seventeenth-century mathematics progressed so opportunely as to satisfy, very largely, all the demands that science made upon it. Only the activity of a number of highly original mathematicians made this possible, and obviously their activity was not wholly fortuitous (it is remarkable, for instance, that there was so little interest in the theory of numbers). It was inspired, to some extent, by knowledge of the usefulness in science of advances in certain directions. But it is generally true, of other periods as of this, that the mathematics required for any particular step in science has been available. Greek science was never near to exhausting the subtlety of the Greek mathematician. In recent times the mathematics of the theory of relativity was in being well before Einstein. Perhaps this empirical fact, more than anything else, justifies the qualification of mathematics as a scientific technique. When a class of facts, or concepts, is first subjected to its use, the mathematics involved will tend to be relatively simple, if only for the reason that the innovator will probably be a scientist, more familiar with the science than with the most recent advances in mathematics itself. Galileo was not a great mathematician. But as the mathematical theory develops, its continuance will increasingly depend upon the activity of men who have been trained as mathematicians to digest and analyse the results of experiment, even to suggest and perhaps carry out the experiments and measurements necessary for the prosecution of the theory. Yet it will rarely happen that the mathematician who applies his skill in this way penetrates to the very frontiers of pure mathematics. While the application is perfected, mathematics itself is not standing still. And the application is made because the mathematical theorist perceives that the particular technique is appropriate, or adaptable, to the problem. The conjunction of the highest mathematical with the highest scientific ability, as in Archimedes or Newton, is extremely rare; the fabrication of scientific theory by the use of well-known procedures has, on the other hand, been a frequent occurrence.

The invention of numerous scientific instruments during the seventeenth century, and their fertile use in many capacities, has long been associated with the revolution in scientific thought and method. The idea of science as a product of the laboratory (in the modern sense of the word) is indeed one of the creations of the scientific revolution. In no previous period had the study of natural philosophy or medicine been particularly linked with the use of specialized techniques or tools of inquiry, and though the

surgeon or the astronomer had been equipped with a limited
range of instruments, little attention was paid to their fitness for
use or to the possibility of extending and perfecting their uses.
More variants of the astrolabe, the most characteristic of all
medieval scientific instruments, were designed during the last
half-century of its use in Europe (*c.* 1575–1625) than in all its
preceding history. The Greeks had known the magnifying power
of a spherical vessel filled with water, but the lens was an invention of the eleventh century, the spectacle glass of the thirteenth,
and the optical instrument of the seventeenth. Navigational
instruments also were extremely crude before the later sixteenth
century. It was not that ingenuity and craftsmanship were wholly
lacking (for many examples of fine metal-work, for artistic and
military purposes, prove the contrary); rather the will to refine
and extend instrumental techniques was absent. On the other
hand, it has justly been pointed out that the early strategic stages
of the scientific revolution were accomplished without the aid of
the new instruments. They were unknown to Copernicus, to
Vesalius, to Harvey, Bacon and Gilbert. They leave no trace in
the most important of Galileo's writings, in which he reveals
himself driven to measure small intervals of time by a device
considerably less accurate than the ancient clepsydra. It is clear
that, great as was the influence of the instrumental ingenuity of
the seventeenth century upon the course of modern science, such
ingenuity was not at all responsible for the original deflection of
science into this new course.

Thus it would seem that the first factor limiting the introduction
into scientific practice of higher standards of observation and
measurement, or of more complex manipulations, lay in the
nature of science itself. Only when the concept of scientific research had changed, as it had by the end of the first quarter of
the seventeenth century, was it possible to pay attention to the
attainment of these higher standards. A number of the new instruments of the seventeenth century were not the product of scientific
invention, but were adopted for scientific purposes because the
new attitude enabled their usefulness to be perceived. The balance
was borrowed from chemical craftsmanship. The telescope was
brought into use by artisans, initially for military service. The
microscope was an amusing toy before it became a serious instrument of research. The air pump in the laboratory was an improved

form of the common well pump—Otto von Guericke, its original inventor, had at the beginning exhausted vessels by pumping out water with which they had been filled. And inevitably the techniques used in the construction of the new scientific instruments were those already in existence; they were not summoned out of nothing by the unprecedented scientific demand. Some instruments were only practicable because methods of lathe-turning and screw-cutting had been gradually perfected during two or three centuries, partly owing to the more ready availability of steel tools, others, because it was possible to produce larger, stronger and smoother sheets or strips of metal. The techniques of glass grinding and blowing which provided lenses and tubes might have been turned to scientific use long before they were. The established skill of the astrolabe-maker could be devoted to the fabrication of other instruments requiring divided circles and engraved lines, that of the watch-maker to various computers and models involving exact wheel-work, and so on. Common craftsmanship held a considerable reservoir of ingenuity, when scientific imagination arrived to draw upon it.

On the other hand, once interest in the kind of result that could be obtained from the employment of specialized instruments had been created, particularly as this employment began to extend under a more disciplined direction towards a greater qualitative depth of information and a greater quantitative accuracy of measurement, the limitations of normal craftsmanship were soon reached. Then it was necessary to begin a more conscious examination of the instruments themselves. Descartes was virtually the founder of the scientific study of the apparatus of science, in his investigation of the causes of the distortions present in the images of crude telescopes. (At the same time, purely empirical measures were also being adopted to remedy their defects.) Descartes concluded that lenses should be ground to a non-spherical curvature, which would introduce greater complexity into their manufacture. Some scientists (including the astronomer Hevelius, and the microscopist Leeuwenhoek) became masters in the art of grinding the lenses they required for their work; others, like Newton, experimented with an alternative form of optical instrument. Towards the end of the century a scientist wishing to have a really good telescope or microscope could no longer simply make use of craftsmanship; he had to direct the work in accordance with a pre-determined specification. Astronomy had already at the end of the sixteenth century reached the point where further advances in precision involved great effort. Devices like the vernier scale and the tangent screw were major steps, and the attachment of telescopes to instruments for measuring angles reduced sighting errors. But the advantages gained by increased complexity in mechanical construction were all dependent on progressive refinement in workmanship, and the forethought and supervision of the scientist. The astronomer, in fact, had to consider his observatory as an exercise in design; he had to build walls, duly orientated, that would not settle; to design quadrants that were rigid, yet light, and true; to ascertain the probable errors of divided scales; to collimate his telescopes and rate his clocks. He had become aware that the limitations to his work were imposed by factors that were, in the main, technical and mechanical. As such, they deserved, and received, increasingly meticulous attention.

From the historical point of view, instruments may be divided into two classes: those which render qualitative information only, and those which permit of the making of measurements. Naturally, these uses of an instrument are not necessarily exclusive, in fact a little consideration makes it obvious that in their evolution most early scientific instruments tended to move into the second class. Thus, devices like the micrometer could be added to telescopes and microscopes so that very small or very distant objects could be measured; or alternatively these optical systems could be added to other measuring instruments to improve their performance. But the *first* use was purely qualitative. Similarly, in the eighteenth century, the electrometer designed in the first place for the detection of charges, was later applied to their comparison and measurement. The balance was first used in chemistry to establish a simple loss or gain in weight; its employment to determine accurately the masses involved in a chemical reaction came much later. It is therefore a natural and plausible proposition that the quantitative potentialities of a new instrument or piece of apparatus are generally appreciated less readily than the qualitative, and this was particularly the case in the seventeenth and eighteenth centuries. The invention of instruments, therefore, did not have that immediate effect of inducing greater rigour, and greater interest in refined measurement, which might be anticipated *a priori.* The barometer, for example, was invented by Torricelli in 1643. It was used originally to demonstrate the existence of atmospheric pressure, and secondly as a means of exhausting a small chamber formed at the top of the tube in which experiments could be made. Only about 1660 was the correlation between barometric pressure and climatic conditions discovered, and only after this were attempts made to improve the readability of the instrument and collect a "history of the weather." Later still, Boyle employed the barometer as a gauge to measure the quality of the vacuum formed by his air-pumps, and the amount of "air" evolved from fermentations. The thermometer has an even longer, and more surprising, history as a merely qualitative instrument. The thermoscope, an instrument in which the expansion of air in a bulb moved a column of water in a narrow tube upwards when heat was applied, was invented by Galileo about 1600. Liquid thermometers were introduced about the middle of the century, and were extensively used by the Accademia del Cimento, but none of these was calibrated. The first suggestions for systematic calibration with the use of two fixed points were made about 1665; Fahrenheit's scale was devised about fifty years later, and the modern centigrade scale only in 1743. Thus the first century of thermometry yielded no quantitative measurements which can now be interpreted with any degree of confidence.[1]

While seventeenth-century astronomers, continuing a long tradition, effected a refinement of angular measure which bore fruit in Flamsteed's *Historia Coelestis Britannica* (1725), in physics and biology qualitative results were far more significant. Even the allied science of terrestrial angular measure (in surveying and geodesy) remained rather crude until vernier scales and telescopic sights were introduced at the close of the century. Consequently it can scarcely be maintained that *technical* limitations to accuracy of measurement were significant in any other branch of science than astronomy before the early part of the eighteenth century. Certainly it has been argued that chemistry would have progressed faster, and the science of heat have been more systematic, if greater attention had been paid to the quantitative aspects of

---

[1] The tubes of the early thermometers were of course divided, so that comparative readings could be taken, but without reference to any standard scale. Thus, extensive observations were rendered useless by the fact that they depended on the arbitrary divisions of a unique instrument.

experiment; but the reasons for the neglect of these aspects are to be sought rather in the nature of scientific activity in the seventeenth century, than in instrumental deficiencies. The importance of accurate measurements was not adequately understood, and therefore they were rarely made; so that it was the texture of science that hindered the effective exploitation of devices already in being, rather than *vice versa*.

On the other hand, with regard to the two qualitative instruments which most strikingly opened up great new fields of activity, the telescope and microscope, it is plain that technical limitations rapidly became serious, and that the nature of these limitations was well understood. Both instruments began in very crude form. Systems of convex lenses replaced the concave-convex combination (the so-called Galilean arrangement) only gradually from about 1640, when rules for working out the appropriate focal lengths and apertures were better understood. The first true compound microscopes date from about this period, and the new (Keplerian) telescope brought more detail of the solar system into visibility. Additional satellites were discovered; the mysterious appearance of Saturn was accounted for; transits and occultations could be observed with higher accuracy. But the great desideratum of seventeenth-century astronomy—an observational proof of the earth's rotation—was not accomplished. To increase magnification without a corresponding vitiating increase of the aberrations the astronomer was compelled to use enormous focal lengths and small apertures. The light-gathering power of such instruments was poor, and a practical limit to length (about 100 feet) was soon reached. Non-spherical curvatures for lenses were theoretically desirable, but technically impracticable. Newton's optical theory explained the nature of chromatic aberration without suggesting an appropriate remedy, for he found that the separate colours into which white light can be resolved could not be brought to a single focus by a simple lens. The reflecting telescope, free from chromatic aberration, was suggested by James Gregory and first constructed by Newton, but it was hardly of serious value to astronomers before the later years of the eighteenth century.

Similar problems were encountered in the microscope. Simple glasses, with a magnification of about ten diameters, were used early in the seventeenth century; by Harvey, who observed the pulsation of the heart in insects, and by Francesco Stelluti who published in 1625 a microscopic study of bees. The small tubular "flea-glass," with the lens mounted at one end, and the object set against a glass plate at the other, became popular among the virtuosi. Towards the middle of the century the compound microscope attracted renewed interest, being now constructed with a bi-convex objective and eye-lens, with a plano-convex field lens placed between to concentrate the rays. In the improved design of Hooke (described in *Micrographia*, 1665) the body, containing extensible draw-tubes, was mounted so that it could be tilted to a convenient angle; a long nose-piece engaged in a large nut, so that the body could be brought to focus on the object by screwing it in or out. The lead-screw and slide method of adjustment was invented later by Hevelius. To illuminate opaque objects Hooke used an oil-lamp and bull's-eye lenses; before the reflecting-mirror was fitted (about 1720), transparent objects were examined by placing a lamp or candle on the floor beneath the instrument, which was often pierced through the base. The compound microscope was complicated and expensive, but it was easy to handle, and in mechanical design became steadily more efficient. Optically it was less satisfactory. Magnifications exceeding 100 diameters could be obtained, but the uncorrected lenses, made of poor glass, gave low resolution. As a result, the point was soon reached where, though the object could be made to appear larger, no finer detail in it could be seen. From 1665 until about 1830, when satisfactory corrected lenses became available, the compound microscope made comparatively slight advance in optical properties. The limitation imposed, on biological research in particular, is obvious.

Nevertheless, the compound microscope was eminently a scientific instrument, and Hooke's *Micrographia* the first treatise on microscopy. Since his objects were fairly coarse (insects, seeds, stones, fabrics, a razor's edge, leaves, wings, feathers, etc.), and since he did not seek to penetrate into anatomical structure by dissection (though he examined the compound insect eye, and discovered the cellular composition of cork) he was able to produce a series of admirable illustrations despite the limitations of his microscope. Most of the discoveries of the time in minute anatomy, associated with the names of Malpighi, Swammerdam and Grew, such as the capillary circulation of the blood, could also be demonstrated with the compound instrument.[1] For the very finest observations, however, another technique was required, in which the Dutch microscopist Antoni van Leeuwenhoek excelled. The compound microscope had stimulated the grinding of very small bi-convex lenses of short focal length for use as objectives. It was found that better results could be obtained by mounting such high-power lenses, or even tiny fused glass spheres, as simple microscopes than by using them as elements in an optical system that multiplied the aberrations. For considerable magnification the lenses had to be less than one-tenth of an inch in diameter; they were proportionately difficult to grind and manipulate, and they imposed severe eye-strain. But Leeuwenhoek reported, in his letters to the Royal Society, observations obtained by this means which were only repeated with the achromatic microscopes of the nineteenth century. His skill as an optician is further shown by the fact that one of his few surviving lenses has been proved, by recent tests, far superior to any other known simple lens; others of his own make are good but not outstanding. This skill enabled him to study more thoroughly than any other observer spermatozoa and the red corpuscles in the blood and to become the first to discern protozoa and bacteria. Despite some contemporary incredulity, aroused by the great number and disparity of Leeuwenhoek's original discoveries, and the difficulty of confirming them, his work was astonishingly accurate. He was also creditably free from the tendency to theorize, or to allow his imagination to play with his microscopic images. At the end of the century Leeuwenhoek was alone in his investigation of microscopic creatures, although others were engaged on the study of the microscopic parts of larger creatures; his results, therefore, remained largely isolated curiosities. In the eighteenth century the description of various animals, visible to the naked eye but capable of being studied only with the aid of the microscope, was taken up both in England (Baker, Ellis) and in France (Réaumur, Bonnet, Lyonet). Trembley, whose monograph on the hydra has become a classic, worked in Holland and was closely associated with both the English and the French groups of naturalists. All these worked with the simple microscope but at a much lower magnification than that frequently employed by Leeuwenhoek. This instrument thus became established in familiar use among zoologists and botanists for much the

[1] See below, p. 288.

same purposes as it serves at the present time, when the higher-powered compound microscope, given a beautiful mechanical construction by the English instrument-makers, was still of little scientific value. The continuation of the sciences of histology and cytology, begun by Malpighi and Leeuwenhoek, depended upon the perfecting of lenses which proceeded swiftly in the early nineteenth century.

It would be possible to develop other, comparable, instances of the way in which, after an initial seventeenth-century invention, a long interval followed before, in a stage of higher proficiency in instrumental techniques, observations or measurements of a different order became practicable. The Newtonian reflecting telescope, with Herschel's improvements, for the first time enabled the astronomer to escape the confines of the solar system. If the "chemical revolution" of the eighteenth century was effected without profound modifications of apparatus, on the other hand the chronology of electrical science was fixed by the discovery of instruments for the creation, and the measurement, of charges and currents. These in turn induced new chemical techniques, such as electrolysis. During that century a considerable literature grew up dealing with the manufacture and use of scientific instruments of all kinds, and teaching the technique of making experiments and observations, while the actual manufacturers strove intelligently to improve their wares. John Dollond, the practical man who solved the problem of making achromatic telescope objectives which had baffled mathematicians, was an instrument-maker. The marine chronometer, in the perfecting of which so much was due to another practical man, John Harrison, imposed a close collaboration between watch-makers and astronomers. Science, therefore, entered into a promising situation with the early nineteenth century, being able to call upon the services of a skilful and progressive specialized craft, and realizing far more pertinently than hitherto its own dependence upon its material equipment. In nearly every respect its progress was involved in that of some instrument, or in that of a variety of laboratory techniques.

Because a three-fold division of function may exist in science, between the instrument-maker, the laboratory worker and the theorist, it has always been possible, and still is, for the strategic thinking in science to take place outside the laboratory, away from the instruments (though it may still be controlled by the accessible mathematical techniques). Thus Tycho's instruments were made by the metal-workers of Augsburg; he himself managed their use with consummate skill; and his results were interpreted by the mathematician Kepler. But the third function without the second, and the second without the first, can clearly yield only diminishing returns. The progress of science demands originality at all three levels; more than this, it may demand the existence of resources of industrial magnitude, of a glass-industry, a gas-industry, of the great plants required to produce antibiotics and radio-active isotopes. If it seems increasingly likely that the major advances of the future will come from large institutes, freely endowed, and as the result of co-operative labours, it is no more than a fresh step in that growth of complexity, and of an increasing reliance on techniques and tools of investigation, which was typical of the scientific revolution. In a sense it is the fulfilment of Bacon's foresight.

CHAPTER IX

THE PRINCIPATE OF NEWTON

W ITH the work of Isaac Newton the scientific revolution reached its climax so far as the physical sciences are concerned. Within their range Galileo's confidence in the mathematical structure of nature was justified, and "mechanical principles" were proved to be a universally sufficient basis for explanation. The unity of nature was made manifest in a grand synthesis revealing the effects of identical physical laws in the heavens and on the earth. Copernicus's revolutions, Kepler's laws, and the discoveries of Galileo and Huygens in mechanics were shown to be necessary consequences of these laws, and Newton also disclosed the mechanical nature of light itself. Indeed, to Newton's mind (though he could not prove his belief) the unity of nature ran through all phenomena, arising from the very composition and properties of matter itself, and from the operation of

> certain forces by which the particles of bodies, by some causes hither-to unknown, are either mutually impelled towards one another, and cohere in regular figures, or are repelled and recede from one another. These forces being unknown, philosophers have hitherto attempted the search of nature in vain.[1]

The *Principia* vindicated the scientific endeavour of the seventeenth century, its reaction against tradition and its search for new methods, above all those of mathematical and experimental investigation. Newton proved that the world was as the "new philosophers" had suspected it to be. Of course, his efforts did not meet with invariable success nor (as the last quotation indicates) did he wholly fulfil his hopes. Inevitably the nature and properties of the ultimate particles of matter escaped him. Even in less recondite tasks Newton's thought sometimes required revision and correction; particularly was this the case with Book II of the *Principia*, where Newton practically initiated the theory of fluids. In this part of mechanics Bernoulli and Euler in the eighteenth century had to traverse the ground again, aiming at greater rigour

[1] Newton, *Principia*, Preface.

244

and accuracy than Newton had attained. But fluid mechanics had not been Newton's central interest; Book II was hurriedly written in 1685–6. His great discoveries in optics and celestial mechanics and his interpretation of them stood unchallenged until the nineteenth century. In its first decade Thomas Young criticized the corpuscular theory of light of which Newton was then regarded as the chief upholder, starting off from Newton's own recognition of the periodicity of light and developing a wave-theory which he took to be essentially Newtonian in character. With the further elaboration of the wave-theory by Fresnel and others, however, this character was completely lost. Serious doubts of the inviolability of Newtonian or classical mechanics only arose at the very end of the nineteenth century.

For the eighteenth century the astronomical and cosmological issues that had so troubled the world since Copernicus's time were regarded as settled for good; it only remained for mathematicians to arrange the details of the Newtonian universe in somewhat more exact order. So, to all intents, was the "mechanical philosophy" that Newton had embraced. Hence, although the labours of eighteenth-century mathematicians to perfect physical science are important the focus of interest in scientific discovery tended to shift to other problems, such as those of heat, electricity, and chemistry, where Newton had not penetrated far. Yet the eighteenth and nineteenth centuries, which admired Newton almost without reservation, did so out of inadequate understanding of Newton's character both as a scientist and as a man. No one cared to explore his unorthodox religious beliefs. No one sought to emphasize the strange legacies from the past to which Newton gave expression on occasion, nor his frequent injustices to those whom he judged rivals in discovery. If Newton was incomparable he was not perfect. Nor was he only one of the three greatest mathematical physicists in human history. While his view of the relations of God, the universe and man was less naïve than that of (say) his English contemporaries Boyle and Ray, his profound religious conviction made him declare that "to discourse of [God] from the appearances of things does certainly belong to natural philosophy," and he doubted that the universe, which demanded a creator, could maintain itself mechanistically without his repeated intervention. A simple, billiard-ball mechanistic philosophy of the Cartesian type could not explain the complexity of nature.

To most of his contemporaries Newton appeared only what indeed he was—an experimental and theoretical physicist of genius. This was quickly recognized after the publication of the *Principia* in 1687, even by those who could not accept its doctrine. Earlier work in optics had brought Newton a name, but not the highest reputation; his private mathematical discoveries had been revealed to very few. With the *Principia* Newton appeared as the possessor of the abilities his age most required, for he was fortunate in the hour of his birth: in 1642, the year of Galileo's death. His adolescence was accompanied by the foundation of scientific societies, the practice of systematic experiment, the gestation of modern mathematics. In 1665–6 the rather unpromising, and certainly ill-grounded student upon whom Barrow had enforced the study of Euclid's *Elements* was in the "prime of his age for invention." How much the fruit of his originality owed to the ground from which it sprang, in mathematics to Barrow himself, and Wallis; in physics to Galileo, Descartes and Boyle; in astronomy to Kepler, Galileo and Borelli—and many others whom he read in those early Cambridge years! How many, and how rich, were the threads which these giants led to the hand of so able a spinner of theorems, so close a weaver of theories! The genius that was frustrated by the dense mysteries of chemistry reaped a splendid harvest in the riper fields of mathematical science in which its fullest powers could be exercised. It is no detraction from Newton's originality to point out that all his discoveries were firmly rooted in science of the time—that like a helmsman he was borne along by the stream—that each of them has a quality of inevitability in its contemporary context. Newton, in fact, won such immediate esteem because he saw clearly the things to which others were groping, because he was so fully in harmony with his age. Not scientists and mathematicians alone, but philosophers and theologians could find in Newton exactly that of which they wished to be assured. Even in his political allegiance to a limited Protestant monarchy, in his ambition to rise in the great world of affairs, and in his shrewd financial sense, Newton was a good and successful citizen of Augustan England, a *bon bourgeois* when the middle classes were rising to luxury and power.

There are no singularities in Newton's early years. As a student in Cambridge, he was fashionably inclined to scepticism of the

Cartesian system, looked favourably upon hypotheses of atoms or corpuscles, and developed a mild academic interest in the empirical method represented by Boyle. Starting rather late, he rapidly assimilated the latest mathematical knowledge, but although his notebooks show him at the age of twenty-three already deeply interested in science, they contain no sketch of an original contribution, no hint of a sudden revelation of the importance of natural knowledge. He was always apt, much later in life, to regard science as an importunate mistress, to work with immense concentration and speed upon a problem when the passion to solve it fell upon him, because a part of himself doubted the problem's cosmic significance. Then, for the better part of two years after he had taken his degree, Newton was forced into retirement at his country home in Lincolnshire by the plague which raged in the towns and villages during 1665 and 1666. He had already begun his optical experiments (which seem to have been prompted by undirected curiosity, rather than by any clearly perceived ambition), and his mind was turning towards new conceptions in mathematics. In the winter of 1664–5 he had found the method of infinite series, and shortly after discovered the binomial theorem; in Lincolnshire during the following summer he calculated a hyperbolic area to fifty-two places of decimals; in November of the same year he had formulated the direct method of fluxions (differentiation), and in May 1666 he began upon the inverse method (integration). During the whole of 1665 he must have been working at his optical experiments, and trying to grind lenses of non-spherical curvature, and early in 1666 he "had the Theory of Colours." This quickly directed him to the construction of his first reflecting telescope. 'And the same year,' wrote Newton long afterwards, 'I began to think of Gravity extending to ye orb of the Moon, & (having found out how to estimate the force with which a globe revolving within a sphere presses the surface of the sphere), from Kepler's rule . . . I deduced that the forces which keep the Planets in their Orbs must [be] reciprocally as the squares of their distances from the centers about which they revolve.' To this period belongs the famous story of Newton and the Apple, told at first hand by Newton's younger friend, William Stukeley:

After dinner, the weather being warm [the date was 15 April 1726], we went into the garden and drank thea, under the shade of some appletrees, only he and myself. Amidst other discourse, he told me, he was just in the same situation, as when formerly, the notion of gravitation came into his mind. It was occasion'd by the fall of an apple, as he sat in a contemplative mood. Why should that apple always descend perpendicularly to the ground, thought he to himself. Why should it not go sideways or upwards, but constantly to the earths centre? Assuredly, the reason is, that the earth draws it. There must be a drawing power in matter: and the sum of the drawing power must be in the earths center, not in any side of the earth. Therefore does this apple fall perpendicularly, or towards the center. If matter thus draws matter, it must be in proportion of its quantity. Therefore the apple draws the earth, as well as the earth draws the apple. That there is a power, like that we here call gravity, which extends itself thro the universe.[1]

When the plague had subsided, Newton returned to Cambridge, presumably in the hope (which was soon fulfilled) of becoming a Fellow of Trinity College; in 1669 he succeeded Isaac Barrow in the Lucasian Professorship of Mathematics. He was now comfortably established in a life that was to be his for nearly thirty years. Professorial pupils were few, though Newton must have done some college teaching; he had ample leisure for his experiments in optics and chemistry, and for his mathematics. In 1672 letters written to Oldenburg, describing the new theory of colours, brought him into the Royal Society but the disputes and criticisms which followed during the next four years (aggravated by the officiousness of Oldenburg) persuaded Newton virtually to cut off this connection. His interest in alchemy and chemistry quickened; the wooden staircase leading from his first-floor rooms to the furnace in the little garden below was shaken by his eager tread. It was Halley who brought Newton back into the scientific movement, who encouraged his interest in mathematical matters, and was the godfather of the *Principia*. The book was begun towards the end of 1684, and was finished by the spring of 1687. Thereafter Newton's life became more full of incident, and more empty of science. He took part in the resistance to the Catholic policy of James II; sat (silently) as a Member of Parliament; and sought

[1] William Stukeley: *Memoirs of Sir Isaac Newton's Life*, ed. by A. Hastings White (London, 1936), pp. 19–20. Stukeley was forty-five years junior to Newton, who was in his eighty-third year in 1726, and did not compose his memoir until 1752. His expressions, therefore, cannot be taken very strictly as interpreting Newton's state of mind in 1665–6.

to become Provost of King's College.[1] His fame, and his new acquaintance with John Locke, made him known personally to men who possessed political power and influence. The prospect of exchanging academic seclusion in a provincial town for an office of honour and profit in London was welcome. The initial frustration of this ambition, working in a mind intolerant of opposition at a time when it was strained by overwork, caused a serious breakdown during 1693. After some months Newton recovered his full intellectual powers, but it is to be doubted whether his judgement of men's actions ever regained a normal balance.[2] This illness brought his creative scientific work to a close. Honours and fame came later; in 1696 Newton left Cambridge to become first Warden, then Master of the Royal Mint; in 1703 he was elected to the Presidency of the Royal Society, which he held for the rest of his life; in 1705 he was knighted. In the Augustan Age Newton and Pope ruled the intellectual world with a sway more absolute than that of the Queen herself.

That branch of science in which Newton had first published original discoveries was also the last to be enriched by his mature reflections. The bulk of the *Opticks* of 1704 had long been written, but the *Queries* with which the book concludes represent Newton's final contribution to science, and perhaps the sum of his perception of nature. In the *Queries* optical phenomena are discussed in order to throw light on the ultimate particulate structure of matter:

> Have not the small Particles of Bodies certain Powers, Virtues, or Forces, by which they act at a distance, not only upon the Rays of Light for reflecting, refracting and inflecting them, but also upon one another for producing a great Part of the Phænomena of Nature?[3]

[1] Newton's extraordinary mastery of detailed argument is well revealed in the legal briefs that he drew up for the royal benefit to confute the contentions of the Fellows of King's. His claim brought him into touch with Christiaan Huygens, whose brother Constantyn had come to England in the service of William III. It is pleasant to think of these two great, and very different, scientists setting off in a coach to petition their common sovereign to promote Newton in the academic hierarchy.

[2] Newton's mental illness was a definite paranoia, in which he accused his closest friends of persecuting and calumniating him. Locke, in 1705, privately described him as 'a little too apt to raise in himself suspicions where there is no ground'; this mental trait, which he always had, was played upon by men of greater ambition than discretion in the quarrel with Leibniz, wherein Newton lost all sense of proportion and probity.

[3] *Opticks*, Bk. III, Qu. 31.

Newton, indeed, for the first time made optics a branch of physics, by his demonstration that the theory of light and the theory of matter were cognate and complementary. Before the mid-seventeenth century the science of optics was almost wholly geometrical, and it is interesting that progress was achieved through renewed attention to its qualitative aspects. Since the characteristic colours of bodies and pigments had long been regarded as true Aristotelean qualities, confusion followed upon the challenge to Aristotle's philosophy in the time of Galileo. The prismatic colours also had long been familiar; in the middle ages an increasingly more perfect account of the formation of the colours in the rainbow had been given by the Latin writers who, following Alhazen, regarded them as produced by the refraction of sunlight in shining through rain-drops. The geometrical optics of the rainbow were further perfected by Descartes, who also put forward a new and more definite corpuscular hypothesis of light. In his view, light was a sensation caused by pressure of the *matière subtile* on the optic nerve, set up by the tendency of this matter to expand in all directions from its concentration in the luminous source; colours were caused by rotations of the particles of the *matière subtile*.

This was, like all the science of the *Principles of Philosophy*, an unsupported hypothesis. Descartes assumed that light travels more rapidly in a dense medium (such as water) than in air, and from this the sine-law of refraction could be deduced; but the same law could also be deduced from the contrary assumption. There was no method of proving either by experiment. The explanation of what happens to "light" itself when a beam of white light is converted into a beam of one or more colours was vague and wholly speculative. In all theories before Newton's, a qualitative change in the nature of the beam was suggested, a modification of the physical entity of light, which at least in imagination could be reversed, so that as red light can be derived from white, white ought to be derivable from red. The same potential reversibility is implicit in the more ingenious theory of colours described by Robert Hooke in *Micrographia* (1665). Hooke was the first investigator of the colours produced by optical interference in the form known, somewhat unjustly, as "Newton's rings," which he observed in the laminations of mica, and in plates of glass pressed together, finding that the manifestation of the colours depended upon the existence of a very thin refracting medium, each colour corresponding to a determinate thickness. According to Hooke, the sensation of light is caused by a very rapid, short vibrating motion in the transmitting medium, every pulse or vibration of the luminous body generating a sphere 'which will continually increase and grow bigger, just after the same manner (though indefinitely swifter) as the waves or rings on the surface of the water do swell into bigger and bigger circles about a point of it, where, by the sinking of a stone the motion was begun.' Therefore, each successive pulse or vibration may be considered as being at right angles to the direction of propagation radially from the centre, and in a narrow beam of light each pulse as a plane surface perpendicular to the beam.[1]

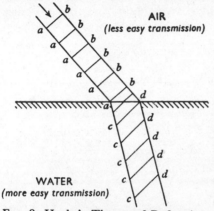

FIG. 8. Hooke's Theory of Refraction. *aaabbb*, incident ray; *cccddd*, refracted ray. *ab*, *ab* perpendicular pulses, *cd*, *cd* oblique pulses.

Hooke next investigated geometrically the results of the passage of the beam from one medium into a second which is capable of transmitting the pulses more, or less, easily than the first. He reasoned that if the light falls obliquely upon the surface separating the two media, one "edge" of the pulse (treated as a plane) would enter the second medium before the other, and would therefore be accelerated or retarded before the latter, in turn, had started to travel through the second medium (Fig. 8). He assumed (without proof) that when the second medium transmitted the pulses more easily the beam would be refracted towards the perpendicular, and consequently that water transmits more easily than air.[2] This refracted ray, moreover, would be distinguished from the incident ray in that the pulses would now be oblique,

---

[1] *Micrographia*, pp. 55–7.

[2] In this Hooke followed Descartes. Fermat had already demonstrated that refraction towards the perpendicular occurs when the velocity of light in the second medium is *less*, i.e. when like water or glass it transmits light *less easily* because it is more dense than air. Huygens, in the *Traité de la Lumière* (1690) wherein he follows Fermat, by introducing the conception of the wave-front into the pulse-theory showed that the sine-law of refraction followed from it.

not perpendicular to the ray, due to the acceleration or retardation of the "leading edge" of each pulse.[1] To this obliquity of the
pulse Hooke traced the sensation of colour, on the grounds that in
a whole beam of light there would be some confusion of the oblique
pulses, and that one "edge" of each pulse would be weakened or
blunted through having to initiate the vibration in a medium at
rest; thus:

> Blue is an impression on the Retina of an oblique and confus'd pulse
> of light whose weakest part precedes, and whose strongest follows.
> Red is an impression on the Retina of an oblique and confus'd pulse
> of light, whose strongest part precedes, and whose weakest follows.

For example, in Fig. 9, towards *a* the leading edge of each oblique
pulse, being adjacent to the un-
disturbed medium, is weakened,
whereas towards *d* the lagging
edge pulse is weakened for the
same reason. Hence the colour
blue will be seen about *a*, and red
about *d*. Hooke thought that
the intermediate colours of the
spectrum 'arise from the composition and dilutings of these
two' produced by the confusion
of the two primary types of

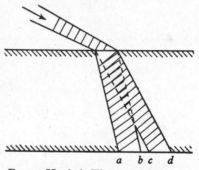

FIG. 9. Hooke's Theory of Colours.

oblique pulse towards the middle of the refracted beam. He
further showed by a very ingenious analysis that when a ray
passes through a very thin medium a similar succession of strong-
weak or weak-strong pulses is created, according to its thickness,
producing colours.

In this theory of light, Hooke devised a subtle mechanism by
which colours might be derived from white light considered as a
train of uniform and homogeneous pulses. It accounted for the
association of heat, light and motion in a crudely sketched kinetic
theory; for the fact that refraction is always accompanied by
coloration; for the order of the colours as produced by refraction
or by interference; and for the fact that the spectrum produced by
one prism can be re-converted into white light by a second. On

---

[1] Huygens (see note 2, above) avoided this supposition of obliquity by
different reasoning on the formation of the refracted wave-front.

the other hand, it was not very clearly conceived in detail, and it explained only a selection of the known facts. Hooke seems to have observed that the two sides of a refracted ray are not parallel, but makes no mention of the fact (cf. Fig. 9).[1] Probably no experiments on homogeneous coloured light were made by him, since he seems not to have known that further refractions have no effect upon such light; this would certainly have been difficult to reconcile with his theory. The most serious objection against its broad form was that it failed to account for the rectilinear propagation of light. If a ray was a train of pulses, why did not these spread out into the surrounding medium, as sound-waves do? This was, indeed, for long a profound objection against any pulse or undulatory theory, having been thoroughly discussed by Newton in the *Opticks*.

So far the theory of light had been illuminated by few new experiments, and none that were decisive. The ideas of Descartes and of Hooke could have been as well propounded in the fourteenth century as in the seventeenth. The new experimental evidence, relating to interference (Hooke) and diffraction (Grimaldi, also 1665) seemed to favour the pulse hypothesis, but though the two opposed theories were equally mechanical in character, they were nevertheless *ad hoc* hypotheses, entirely lacking in the demonstrative solidity on which the "new philosophy" was supposed to be founded. By introducing imaginary mechanical qualities of pellucid matter, the old Aristotelean theory of colour as a quality had only been pushed back to a further remove. Newton's experiments began where the theories of Descartes and Hooke, with which he was certainly well acquainted, ceased—in new experiments. His papers in the *Philosophical Transactions*, containing his new theory of prismatic colours, are hardly more than the statement of experimental results. The deductions he drew were simply those enforced by the new evidence.

His first paper begins with the observation that a parallel beam of white light diverges into the spectrum when it is refracted by a prism. The spectrum is longer in one direction, not of the same shape as the aperture through which light is admitted. This must have been noticed often, and disregarded. Both in the *Opticks*, and in his note-book, however, Newton first describes another

[1] This is shown in Fig. 2 of Schem. VI of *Micrographia*, where, however, the incident ray is shown as convergent though Hooke purports to be using sunlight.

experiment, in which he observed a continuous line, painted in two colours, through the prism and found that the image of the parti-coloured line was no longer rectilinear. Perhaps this gave him the first clue to what followed. These simple experiments taught Newton that 'ye rays which make blew are refracted more y^n ye rays which make red,'[1] or as he phrased it after many more experiments described in the *Opticks*, 'the Sun's light is an heterogeneous Mixture of Rays, some of which are constantly more refrangible than others.' Any refracting agent, such as a prism, simply acted like a filter to distinguish the infinite components of white light according to their refrangibility. Therefore, he thought, it was necessary to give up all hope of contriving a lens which would refract all colours equally, and so yield an achromatic image. Newton was careful to prove that monochromatic light has all the optical properties that were "vulgarly" attributed to white light, and that there could be no question of the white ray being actually shattered by refraction—it was merely divided, and it could be reassembled. Each pure coloured constituent, however, was homogeneous and indivisible.

This doctrine appeared very shocking to contemporaries. The simple nature of white light had always been accepted as axiomatic. The proof that all the colours of the spectrum are equally primary and necessary in white light cut directly across notions which supposed them the product of mixtures of red and blue, or blue and yellow. While some theorists of empiricism welcomed Newton's Baconian use of experiment, many who prided themselves on their knowledge of optics were hostile. The reactions of the critics, Hooke and Huygens amongst them, are interesting. When Newton's first paper appeared in the *Philosophical Transactions*, they judged initially that he was merely speculating, and tried to answer with irrelevant arguments. Then they denied that the experiments gave the results described by Newton, or maintained that if the experiments were correct the conclusions drawn from them were false. Finally, it was alleged that if Newton's ideas were justifiable, they were not original. Neglecting the minor critics, it may be doubted whether either Hooke or Huygens— two of the leaders of the scientific movement—ever succeeded entirely in adjusting their thinking in accordance with the evidence

[1] Cambridge MS. Add. 3996: cf. the author's note in *Cambridge Historical Journal*, vol. IX (1948), pp. 239–50.

of Newton's experiments. The former never understood that his own pulse-theory could not account for them. The latter, in his *Traité de la Lumière* (1690), tactfully omitted the subject of colour completely. Their failure is not more surprising than that of other scientists in more recent periods who have equally resisted an innovation which has inevitably overwhelmed their criticism. Newton's propositions were revolutionary, not only in their content, but because they were founded straightforwardly on new experimental evidence. It is, in one sense, an indication of the superficiality of the change in spirit effected by the scientific revolution that such obvious conclusions drawn from such easily repeatable experiments should have been treated as matters for argument. Empiricism—that is, in this instance, Newton's reticence in his original papers on the question of what light is, and the reasons for the properties exhibited by a lens or prism—was still obstructed by the inertia of established theory, which could prevent an accurate factual description being estimated at its true worth. In another sense, the dispute carried on in the pages of the *Philosophical Transactions* was the product of a confusion between facts and their interpretation which is perhaps inevitable in science at its growing point. Of this there are many examples, in the incredulity with which Lavoisier, Darwin, Joule or Pasteur was heard. Major scientific advances have not been made simply by uttering the statement that, in certain conditions, "The red line is higher than the blue," or "the thermometer reading was $x°$." Newton did not make such a purely descriptive statement, for he added that the red was higher than the blue *because* the constituent coloured rays of white light are not equally refrangible. If uncomfortable experiments have stuck in the throats of those wedded to established ideas, it is owing to their indissoluble connection with a heterodox theory. In such circumstances, the new theoretical attitude may seem to leave incomprehensible more than it seeks to explain, and then the question arises: "Is it worth while to alter the balance which is already struck between the mysteriousness of nature and human understanding?"

The most important of Newton's discoveries, because they were most unconventional, aroused in many of his contemporaries the feeling that the accepted conception of natural processes was being wantonly disturbed in order to account for phenomena that could equally well be treated without such intellectual upheavals.

It is, in the last resort, the function of cumulative experiment and observation to prove that such supposed alternatives in explanation are unreal; triumphantly to defend the old, or vindicate the new. In the course of the scientific revolution, the dominant antithesis had been between old and new methods and ideas, broadly divisions among the "moderns" were trivial. Towards the end of the seventeenth century this antithesis was becoming lifeless, and other divisions between the moderns themselves became far more critical. The appeal to experiment and observation had played a useful part in resolving the former antithesis, but the appeal to a new conception of what science should be, to a new image of nature, to a new mathematics and structure of reasoning, in short to a new appraisal of familiar facts, had achieved far more sweeping results. In many ways, genuine observation and experiment had been the pivots of seventeenth-century biology far more than they had been of its mechanics and physics. With the empiricist reaction against Cartesian science, which had seemed for a moment almost to sum up the whole revolt against tradition, and especially with the discoveries of Newton, came the test of the ability of the heirs of Copernicus and Galileo to resolve their own internal contradictions. If these were not in turn to lead to endless debate, such as had embroiled the heirs of Aristotle, it could only be by a more rigorous attention to the criteria of experiment. The fundamental significance of Newton's scientific method was that it achieved this exactly; it did not show merely that a theory roughly agreed with a selected group of facts, but that a group—however limited and restricted—of theoretical propositions could be associated with a range of experimental facts, carefully checked and often repeated. Confidence could be granted to such a group of propositions because it was unique, and the minimum necessary; because it claimed only to comprehend a limited range of phenomena that had been exactly studied, and not to extrapolate from a few particulars to universal truths. Against such a method the kind of criticism directed towards Newton's theory of colour was unavailing in the long run; as was that directed against Lavoisier or Joule. If, with regard to Newton, expressions of incredulity were less well founded, it was perhaps because his precise use of the experimental method was not yet understood.

Of course Newton did not restrict himself entirely to such

theoretical propositions as were firmly based on experiment. Too much has been made of his celebrated *obiter dictum*, "I do not frame hypotheses." Scattered through his works, and more freely through his unpublished papers, are numerous comments suggestive of a deeper penetration into the mysteriousness of nature than exact science could yet achieve. Newton framed many hypotheses which, he was inclined to think, might account for natural phenomena, like gravitation or refraction, hardly as yet illuminated by experimental inquiry; but he was always careful to distinguish between such hypotheses and an experimentally established theory. With him this distinction—totally ignored by Descartes—becomes fully self-conscious; hence the form of the *Queries* in the *Opticks*. In these *Queries* Newton sketched out an interlocking but not wholly consistent body of ideas relating to the nature of light and matter which he regarded as possible, or probable, but unproven. The most influential portions of this discussion were those in which Newton considered the relative merits of the undulatory (or pulse) and the corpuscular theories of light. He has been generally regarded, with some truth, as the classical advocate of the latter. But he was not wholly decided. For example, in Query 13 the suggestion is made that the colour of a ray depends upon the "bigness" of its vibration, the shortest vibrations corresponding to violet, and the largest to red, just as the pitch of sound corresponds to the "bigness" of the vibration in the medium.[1] Again, in Query 16, Newton asks, 'considering the lastingness of the Motions excited in the bottom of the Eye by Light, are they not of a vibrating nature?'; and in Query 17 he suggests that when a ray of light is reflected or refracted, vibrations are set up in the medium which, radiating from the point of incidence, 'overtake the Rays of Light, and by overtaking them successively, do they not put them into the Fits of easy Reflexion and easy Transmission described above?' As for the medium whose vibrations should transmit light, Newton could conceive of an æther, less dense than air but far more elastic, pervading all bodies and expanded throughout the universe, transmitting heat by its vibrations as well as light,[2] acting as the agent of reflection and

---

[1] Newton's terminology is obscure: it is not quite clear whether by "bigness" he means amplitude or wave-length.

[2] The existence of invisible (infra-red) heat rays was first demonstrated by Sir William Herschel (1800).

refraction. Conversely, in Query 29 Newton declares: 'Are not the Rays of Light very small Bodies emitted from shining Substances? For such Bodies will pass through uniform Mediums in right Lines without bending into the Shadow, which is the Nature of Rays of Light.'

Colour might be accounted for by supposing that the smallest corpuscles of light gave the sensation violet, and the largest, being less refracted, the sensation red. From his study of the double refraction in Icelandic spar (which Huygens, in his *Traité de la Lumière*, had failed to explain completely in terms of the wave-theory) Newton concluded that the "sides" of a ray of light might have different properties, and he did not see how these could be associated with wave-propagation. These two points were, for him, decisive objections against the ideas of Hooke and Huygens, yet he himself found it necessary to conceive of a periodicity in the "Fits of Easy Reflection and Transmission" which he introduced in Book II in order to deal with the phenomena now ascribed to optical interference.[1] Newton's corpuscular theory, therefore, did not render an æther unnecessary, nor did it dispense entirely with the concept of ætherial vibrations. As he was evidently not satisfied with its completeness, it was unfortunate that some later physicists paid insufficient attention to Newton's caution: 'Since I have not finish'd this part of my Design, I shall conclude with proposing only some Queries, in order to a farther search to be made by others.'

Not only does the *Principia* differ from the *Opticks* in form—a mathematical, as compared with an experimental, treatise—but its relationship to the contemporary scientific background is different also. It presented no less of a puzzle to contemporaries than the optical discoveries of fifteen years earlier, partly for the same reason that Newton chose to propound even apparently absurd theories if they alone suited the facts described, partly for the additional reason that the propositions of the *Principia* seemed to contravene the mechanistic philosophy which science had so

[1] In each regularly successive "Fit of Easy Reflection" the ray tended to be reflected upon meeting a surface; in between in the "Fits of Easy Transmission" to penetrate into it. Thus the behaviour of a ray encountering the two surfaces of a thin medium (such as the air gap between two glass plates) depended upon the dimensions of the intervals between the surfaces and of the intervals between the "Fits."

recently adopted with confidence. The theory of gravitation stated in terms of action at a distance was regarded by Cartesians as no better than a revival of the occult forces from which science had been liberated by mechanical hypotheses. Nevertheless, the *Principia* was far more in keeping with a certain scientific tradition —one overshadowed by the triumph of Cartesian science—than was the *Opticks*; its thesis was more clearly anticipated in earlier notions, due to Gilbert, Kepler, Borelli and Hooke; its materials had largely been prepared by other hands than those of Newton himself.

Descartes had flatly denied the validity of attraction as a scientific concept. Since all forces were mechanical in his view, a body could move only because other bodies impacted upon it, and so impelled it by pushing. Therefore he had described probable ways in which gravitational, magnetic, or electric "attractions" might be caused by such impacts of the *matière subtile* upon solid bodies. His system implied that the various motions of the planets around the sun were conditioned by the structure of the solar system as a whole, and particularly by the properties of the ætherial matter, rotating like a vortex, with which it was filled. They were not determined by separate relations between the individual celestial bodies; this perhaps explains why Kepler's empirical laws of planetary motion were of little interest to him. The earlier notions of attraction, however, had conceived rather of each celestial body, regarded as a physical entity, exerting an influence directly upon cognate matter. In Aristotle's theory of gravity and levity the matter of the universe was drawn, or attracted, to its natural place, the attraction being specific to each kind of matter, since earth could not tend naturally to the place of fire, nor *vice versa*. Thus the same kind of matter tended to collect together in the same place. In the Copernican system, this concept of attraction to specific spherical layers about the centre of the universe no longer had meaning, but the cosmic order could be preserved by transferring the attraction from the place to the matter which had, in Aristotle's cosmology, occupied that place. The basic principle—that matter tends to coalesce with like matter—implied by Aristotle, could now emerge more clearly. For Gilbert this principle, justified quite naïvely by teleological reasoning, enabled bodies to preserve their integrity. 'Cohesion of parts, and aggregation of matter, exist in the Sun, in the Moon, in the planets, in the fixed stars,' so that in all these bodies the parts tend to unite with the whole 'with which they connect themselves with the same appetence as terrestrial things, which we call heavy, with the Earth.'[1] This means that gravitation is a universal property of matter, but peculiar to each body; the same gravity is not common to all, in Gilbert's view, because a piece of lunar matter would always tend towards the moon, and never adhere to the earth.

It might seem that after Galileo's contention that the matter of earth, moon and planets—the non-luminous heavenly bodies— was of the same kind, it would have been a straightforward step to argue that all this earthy matter shared a common attraction, like drawing like. But against such an argument the teleology of the theory of attraction—which was in no way required to explain the known behaviour of lunar or solar matter, apart from its cohesion—was doubly effective. In the first place, if the matter of the moon, for example, were attracted towards the earth, the theory would cease to explain the cohesion of the parts of the moon. Secondly, a common gravitational attraction would suggest that all the earthy matter in the universe would collect in one mass—this was Aristotle's view, which Galileo opposed. Thirdly, Galileo and his followers were reluctant to introduce into astronomy the esoteric principle of attraction, which would disturb the perfect inertial revolution of the heavenly bodies. The theory of specific attractions remained far more plausible.

This theory had been used by Gilbert, and by Copernicus before him, as an alternative to the Aristotelean causation of the motions of heavy terrestrial bodies. It was less a new cosmological principle, than a new physical principle applied to cosmology. As such it is also used by Kepler:

A mathematical point, whether it be the centre of the universe or not, cannot move heavy bodies either effectively or objectively so that they approach itself. . . . It is impossible, that the form of a stone, moving its mass [corpus], should seek a mathematical point or the centre of the world, except with respect to the body in which that point resides. . . . Gravity is a mutual corporeal affection between cognate bodies towards their union or conjunction (of which kind the magnetic faculty is also), so that the Earth draws a stone much more than the stone seeks the Earth. Supposing the Earth to be in the

[1] *On the Magnet*, trans. by S. P. Thompson (London, 1900), pp. 219, 229.

centre of the Universe, heavy bodies would not be borne to the centre of the Universe as such, but to the centre of a cognate spherical body, to wit the Earth. And thus wherever the Earth is assumed to be carried by its animal faculty, heavy bodies will always tend towards it.[1]

So far Kepler has said nothing very new. He has repeated that the concept of attraction of like to like can replace the Aristotelean concept of matter being attracted to specific places, and he has limited his use of this concept to heavy bodies cognate with the earth. But he has stated, for the first time, that the attraction is mutual (the analogy between gravity and magnetism, so fruit-fully begun by Gilbert, is now being extended), and this point he amplified further:

> If two stones were placed close together in any place in the Universe outside the sphere of the virtue of a third cognate body, they would like two magnetic bodies come together at an intermediate point, each moving such a distance towards the other, as the mass of the other is in proportion to its own.

Here an original conception of the magnitude of the motion due to gravitational attraction was introduced $\left(\dfrac{d_1}{d_2} = \dfrac{m_2}{m_1}\right)$, in which it was related to the ratio of the masses of the two bodies. Kepler, therefore, began the investiture of the theory of attraction with a definite dynamical form. Further, he postulated that the earth and the moon were cognate matter, like the two stones:

> If the Moon and the Earth were not retained, each in its orbit, by their animal or other equivalent forces, the Earth would ascend towards the Moon one fifty-fourth part of the distance between them, and the Moon descend towards the Earth about fifty-three parts; and they would there join together; assuming, however, that the substance of each is of one and the same density.[2]

Kepler then went on to demonstrate, from the ebbing and flowing of the tides, that this attractive force in the moon does actually extend to the earth, pulling the waters of the seas towards itself; much more likely was it that the far greater attractive force of the earth would reach to the moon, and greatly beyond it, so that no kind of earthy matter could escape from it.[3]

---

[1] *Astronomia Nova; Gesammelte Werke*, vol. III, pp. 24–5.

[2] Assuming also that the diameter of the earth is about $3\frac{3}{4}$ $\left(\sqrt[3]{53}\right)$ that of the moon, which is a little too large.

[3] *Loc. cit.*, pp. 25–7.

Clearly no one invented the theory of gravitational attraction; it grew through many diverse stages. And clearly also the genesis of the theory of *universal* gravitation is found in Kepler. Newton's hasty calculation of 1666, his later theory of the moon, and his theory of the tides, are all embryonically sketched in the *Astronomia Nova*. But the attraction was still specific, applicable only to heavy, earthy matter; Kepler himself did not go so far as to suppose that the sun and planets were also mutually attracting masses, or that the dynamical balance he indicated as retaining the earth and moon in their orbits with respect to each other also preserved the stability of the planetary orbits with respect to the sun. He failed, as Copernicus, Gilbert and Galileo failed, to see the full power of gravitational attraction as a cosmological concept.

Nevertheless, Kepler's idea that the satellite revolving round a central body is maintained in its path by two forces, one of which is an attraction towards the central body, although applied only to the earth-moon system, holds the key to all that followed and to the *Principia* itself. Galileo, like Copernicus, had believed the planetary revolutions to be "natural," i.e. inertial; the celestial bodies were subject to no forces. Kepler, however, believed that the motive force of the universe resided in the sun which, rotating upon its own axis, 'emits from itself through the extent of the Universe an immaterial image [species] of its body, analogous to the immaterial image [species] of its light, which image is itself rotated also like a most swift whirlpool and carries round with itself the bodies of the planets.'[1] Each planet, moreover, was endowed with its own "soul" which influenced its motions.[2] Such notions confused the dynamical elements of the situation for Kepler—since the sun's force operated tangentially upon the planet, he did not imagine that a centripetal force was necessary to retain it in the orbit. In the singular case of the earth and moon, it was necessary for him to suppose that the "animal or other equivalent force" of the moon was sufficient to overcome the attraction towards the earth which would have distorted its path. This physical, attractive property of heavy matter could not as yet be made the basis of the stability of the celestial system; rather it was a disturbing feature which the cosmological properties of the heavenly bodies had to overcome.

[1] *Loc. cit.*, p. 34.
[2] Cf. *Harmonices Mundi* (1619), *Gesammelte Werke*, vol. VI, p. 264 *et seq.*

With Descartes the position was altogether reversed. He knew that bodies in free motion move in straight lines. He knew that his planets, swirled like Kepler's in a solar vortex, would if unconstrained travel in straight lines outside its limits. He knew, therefore, that some centripetal force must bend these straight lines into the closed curves of the orbits. Rejecting Kepler's mysterious attraction, he supposed this force to be provided by the varying density of the solar vortex, which resisted the planets' natural tendency to recede towards its periphery.

After the publication of Descartes' *Principia Philosophiæ* (1644) which for the first time applied the law of inertia systematically to the planetary motions, the elements of the problem of universal gravitation were completely assembled. The essential step was to replace Descartes' conception of the nature of the centripetal force required to hold the universe together by the Keplerian idea of attraction, with the sun taken as the central body. Kepler's problem could now be approached from a completely fresh aspect: knowing that the moon must be, as it were, chained to the earth to prevent it flying off into space, might not this bond be that "corporeal affection between cognate bodies towards their union" described by Kepler? Three men, all about the year 1665, formulated this question in similar terms, and attempted to answer it: Alphonso Borelli, Robert Hooke and Isaac Newton.

Borelli, who was a member of the Accademia del Cimento, tried to find in Kepler's ideas the basis for a complete mathematico-mechanical system of the universe.[1] He regarded the light-rays radiating from the sun as levers pressing upon the planets, revolving because the sun revolved, and able to exert a pressure because they were themselves material emanations. He explained that the least force would impart some motion to the greatest mass, and that therefore (in the absence of resistance) the planets would move with a speed proportionate to the force impressed.[2] This, like Kepler, he supposed to become more feeble as the distance from the sun was greater, so that the outer planets would move more slowly than the inner. Instructed by Descartes, Borelli knew that under such circumstances a centripetal force was necessary to

---

[1] *Theoricæ Mediceorum Planetarum ex causis physicis deductæ* (Florence, 1666).

[2] As Prof. Koyré has pointed out, Borelli has unwittingly formulated the Aristotelean doctrine. Acceleration, not velocity, is proportional to force applied: hence Borelli's planets could never maintain a uniform speed.

maintain the planets in their orbits, but he carefully avoided speaking of this as an *attraction* since the word was banned from the phraseology of mechanism. Nor did he identify this force with that which in the earth is called gravity. Instead, he postulated that all satellites in the celestial machine had a natural tendency or appetite to approach the central body about which they revolve—thus the planets sought the sun—an appetite constant at all distances, and not at all affecting the central body itself. The stability of the planet in its orbit was therefore conditioned by the perfect balance of the centrifugal and centripetal forces to which it was subject, and this he was able to illustrate experimentally; but Borelli was further required to explain why these orbits are elliptical, not circular. The answer is highly ingenious: Borelli imagined that each planet was created *outside* its circular orbit. In this position the excess of centripetal over centrifugal force would urge the planet to its proper distance from the sun, but the momentum acquired would carry it beyond, to a point *inside* the circular orbit. Here the centrifugal would exceed the centripetal force, and the planet would again be pressed outwards, and again carried by momentum to its former station.[1] Then the cycle would be repeated. Thus the ellipse was a result of a slow oscillation about a stable position—a circle round the central body—compared by Borelli to the oscillation of a pendulum, to and fro, passing through a stable position at the perpendicular. This hypothesis was in accord with the observed fact that the velocity of the planet is greatest at its nearest approach to the sun.

Borelli's theory, an amalgamation of those of Kepler and Descartes, in regarding the planets as impelled by a sort of vortex centred upon the sun conceived of the universe as a driven, not a free-spinning, machine. To it the law of inertia could not be directly applied—hence Borelli's confusion concerning force and momentum. Both Hooke and Newton assumed that the planets' inertial motions were pulled into closed curves by a centripetal force in the sun, and hence that there was a universal, mutual attraction between masses of matter. Newton, moreover, remedying the mistakes in the principles of dynamics which permeate

---

[1] Nearly two hundred years later, Clerk-Maxwell accounted for the stability of the innumerable small satellites which compose Saturn's rings in a mathematical analysis, the principle of which was dimly anticipated in this hypothesis of Borelli.

Borelli's treatise, effected a meticulous analysis of the forces which the latter (and Hooke) had described so loosely. Hooke's tragedy was to fail in mathematics.

There is ample evidence that by 1685 Robert Hooke had a very complete picture of a mechanical system of the universe founded on universal gravitation. In the early days of the Royal Society he performed unsuccessful experiments to discover whether gravity varies above and below the earth's surface. In *Micrographia* (1665) he conjectured that the moon might have a 'gravitating principle' like the earth. In a discourse read to the Royal Society in 1666 Hooke improved on Borelli with the supposition that a 'direct motion' might be inflected into a curve by 'an attractive property of the body placed at the centre.'[1] Like earlier writers he compared this centripetal attraction to the tension in the string of a conical pendulum, which retains the bob in its circular path. In 1678 he wrote: 'I suppose the gravitating power of the Sun in the center of this part of the Heaven in which we are, hath an attractive power upon all the planets, . . . and that those again have a respect answerable.'[2] This is the first enunciation of the true theory of universal gravitation—of gravity as a universal principle that binds all the bodies of the solar system together. The same force whereby the heavenly bodies 'attract their own parts, and keep them from flying from them,' also attracts 'all the other celestial bodies within the sphere of this activity.' It is this force which, in the sun, bends the rectilinear motions of the planets into closed curves. And this force is 'the more powerful in operating, by how much nearer the body wrought upon is' to the attracting body.[3]

These ideas, Hooke claimed, he had expounded as early as 1670. But it was not until 1679 that he hit upon a hypothesis to describe the rate at which the gravitational attraction should decrease with distance. In that year he renewed his correspondence with Newton, discussing an experiment to detect the earth's rotation through the deviation of falling bodies. This in turn led to a debate on the nature of the curve which a heavy body would describe if it were supposed to be able to fall freely towards the centre of the earth, during which (in a letter to Newton dated

[1] R. T. Gunther: *Early Science in Oxford*, vol. VI (Oxford, 1930), p. 266.
[2] *Ibid.*, vol. VIII, p. 228.
[3] *Ibid.*, vol. VIII, pp. 27–8, 229–30, etc.

6 January 1680) Hooke stated the proposition that the force of gravity is inversely proportional to the square of the distance, measured from the centre of the gravitating mass.[1] He was convinced that this "inverse square law" of attraction, combined with the ideas he had already sketched out, would be sufficient to explain all the planetary motions.

Hooke's scientific intuition was certainly brilliant. Of all the early Fellows of the Royal Society, in a generation richly endowed with genius, his was the mind most spontaneously, and sanely, imaginative; schemes for new experiments and observations occurred to him so readily that each day was divided between a multiplicity of investigations; physiology, microscopy, astronomy, chemistry, mechanics, optics, were each in rapid succession subjects for his insight and ingenuity. No topic could ever be broached without Hooke rising to make a number of pertinent points and to suggest fruitful methods of inquiry. With regard to celestial mechanics, Hooke's conception was as far-reaching as Newton's; but it was not prior and it was not proven. The publication of the *Principia*, accompanied by Hooke's charge of plagiary against Newton, inflamed the suspicion between the two men into outraged anger. Both suffered from a touchy pride; neither would recognize the true merits of the other. In Newton's eyes, Hooke grasped after other men's achievements, having merely patched together some notions borrowed from Kepler, Borelli and Huygens. To Hooke, Newton had merely turned into mathematical symbols ideas that he had himself already expressed without attracting notice or reward. He was ever unwilling to admit the supreme advantage that Newton held over himself, of being mathematician enough to demonstrate as a theory, confirmed by observation, that which Hooke himself had only been able to assert as a hypothesis. In fact the different status which attaches to a scientific *theory* and a scientific *hypothesis*—a difference which

[1] Hooke thought that this same law would apply to the force of gravity below the earth's surface. Newton later proved (*Principia*, Bk. I, Prop. LXXIII) that within a sphere the centripetal force is proportional to the distance from the centre, not inversely to the square of the distance. Ismael Bouillau, in *Astronomia Philolaica* (1645), had argued that the intensity of Kepler's "moving virtue" resident in the sun would decrease, like that of light, as $\frac{1}{d^2}$. Therefore, he said, since the velocities of the planets are not in this proportion to their distances from the sun, they could not be impelled by such a force emanating from it.

Newton emphasized more than once—was something to which Hooke proved himself insensitive by a number of episodes in his career. As Newton pointed out, Hooke did not invent the theory of attractive forces, as such; and after Huygens' theorems on the centrifugal acceleration of rotating masses had been published, the inverse square law could easily be deduced. Granting the highest merit to Hooke's scientific intuition, it is quite clear from certain confusions inherent in the development of his hypothesis between 1666 and 1685 that his mastery of the principles of dynamics was never completely confident, and that his thought was never safeguarded by the precision of mathematical analysis.

Newton complained, in a letter to Halley at the time when Hooke was voicing his protests before the Royal Society, that the latter wished to assign to him the status of a mathematical drudge and claim for himself the sole invention of a new system of celestial mechanics. The truth is far otherwise. By 1666 Newton was already able to calculate centrifugal accelerations. This calculation, applied to Kepler's laws, gave him the inverse square law of attraction.[1] If the earth's gravitational attraction was assumed to act upon the moon, he computed that in accordance with this law, at the distance of the moon from the earth this centripetal force would be "pretty nearly" equal to the centrifugal force created by the moon's own revolution about the earth. Gravity would be precisely the chain required to bind the moon in its orbit.[2] But such a calculation was not strictly appropriate: for a planet's (and the moon's) orbit is not circular but elliptical, with the central body (sun or earth) at one focus of the ellipse. Moreover, in calculating the distances and relative forces, Newton had proceeded as though the earth and moon were points, that is, reckoned that if

the force of gravity at the surface of the earth was $\frac{g}{1}$, then at the

[1] In a circle of radius r, if T is the time taken by a body to complete one revolution, the centripetal force required to hold it in its path is $F = \frac{kr}{T^2}$. But according to Kepler's Third Law, in the solar system $T^2 = k'r^3$. Substituting, $F = \frac{kr}{k'r^3}$ whence, omitting the constants, F is proportional to $\frac{1}{r^2}$.

[2] Newton's calculation in 1666 did not give a perfect confirmation of the inverse square law, because the value he took for the earth's radius (and hence the distance of the moon from the earth) was too small. It used to be supposed that this discrepancy induced him to lay the matter aside. The view given below is now generally accepted.

moon it was $\frac{g}{(60)^2}$. He had not *proved* that the external gravitational attraction of a sphere could be computed as though its mass
were concentrated in a point at the centre. The difficulties involved in perfecting the hypothesis upon which his first casual
trial was founded were mainly mathematical, and at this time
beyond Newton's skill. So great were they that Halley was astonished to learn (in 1684) that Newton had overcome them; had
proved, in fact, that the path followed by a body, moving obliquely
in relation to a second which exerts a centripetal force upon it,
must correspond to one of the conic sections.[1]

In 1666, therefore, Newton was not satisfied that the inverse
square law represented more than an approximation to the truth.
Impressed by the mathematical difficulties involved, he laid the
hypothesis aside. For about thirteen years (1666–79) there is not
the least evidence that he paid any attention to dynamics, universal gravitation or celestial mechanics. Optics, mathematics,
chemistry and perhaps already theology, filled his mind. So much,
at least, Newton owed to Hooke: that he was compelled to return
to his former hypothesis. Even when driven, by Hooke's unwel-
come letters, to review the mystery of gravity, he was guilty of
blunders and misapprehensions. Even when, pricked by Hooke's
corrections, he had solved the problem of the inverse square law
and the elliptical orbit, Newton once more set his success to cool
for a further five years. Only Halley's visit to Cambridge in 1684,
and the warmth of his admiration and offers of assistance, set the
Principia in train. Two and a half years later it was finished.

The main, though not the sole, function of the book was to
establish what has been called the Newtonian system of the
universe. This is, indeed, a system of physics in the widest possible
sense, by no means limited to celestial mechanics. Book I sets out
the general theory of dynamics, especially the theory of the
motions of bodies under the action of a centripetal force directed
towards some point or body; Book III applies this general theory
to the inverse-square law of gravitation (as an attractive force
between bodies proportionate to the product of their masses) and
demonstrates from the phenomena of the planets and their

[1] Newton established (*Principia*, Bk. III, Prop. XL) that the path of a comet
is a parabola, and that the orbit of a returning comet, such as that examined
by Halley (*ibid.*, Prop. XLI), is an extended ellipse of which the portion near
the sun is scarcely distinguishable from a parabola.

satellites, the moon and the tides, that this law and the general theory do hold for the solar system.[1] In these Books Newton gave and used the proof that the inverse-square law accounts for the elliptical orbit defined by Kepler; provided the general dynamical theory of Kepler's laws of planetary motion, of Galileo's law of acceleration, and of Huygens's theorems on centrifugal force; described a method of determining orbits from observations of position; offered an approximate solution of the problem (essential to the dynamical theory of the moon's orbit) of determining the motions of three mutually gravitating bodies; established the broad theory of the tides, and determined the degree of asphericity of the earth caused by its rotation. Thus he solved at least a dozen problems in contemporary science by showing the hitherto unsuspected connection between them—the theory of gravitation.

The outline of these achievements by no means exhausts the importance of Books I and III, quite apart from their contributions to pure mathematics and astronomical computation. Newton also laid down the principles of theoretical physics as a mathematical science in the form which it preserved until the end of the nineteenth century: "since . . . the moderns, rejecting substantial forms and occult qualities (he wrote) have endeavoured to subject the phenomena of nature to the laws of mathematics, I have in this treatise cultivated mathematics as far as it relates to philosophy." It might nearly as well be added that he cultivated philosophy in so far as it related to mathematical science. Book I opens with *Definitions* of the fundamental concepts of mechanics— mass, quantity of motion, inertia, impressed force and centripetal force. The last, and most fundamental, of these definitions concerns Newton's distinction between absolute and relative space, time and motion; for he held that reason and the stability of scientific theory alike required the existence of dimensions that are universal and unchanging: the space and time that would exist if no material bodies existed.[2] It was this distinction that was overthrown by Einstein.

---

[1] It should be noted that already in the first edition (and more firmly in later ones) Newton attempted to dissociate the word "attraction" from its esoteric, Keplerian connotations. It signified for him an effect, not a cause, which might possibly be impulse or something else.

[2] The distinction between absolute and relative motion furnishes Newton's proof of the absolute motion of the earth and planets and the (relative) fixity of the sun; that is, a proof of the Copernican hypothesis. For, as the planets exhibit a centrifugal force from the sun, they must be in absolute rotation about it.

After the Definitions follow the Laws of Motion and their corollaries, the contingent principles of nature upon which the consequent mathematical theory is based.[1] In both Book I and Book III (in the 'Rules of Reasoning') Newton stated explicit principles of scientific method, but not less influential was his implicit exemplification of a way of proceeding in science that was at once theoretical and experimental, mathematical and mechanical. Newtonian theoretical science was weakened neither by the loose articulation of Cartesian natural philosophy (since it was cemented step by step by geometrical demonstrations) nor the latter's arbitrariness (since its conclusions were verified by experiment or observation). It proceeded with almost as complete mathematical continuity from the Laws of Motion as the geometry of Euclid did from his Axioms;[2] in accordance with his dictum that the first stage in science was to investigate the forces of nature from the phenomena, Newton created the whole structure of dynamics in a formal manner, without needing to justify any proposition on merely empirical grounds. And this mathematical theory was also a mechanical theory because (as was remarked before) Newton believed that the phenomena of nature are caused by the play of natural forces upon the particles of which all matter is composed, giving them various degrees and types of motion. Though in the *Principia* he analysed mathematically the effects on phenomena of only one force, the force of gravity, there can be no doubt that he believed that there were other forces, acting on the material particles to cause the cohesion and dissolution of matter, chemical and electrical phenomena and so forth. Once the laws of these other forces were known their mathematical theory also could be formulated; thus the *Principia* was no more than a part of a vast unfolding of mechanistic, mathematical science that Newton envisaged but could never complete. To his contemporaries he offered only the hints for the continuation of such a work that are found mainly in the *Quaeries* in *Opticks*. The *Principia* is emphatically the "mechanical philosophy" already developed by Galileo, Descartes and especially Robert Boyle set in a strict

---

[1] Law I is the law of inertia; Law II states the proportionality of force and motion; Law III states the equivalence of action and reaction.

[2] The continuity is in fact broken by the need to introduce subsidiary physical assumptions about the particulate structure of gross bodies, and about the properties of fluids, etc.

mathematical framework, whereby Newton succeeded in rendering exact what his predecessors had left loose and qualitative. Book I deals, firstly, with the dynamics of particles considered as mass-points, and then with physical bodies made up from such particles; in Book III it is shown that the theory for such bodies applies in astronomy. And Book II is of course the theory of fluids considered as particulate in structure.

Newton's object in writing the *Principia* was not limited to verifying the Copernican system, nor even proving the truth of the theory of universal gravitation by bringing about that synthesis of astronomy and mechanics towards which the seventeenth century had been groping since the time of Galileo and Kepler. For Newton the mathematical, mechanical principles of natural science were not merely made manifest by the kinematics of Galileo, nor the descriptive laws of Kepler, nor even by his own majestic theory of the planetary motions: they were to be traced in the whole of physical science, in the structure of matter and the operations of chemistry. This explains the inclusion in the *Principia* of the groups of propositions on the elasticity of gases and the refraction of light, which have nothing to do with the main course of its argument, and more generally of Book II which (except at its close, where the Cartesian theory of vortices is refuted) appears at first sight as a mere interruption between the general theory of Book I and its application in Book III. Indeed, Newton at first intended his work to consist only of Book I, and a shorter, less technical section corresponding to the present Book III. But he finally resolved to write not a treatise on celestial mechanics but *The Mathematical Principles of Natural Philosophy*. Book II is the justification of this title, apart from celestial mechanics. Newton chose to illustrate the mathematical principles in a second field of mechanics, the theory of fluids, for the very obvious reason that celestial mechanics is appropriate to a world of vacuity, while in Book II Newton entered the world of terrestrial experience, a world filled with fluids such as air and water where phenomena occur not in a vacuum but in a plenum. If he was less successful in this Book than in the rest of the *Principia* it was because of the novelty and intrinsic mathematical difficulty of the subject.

Thus the *Principia* is a treatise on physics in a broad sense, in which moreover Newton made meticulous use of the quantitative

experimental method, as in his earlier optical investigations
(similarly, a decade later in *Opticks*, he was to emphasize that the
theory of light is a mathematical theory). He showed that the
theory of motion in resisting, fluid mediums could be made to
square precisely with experiments on the oscillation of pendulums
and the fall of heavy bodies in air and water.[1] Here Newton took
an important step forward in indicating how, mathematically,
the over-simple, highly abstract mathematical physics of Galileo
could be adjusted to the more complex real world; he proved that
with number and measure science could reach beyond the uncontrolled imagination of a Descartes, or even the idealism of a
Galileo. In Newton's eyes, scientific comprehension was not
limited to vague qualitative theories on the one hand, or definite
statements about a state of affairs much simpler than that which is
actually experienced on the other; it could proceed, by due
techniques, to definite ideas about all that is physical, down to the
properties of each constituent corpuscle. To illustrate this conception of science is the purpose of the *Principia*.

Otherwise the framework of celestial mechanics would have
been without foundation. Here Newton and Descartes were more
alike than the crude antithesis of their cosmologies would allow.
Descartes had proceeded, in the *Principles of Philosophy*, from his
clear ideas of what must be, through the laws of motion and the
properties of moving bodies, to his celestial mechanism. Newton
likewise: developing his mathematical method from the definitions
and the laws of motion, through the long analyses of the motions
of bodies in many different circumstances until he could discern
in the heavenly motions special cases of those principles of motion
that he had already elucidated. Newton perceived, as Descartes
had done, as Huygens did when he all but abandoned Descartes,
that a theory which attributed the most noble and enduring
phenomena in nature to the play of mechanical forces could not
stand on a handful of assumptions, two or three happy computations and the vague, undemonstrative, mechanistic philosophy
that had become fashionable.[2] Like Descartes, but with an

[1] In fairness it must also be pointed out that some of the more intricate
investigations—e.g. into the speed of sound in air and the flow of liquids under
pressure—did not (at least in the first edition) conform to experiment.

[2] In this, of course, Newton powerfully revealed his immense superiority
over Hooke, who could never have constructed the laborious scaffolding upon
which Book III of the *Principia* is raised.

infinitely more subtle logic, with all the rigour of mathematics and with cautious appeal to observation and experiment, Newton displayed the whole science of matter-in-motion before he turned to the solar system specifically. An unfinished treatise, *De Motu Corporum*, preceded the *Principia*; the study of a particle in motion must precede that of a circling planet. And if Newton found it necessary to investigate the solid of least resistance, or the flow of liquids, it was to prove the universality of that science of moving particles that he proposed to apply, not to a minute part, but to the whole of man's physical environment. The *Principia*, in fact, does not expound a particular scientific theory to account for the motions of the heavens: it develops a theory of physical nature which embraces these phenomena, and all phenomena of matter-in-motion, within its compass.

Nevertheless, Newton did not exclude God from the universe:

> This most beautiful system of the sun, planets, and comets, could only proceed from the counsel and dominion of an intelligent and powerful Being. . . . He endures forever, and is everywhere present; and by existing always and everywhere, he constitutes duration and space.[1]

God was for Newton the Final Cause of things, but, excellent and laborious theologian as he was, he made no confusion between physics and theology.[2] That Newton seemed, by the theory of universal gravitation, to contravene the principles of mechanism, was due to misapprehension. Though certain phrases in the *Principia* might seem to indicate the contrary, he did not believe that gravity was an innate property of matter, nor that two masses could attract each other at a distance without some relationship. To Bentley, Newton wrote:

> It is inconceivable that inanimate brute matter should, without the mediation of something else, which is not material, operate upon and affect other matter without mutual contact. . . . That gravity should be innate, inherent, and essential to matter, so that one body

[1] *General Scholium*, concluding the *Principia*, Cf. Florian Cajori: *Sir Isaac Newton's Mathematical Principles of Natural Philosophy* (Berkeley, 1946), pp. 544–6.

[2] Newton, however, expressed a sentiment typical of the age in a letter to Bentley (1692): 'When I wrote my treatise about our system, I had an eye on such principles as might work with considering men for the belief of a Deity; and nothing can rejoice me more than to find it useful for that purpose.' (*Ibid.*, p. 669).

may act upon another at a distance through a vacuum, without the mediation of anything else, by and through which their action and force may be conveyed from one to another, is to me so great an absurdity that I believe no man who has in philosophical matters a competent faculty of thinking, can ever fall into it.[1]

Of the actual cause of gravity Newton always professed himself ignorant. It was sufficient to infer that the phenomenon existed and was universal; like Galileo, Newton regarded the effect as established if it could be described, though the cause were hidden. To suppose that particles or masses exert a gravitional attraction was not, therefore, in Newton's language to postulate an occult quality in matter but to describe a fact—a fact that was to be demonstrated in the laboratory by Henry Cavendish seventy years after Newton's death. Nor did Newton believe that the celestial spaces across which the sun's attraction holds the planets in their orbits were necessarily empty of all matter, though it is hard to see how such extremely tenuous matter could exert the force of gravity or transmit light.

In fact Newton on several occasions (most notably, once more, in the *Quaeries*) framed hypotheses explaining gravity and light by aetherial mechanisms, postulating for example the existence of a mutual repulsion among the particles of aether. The concluding paragraph of the General Scholium added to the second edition of the *Principia* (1713) was Newton's last and most puzzling suggestion of this kind:

Something might now be added about a certain very subtle spirit that pervades all dense bodies and is concealed in them, by whose force and actions the particles of bodies attract each other when separated by very small intervals, or cohere when contiguous; and by which electric bodies act at greater distances, both repelling and attracting neighbouring corpuscles; and by which light is emitted, reflected, refracted and inflected, and heats bodies; and by which all sensation is stimulated, and the limbs of animals are moved at will—for this is done by the vibrations of this spirit transmitted through the solid capillaments of the nerves from the external organs of sensation to the brain, and from the brain to the muscles. But these things cannot be explained in a few words, nor have we at hand a sufficient number of experiments by which to determine and demonstrate the laws of action of this spirit accurately, as ought to be done.[2]

[1] Quoted by Cajori, *op. cit.*, p. 634.
[2] Newton later added the words "electric and elastic" before "spirit", which will be found in the English translations.

Newton's scientific imagination was not less fertile than Descartes's. But he probably realized that the existence of forces between aetherial particles would be just as mysterious as their existence between material particles, for he seems to have regarded these hypotheses as no more than devices to satisfy those who demanded mechanical explanations of everything. Besides, he had no ambition to make the universe so perfectly mechanical that there was no room for God left in it—though that was how the Newtonian universe ultimately came to be. In the last resort Newton concluded (apparently) that the physical forces such as gravity and magnetism were natural but not mechanical; not innate attributes of matter because God willed them, and yet not miraculous because they were part of the normal order of things. As such they seemed to him inscrutable.

His refusal to ascribe a cause or mechanism to universal gravitation was indeed one of Newton's principal advantages in celestial mechanics. He was free, as the Cartesians were not, simply to state and analyse the observable facts, and the inferences necessarily drawn from them. He did not seek to construct a model which would be rendered clumsy and contradictory by the very attempt to explain everything in nature by corpuscular mechanisms.

> Hypotheses [wrote Newton], whether metaphysical or physical, whether of occult qualities or mechanical, have no place in experimental philosophy. . . .[1] And to us it is enough that gravity does really exist, and act according to laws which we have explained, and abundantly serves to account for all the motions of the celestial bodies, and of our sea.[2]

Nevertheless, his attitude (less plain certainly in the first edition of the *Principia* than it has since become) was widely misunderstood. Newtonian mechanics was rapidly accepted by his own countrymen: probably no scientist has received a more immediate or a warmer acclaim from the intellectuals as well as the professed scientists of his race. Abroad it was distrusted and Newton was accused of introducing esoteric explanations into science because he did not furnish a mechanical cause of gravity Neither Huygens

---

[1] Newton, of course, did not mean that tentative hypotheses have no use in an investigation; he framed many such himself. He meant that an unconfirmed and undemonstrated hypothesis should not be taught as an adequate theory.

[2] Cajori: *op cit.*, p. 547.

nor Leibniz (who set the tone for many lesser men) could stomach the downright statement of Proposition VII, Book III.[1] Attempts to reconcile the Cartesian mechanical theory of celestial vortices with Newtonian mathematical laws were prolonged into the mid-eighteenth century. Not until fifty years after the publication of the *Principia* did Voltaire's proclamation of his admiration for the profound English geniuses, Newton and Locke, begin to win adherents. The essential truth, that Newton and Descartes shared the same idea of nature, was thus long obscured; and Newton has perhaps been too often praised for being other than he was. The idol of perfection who was endowed by the nineteenth century with every attribute of scientific insight and vigour, with abhorrence of hypothesis and mystery, with serene temper and conventional religion, was not the genuine Newton. It is perhaps paradoxical—but not unjust—that his greatest successor was to arise not from the crowd of reverend English gentlemen who were to claim Newton as their own, but in the person of the sceptical French mathematician, the Marquis de Laplace, whose *Mécanique Céleste* (1799–1825) extended in time the laws that Newton had traced in space.

[1] "That there is a power of gravity pertaining to all bodies, proportional to the several quantities of matter which they contain."

CHAPTER X

DESCRIPTIVE BIOLOGY AND SYSTEMATICS

BIOLOGICAL study, as it is practised today in laboratories and field stations, is essentially a creation of the nineteenth century. The work of Darwin on evolution, of Mendel on genetics, of Schleiden and others on the cell theory, so transformed the texture of the biologist's thought that it would be appropriate to attribute to the period 1830–70, rather than to any earlier age, the "biological revolution" which completed the modern scientific outlook. The belief in the fixity of species was no less respectable than the belief in the fixity of the earth; the belief that the Creator must have personally attended to the fabrication of every kind of diatom and bramble was no less primitively animistic than the belief that His angels governed the revolutions of the planetary orbs. Exactly as the mechanistic philosophy of the seventeenth century was accused of encouraging scepticism and irreligion, on a greater scale (because the issue was more clear and more decisive) the mechanistic biologists of the nineteenth met the full force of ecclesiastical wrath. The liberty of the scientist to direct his theories in accordance with the scientific evidence alone was equally at stake. But there is this difference. Biology was certainly "modern"—in some respects if not all—before the nineteenth century. A great renaissance had already occurred, which itself far surpassed all that had gone before. Materials had been heaped up from which a great generalization such as evolution could be drawn. Above all, the scientific method of biology was already in existence—that was not the creation of the nineteenth century. The researches of Leeuwenhoek and Malpighi, the systematics of Ray and Linnæus, were preliminaries as essential to the syntheses which introduced the truly modern outlook as the work of Copernicus and Galileo was to that of Newton.

None of the ancient founders of biology was primarily interested in collection, description and classification as ends in themselves. Aristotle the zoologist and Theophrastus the botanist were always philosophers—their purpose was to investigate the functioning of living organisms; Dioscorides studied botany as the servant of medicine. Partly, perhaps, because the range of species examined was comparatively small—neither Aristotle nor Theophrastus knew more than about five hundred distinct kinds of animals or plants—the problem of cataloguing them did not become of overriding importance, though much thought was given to order and arrangement. Since the Greek empire extended into India, exotic species were available, but they did not attract great attention.[1] To the Greek mind, the attempt to answer the questions that living nature posed was more important than the compilation of information, and for this the materials close at hand were sufficient. Over-leaping a great space of time, in the last century collection and taxonomy have again become no more than specialized branches of biology. The study of function, of the processes of growth and differentiation, has assumed a more fundamental importance. The experimental has replaced the encyclopædic method, so that a modern zoologist may find a greater interest in the works of Aristotle than in those of any natural historian of the pre-Darwinian age.

The intervening period has, indeed, very special characteristics. For long there were no adequate successors to the Greek botanists of the fourth century B.C. The Romans were competent writers on agriculture, but such an author as Pliny added nothing beyond the cult of marvels to the existing texts which he pillaged. The philosophic spirit of the Greeks almost perished, and was only revived in the botanical work of Albert the Great (*De Vegetabilibus et Plantis*, *c.* 1250). Albert was an Aristotelean botanist—at least, his main authority was a translation of two books on plants then attributed to Aristotle.[2] He was interested in the philosophy of plant growth, in the variety of their structures and (as he believed) in their constant mutations. Care in the morphological analysis of plants for purposes of description and identification was combined with renewed attention to the problem of classification, but Albert was not greatly impressed by the importance of cataloguing. Such an emphasis was then unusual, for in his time herbalism—medical botany—had already become a principal interest. It is a strange paradox that while the learning of the

---

[1] One important exception was the date-palm, which provided the only example of sexuality in plants known before the late seventeenth century.

[2] But now assigned to Nicholas of Damascus (first century B.C.).

later middle ages turned so naturally to argument in metaphysics and philosophy, and thence to logic, cosmology and physics, the intellectual problems posed by the living state were so often ignored. Only in its relations with medicine can medieval biology be generally said to have had a serious intellectual content, to have attempted to answer questions.

Herbalism looked to Dioscorides, rather than to Aristotle and Theophrastus. Before the fall of Rome the tradition he founded had already suffered debasement and the decline, both in matter and in illustration, continued throughout the early middle ages. In the thirteenth century, however, there were already skilful herbalists with a good knowledge of Dioscorides and his commentators, some familiarity with exotic drugs, and an interest in description and identification. The herbal of one of them, Rufinus, serves to show that he, at least, did not scruple to add remarks of his own to the literary tradition, and that he was aware of distinctions in kind unknown to the more famous compilers of the sixteenth century.[1] Rufinus was clearly well acquainted with drug plants and druggists, but he made no attempt to classify, merely arranging his notes in alphabetical order. The greater part of his text was made up of quotations from earlier pharmacological authorities (Dioscorides, the *Circa instans* of about 1150, the *Tables* of Salerno, and others), but Rufinus' own additions were mostly botanical, such as this description of *Aaron's Beard*:

> Aaron's Beard has leaves which are thick in substance, nearly a span broad and long, and it has two little beards to each leaf. The leaves are divided down to the root. It has a tuberous root in the ground, from which a cosmetic ointment is made, and it sometimes has blotched leaves. It forms its flower in a capsule, contrived by a marvellous artifice and having this yellowish capsule around it, in the centre of which is a sort of finger, with two little "apples" below it, wonderfully contrived. The plant which has blotched leaves is masculine and that without blotches on the leaves is feminine.[2]

His manuscript was apparently unillustrated, so that the identification of uncommon plants from it would have been very doubtful.

The herbal flourished, to become enormously popular soon after the invention of printing. But the herbalist's interest in the plant was always in knowledge of means to an end. Some of his

[1] Lynn Thorndike: *The Herbal of Rufinus* (Chicago, 1946).
[2] *Ibid.*, p. 54.

medicaments were minerals, or derived from animal sources, and it was only because such a large proportion of medieval physic was derived from vegetables, that the pharmacopœia assumed a preponderantly botanical form. Thus descriptive zoology was a poor relation of herbalism, though animals were also described as the immediate companions and servants of man, because they offered useful moral lessons, and because some of them had an exotic or symbolic fascination. Conceiving that the world was created for the use and instruction of man in working out his own salvation, the medieval mind naturally adopted a somewhat functional approach to the living state. The task of the naturalist was simply to describe living things, with their particular uses (or wonders, or edifying properties) so that other men might use them (or wonder at them, or be edified). Despite the occasional philosophic questioning of an Albert, there was no powerful motive to elevate him above a lexicographical mentality. And the naturalist was less interested in collecting facts about creatures that might form the material of a science, than in human reactions to this and that, in the diseases against which a given plant was supposed to be beneficial, in the moral to be drawn from the habits of the ant-lion.

Thus the origins of natural history were essentially anthropocentric, in the Roman Pliny, in the early Christian compilers like Isidore of Seville, in the thirteenth-century encyclopædia of Bartholomew the Englishman, in the late medieval herbalists. Human interest in nature was limited to the production of a *catalogue raisonné*.

The early stages of the renaissance brought no important reorientation. Occasionally the representational art of a "Gothic" stone mason or wood carver had enriched a cathedral with a recognizable likeness of a living species. About the beginning of the fifteenth century the graphic artist began to realize the æsthetic possibilities of exact imitation of nature in the illumination of manuscripts—here were the roots of both the naturalistic art of a Dürer, and biological illustration. By 1550 the technique of life-like illustration had been mastered, with greatest distinction in the herbals of Brunfels (1530) and Fuchs (1542). This technique was ultimately as necessary to botany and zoology as to human anatomy, but it did not occasion any immediate enhancement of the level of botanical knowledge, for the texts of both Brunfels' and Fuchs' books were poor, and excelled in description by unillustrated works, such as that of Valerius Cordus. Brunfels, indeed, tried to find some more natural arrangement than that of an alphabetical list, but the latter was by no means abandoned as yet. The botanists of the sixteenth century, with the exception of Cesalpino and Gesner, were still herbalists, and the herbal was still an adjunct to the pharmacopœia, enabling the apothecary to identify such medicinal plants as Swallow-wort and Fennel, Sage and Fumitory, whose names are perpetuated on the delightful majolica drug-pots of the time.

Humanism had its effect upon biology, as upon all branches of science, without challenging the main emphasis on collection and classification. The authority of Dioscorides and Theophrastus was reinforced rather than weakened; their texts were better under-stood, but did not encourage originality in ideas. Mediterranean botanists particularly took up the task of identifying more exactly the species described by Dioscorides; some, like Mattiolo, Cordus and Conrad Gesner, were content to put forward their own work as expansions of his, with considerable display of philological learning. Gradually it was learnt that Greek names had been abused by application to species quite different from those known to the Greeks themselves; and that, moreover, the name often covered a whole group of similar plants, not a specific type. Northern botanists, on the other hand, acquired knowledge of plants not included in the traditional Mediterranean flora; Charles de l'Écluse alone is reputed to have found two hundred new species in Spain and Portugal (1576), and later he was equally successful in Austria and Hungary. Cataloguing and description were extended far beyond the range of the merely useful. Decora-tive plants, like the daffodil and horse-chestnut—this last one of many importations into western Europe at this period—were noticed as well as the medicinal, along with many new species reported by the explorers to the Far East and the Americas. The common and uncommon plants of hedgerow, pasture and upland were no longer neglected. A garden was now judged by the multitude, rarity and beauty of the species represented in it, while the *Hortus Siccus* became a repository of trophies exchanged among collectors. For the men of the renaissance collected plants, plumages and skins as they amassed coins, antique statuary and manuscripts, since greater wealth and leisure permitted such costly and learned ostentations. The plants in some renaissance gardens, like that of the Venetian patrician Michieli (c. 1565), were commemorated in water-colour and written description.

Though the character of the product of this vastly increased activity in botany, or herbalism, was not greatly changed in the sixteenth century, the character of the new herbalist was certainly modified. As he attached less importance to medicinal value, he became more keenly interested in fine distinctions; whereas the ancient and medieval herbalist had hardly been concerned with a unit smaller than the genus, their successors began to discriminate between different species within the genera, and even between varieties of the same species. Again, the new naturalists were often scholars and gentlemen, they had therefore greater opportunities for botanizing over wide areas, even despatching emissaries for this purpose; they could acquire a more extensive literary knowledge and employ the best draughtsmen. Doubtless such men felt the æsthetic appeal of nature more keenly than the apothecary or peasant. As Fuchs wrote:

> There is no reason why I should expatiate on the pleasure and delight of acquiring knowledge of plants, since there is no one who does not know that there is nothing in this life more pleasant and delightful than to wander over mountains, woods and fields garlanded and adorned with most exquisite little flowers and plants of various sorts. . . . But it increases that pleasure and delight not a little, if there be added an acquaintance with the virtues and powers of these plants.[1]

Fuchs' observation ends with a touch of that pedantry which very often divides the scientist from the artist; the scientific tendency is, after all, to dissect and destroy the thing of beauty, but there is no reason to doubt that the intellectual inquisitiveness which leads *via* microscope and herbarium to the unreadability of a *Flora*, may have its æsthetic foundation. This also links naturally with the urge to collect and preserve, the emphasis upon the rare and the expensive, which are so typical of biology from the sixteenth to the nineteenth century. Collector's mania has often been derided, yet it may yield genuine scholarship, as in (for example) the study of ceramics or bibliography. The botanist's character was more complex. He could claim that his activities were useful to man, and contributed to the worship of God. If Sir Joseph Banks'

[1] *De Historia Stirpium* (Basel, 1542), Preface, sig. α2v. Quoted by A. Arber: *Herbals* (Cambridge, 1953), p. 67.

attempts to transplant the breadfruit to the West Indies were vain, naturalists had great success with tobacco, the potato, maize and innumerable ornamental species. In nature they saw abundant evidence of Design, and so created the tradition which led through Ray's *Wisdom of God* to Paley's *Natural Theology, or Evidence of the Existence and Attributes of the Deity, collected from the Appearances of Nature.* There was thus a variety of arguments for commending biology to the attention of a serious and devout mind, of which medical utility was not the least important. Few naturalists in this period would have given whole-hearted support to the views of the Bohemian, Adam Zaluzian (1592):

> It is customary to connect Medicine with Botany, yet scientific treatment demands that we should consider each separately. For the fact is that in every art, theory must be disconnected and separated from practice, and the two must be dealt with singly and individually in their proper order before they are united. And for that reason, in order that Botany, which is (as it were) a special branch of Natural Philosophy, may form a unit by itself before it can be brought into connection with other sciences, it must be divided and unyoked from Medicine.[1]

The task of the descriptive biologist was also far more complex than that of the cataloguer of human artifacts, indeed, it was this complexity which enforced the development of systematics. Problems of nomenclature, identification and classification rather suddenly became acute between 1550 and 1650, and constituted one of the main theoretical topics in biology for nearly three hundred years. Naturalists tried to follow a "natural" order of groupings—which meant that they were long deceived by superficial characteristics. Aristotle had distinguished, in zoology, between viviparous and oviparous creatures, between the cephalopodia and other molluscs; Dioscorides had distributed plants among the four rough groups of trees, shrubs, bushes and herbs. Lesser distinctions, between eggs-with-shells and eggs-without-shells, between deciduous and non-deciduous, flowering and non-flowering, were also very ancient. In the main such distinctions were preserved as the basis of arrangement until late in the seventeenth century. Nomenclature was equally in need of reform, if standardization was to be obtained, and the name to have a logical connection with the system. Description was the very basis of a communion

[1] *Methodi Herbariæ Libri Tres*, quoted in Arber, *op. cit.*, p. 144.

of understanding in biology, for on it depended the hope of arriving at a single comprehensive *Flora* which would enable all men to agree upon the identity of any given specimen. Here the classical tradition was very frail, partly owing to the defects of its language in referring to the parts of animals and flowers.

No consistent answers to the problems of taxonomy were produced before the eighteenth century; even today the concept "species" cannot be exactly defined, and many systems of classification have succeeded that of Linnæus. Nevertheless, the great compilers of the sixteenth century, in their attempts to make an encyclopædic survey of all living things, more than mastered their Greek inheritance and demonstrated the fruits of exact observation. Their view of their undertaking was of course far from strictly biological. Thus Conrad Gesner, in his enormous *Historiæ Animalium* (published 1551–1621), besides naming and describing the animal, discussed its natural functions, the quality of its soul, its use to man in general and as food or medicine in particular, and gave a concordance of literary references to it. The Italian naturalist Ulissi Aldrovandi strove for even deeper omniscience when (for example) writing of the Lion he noted at length its significance in dreams, its appearance in symbolism and mythology, and its use in hunting and tortures. But Aldrovandi was also one of the first zoologists to give a skeletal representation of his subjects where possible. Along with the spirit of sheer compilation there developed a growing tendency to specialize, exemplified in Rondelet's book on Fishes (1554), in Aldrovandi's treatise on the different breeds of dog, in the Englishman Thomas Moufet's *Theatre of Insects* (1634).[1] All these works, and some portions of the vast encyclopædias, were written with conspicuous attention to the kind of detail that could only be obtained through systematic personal observation. Most of the old fables debasing natural history—the birth of bees from the flesh of a dead calf, and of geese from barnacles, the inability of the elephant to bend its legs, and the tearfulness of crocodiles—were at least doubted, though they lingered long in popular books.

The classification of animals in accordance with Aristotle's scheme presented no great difficulties. The Latin names gave sufficient identification, superficial distinctions were marked. In the

----

[1] This, written about forty years earlier, was compiled from the work of Thomas Penny (*c.* 1530–88).

group of oviparous quadrupeds, for instance, Gesner had only a few divisions—frogs, lizards, tortoises—and he knew only three or four different kinds in each. Plants were more recalcitrant. Alphabetical lists had their uses, and so had others in which the groups consisted of plants having a similar habitat or function. When the attempt was made to render identification easier by adopting arrangements based on form and structure, more profound difficulties were encountered. In general, it seemed desirable to make the arrangement as natural as possible, by taking into consideration the maximum number of characteristics, but it was difficult to decide what the most important of these were. Reliance on superficial features, like the possession of prickles, or habits like climbing, was apt to prove very deceptive. The early systematists consequently tended to make increasing use of a single characteristic of the plant as a determinant—de l'Obel chose the leaf, and Cesalpino the fruit. One advantage of this method was that it led to the more intensive study of particular parts of the plant, especially the flower, and to the improvement of descriptive terminology. Such systems, of which Linnæus' was the logical and highly successful climax, were artificial, convenient indices to the prodigality of nature; but they did promote conscious study of the problems of taxonomy. Before 1550 there were hardly any firm principles by which species were distinguished, while the arrangement of the species was a matter for the discretion of each author. By 1650 there was a great measure of agreement on specific identities, and it was gradually becoming clear that there was a difference between a search for a *method*, which would make identification easy, and an endeavour to trace the natural affinities between species and larger groupings.

Attention to systematics was partly enforced by the sheer multiplicity of species. Some six thousand distinct plants had already been described by 1600, and the number trebled during the following century. Since it was the pride of the good botanist to be able to identify every plant presented to him, or if it were a new species to indicate its relationship to known ones, there were strong reasons for correlating identification and arrangement with one or more morphological characteristics. Caspar Bauhin, in 1623, outlined the natural groupings of botanical species more clearly than any of his predecessors, and made more extensive use of the binomial nomenclature in which one element of the name was shared by the *genus*, or group of closely related species. A little later Jung, at Hamburg, greatly improved the technical description of the disposition and shape of leaves, and of the various parts of the flower. A younger contemporary, the Englishman John Ray (1627–1705), laid the foundations of modern descriptive and systematic biology, in botany at least owing something to Jung's methods. Ray had some experience of dissection, but he was not an experimenter, nor a microscopist. Though his interests extended to the ecology, life-history and physiology of his subjects—thus he was much more than a plain cataloguer—he did not himself do much to advance the newer branches of biology growing up in his time. On the other hand, his philosophic and general scientific outlook was wider than that of most succeeding naturalists; like many other Fellows of the Royal Society, he was fascinated by technological progress, accepted the broad picture of a mechanistic universe under divine surveillance, and joined in the expulsion from biology of myth and mystery.

Ray was perhaps the first biologist to write separate treatises on the principles of taxonomy.[1] These were exemplified in his great series of descriptive volumes, the *Historia generalis plantarum* (1686–1704) and *Historia insectorum* (1710), with the *Ornithologia* (1676) and *Historia Piscium* (1686) in which he collaborated with his patron, Francis Willughby. Taken together—for all these books were actually finished and published by Ray—they represented by far the most complete and best arranged survey of living nature that had ever been attempted. Ray had exercised his keen faculty for observation intensively over the whole of England, and extensively over much of western Europe; he was deeply learned in the writings of ancient and modern naturalists; above all, he welcomed new ideas. From Grew he accepted as probable the sexual reproduction of plants; from Redi and Malpighi the experimental disproof of spontaneous generation; and he himself taught that fossils were the true remains of extinct species, not mere "sports" of nature nor God-implanted tests of man's faith in the truth of the Genesis story. If the enumeration of species was his principal task —which still left him room for his *Collection of English Proverbs*, *Topographical Observations*, and *Wisdom of God*—Ray was very far from supposing that classification was the end of biology.

In botanical systematics Ray favoured a "method" which was

[1] *Methodus plantarum nova* (1682); *Synopsis methodica animalium quadrupedum et serpentini generis* (1693); *Methodus insectorum* (1704).

more natural than those of his contemporary Tournefort and his successor Linnæus. He admitted that the familiar triple distinction between trees, shrubs and herbs was popular rather than scientific, though he continued to use it while also making the far more fundamental distinction between mono- and dicotyledonous plants. For finer discrimination he relied upon no single characteristic but appealed to the forms of root, leaf, flower and fruit. The necessity for a formal method of classification was fully apparent to him—it was particularly required by beginners in botany—

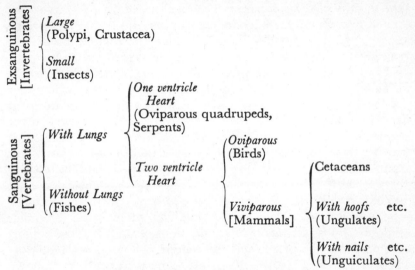

Fig. 10. Ray's Classification of Animals

but he did not expect that all living forms could be perfectly accommodated within it. Taxonomists would always have difficulty with 'species of doubtful classification linking one type with another and having something in common with both.'[1] In zoological classification Ray was perhaps even more successful, through basing his groups upon decisive anatomical features. He was the first taxonomist to make full use of the findings of comparative anatomy, particularly among mammals[2] and with regard to such characteristic features as feet and teeth, thereby discerning such groups as the Ungulates, Rodents, Ruminants, etc. (Fig. 10). This part of his work was freely adopted by Linnæus.

[1] Cf. the Preface to *Methodus Plantarum*, and C. E. Raven: *John Ray, Naturalist* (Cambridge, 1950), Ch. viii.
[2] This class was recognized by Ray, though not given this name.

Meanwhile, the naturalist's range of observation was being vastly extended by the microscope (Chapter 8). In the use of this instrument the primary emphasis was still on description; at this stage, attempts to construct elaborate theories upon the new evidence were infrequent and misleading. There was opportunity for the ramification of activity, and it was not neglected. The study of plant anatomy, originally enforced by the need for classificatory systems, could now proceed to the structure of tissues and reproductive mechanisms; zoological anatomy, likewise, stimulated by the fertility of the comparative method as shown by Harvey and many before him, was extended to strange creatures like the "orang-outang" (dissected by Dr Edward Tyson),[1] and, with the aid of the microscope, to levels of detail inaccessible to the naked eye.

Most of this new work prolonged existing tendencies. Marcello Malpighi (1628–94), for example, completed Harvey's discovery of the circulation of the blood by following its passage from the arterial to the venous system through the capillary vessels, at the same time observing its red corpuscles. He was also able to go farther than Harvey and Fabricius in examining the microscopic foetus of the chick within the first hours of incubation, from which he was led to believe that growth was a process of enlargement or unfolding only: the foetus was "pre-formed" in the unfertilized egg.[2] As a pioneer of histology Malpighi entered on less familiar ground, in his microscopic examinations of the liver, the kidney, the cortex of the brain, and the tongue, whose "taste buds" he discovered. In the study of insects—where Aristotle had shown wonderful insight—the serious scientific curiosity in which Malpighi was joined by Jan Swammerdam (1637–80) had already been anticipated by Hooke in *Micrographia*, and by even earlier virtuosi with their "flea-glasses." These two naturalists, however, were the first to explore fully the internal anatomy of minute creatures, demonstrating that their organs are as highly differentiated as those of large animals. Malpighi's treatise on the silkworm has been described as the earliest monograph on an invertebrate; in it he indicated the function of the *tracheæ* first observed by him, which distribute air about the insect's body, and of other tubes by

---

[1] Cf. M. F. A. Montagu, *Edward Tyson, M.D., F.R.S. 1650–1708* (Memoir XX, Amer. Phil. Soc., Philadelphia, 1943). The creature was in fact a chimpanzee. Tyson also published monographs on the "porpess", rattlesnake, opossum etc.

[2] Joseph Needham: *History of Embryology* (Cambridge, 1934), pp. 144 *et seq.*

which the products of metabolism are excreted. He did much work on the anatomy of the larval stages of insects, and observed their evolution to maturity, but here he was excelled by Swammerdam, who also denied that there was any true transformation, even in the emergence of the butterfly from the caterpillar, or of the frog from the tadpole—processes which he studied with enormous care. In sheer technical skill—exemplified in the quality of his drawings as well as in the fineness of his dissection under the lens and his unique methods of injection—Swammerdam foreshadowed the greater manipulative resources of the mid-nineteenth century. Leeuwenhoek is chiefly remarkable for his work at much higher magnifications and the discovery of a new world inhabited by Infusoria and Bacteria (p. 241), but the ubiquitous curiosity which led him to examine hairs, nerves, the bile, parts of plants, crystals —indeed almost everything that could be brought before his lenses—induced him to make some observations comparable to those of Malpighi and Swammerdam, among which those on the compound insect eye and on ants were particularly novel. From observations on aphids he discovered parthenogenesis in animals —reproduction by the female parent alone.

In the plant kingdom, the microscope could not reveal a new order of magnitude within the living state, as it did in the animal; on the other hand, a much clearer idea of the structure of plant tissues emerged—including the description of their minute components, the cells—than was yet obtained in zoology. The presumed anatomical and physiological analogies between animals and plants were indeed powerful incentives to inquiry at this time. Sometimes analogy was wholly misleading, as with the theory (popular until disproved by repeated experiments) that the sap in plants circulates like the blood in animals, but in other aspects, as when the "breathing" of plants was compared with that of animals by Malpighi and later by Stephen Hales (1679–1761), it led towards a more correct understanding. Malpighi, despite the excellence of his descriptions of the differing structures found in wood, pith, leaf and flower under the microscope, and of the germination of seedlings, thought too exclusively in terms of the animal form. Thus he wrongly identified the function of the spiral vessels that he observed in plant tissue with that of the tracheæ in insects, and erected upon this identification a broad theory of the increasing specialization of the respiratory organs, reaching its climax in mammals. He also tried to find in plants the reproductive organs familiar from vertebrate anatomy. The Englishman, Nehemiah Grew (1641–1712), whose independent work is closely parallel to that of Malpighi, and of equal quality, was a more restrained observer, though he believed (as he quaintly wrote) 'that a *Plant* is, as it were, an *Animal* in Quires, as an *Animal* is a *Plant*, or rather several *Plants*, bound up into one volume'—a remark which, however strange the metaphor, expresses profound intuition.

Grew was well aware, not only that the attainments of ordinary naturalists still fell far short of their aims, but that these aims by no means amounted to a true "*Knowledge* of *Nature*." His *Philosophical History of Plants* (1672)[1] sketched a new and more ambitious programme. Many of the problems he proposed remain unsolved:

> First, by what means it is that a *Plant*, or any *Part* of it, comes to *Grow*, a *Seed* to put forth a *Root* and *Trunk*. . . . How the Aliment by which a *Plant* is fed, is duly prepared in its several *Parts*. . . . How not only their *Sizes*, but also their *Shapes* are so exceeding various. . . . Then to inquire, What should be the reason of their various *Motions*; that the *Root* should *descend*; that its descent should sometimes be *perpendicular*, sometimes more *level*: That the *Trunk* doth *ascend*, and that the ascent thereof, as to the space of *Time* wherein it is made, is of different *measures*. . . . Further, what may be the Causes as of the *Seasons* of their *Growth*; so of the *Periods* of their *Lives*; some being *Annual*, others *Biennial*, others *Perennial* . . . and lastly in what manner the *Seed* is prepared, formed and fitted for *Propagation*.

Some of these questions Grew himself tried to elucidate, most brilliantly deducing that plants reproduce sexually, the flowers being hermaphrodite like snails, with the stamens acting as the male organs.[2] Nor did he neglect the possibility of examining the plant substance by combustion, calcination, distillation and other experimental methods of chemistry, though these were as yet too primitive to be of real service. In this way he showed that the matter of the pithy or starchy part of the plant was quite distinct from that of the woody or fibrous part. Like Ray and other naturalists Grew saw no reason to reject mechanism as a working

[1] Reprinted in *The Anatomy of Plants* (London, 1682).
[2] *Op. cit.*, pp. 171–3. Hermaphroditism is not, of course, universal among plants, as Grew thought.

hypothesis which he developed (for example) in his account of plant nutrition; as he put it, with a familiar simile:

[We need not think] that there is any Contradiction, when *Philosophy* teaches that to be done by *Nature*; which *Religion*, and the Sacred *Scriptures*, teach us to be done by *God*: no more, than to say, That the Ballance of a *Watch* is moved by the next *Wheel*, is to deny that *Wheel*, and the rest, to be moved by the *Spring*; and that both the *Spring*, and all the other *Parts*, are caused to move together by the *Maker* of them. So *God* may be truly the *Cause* of *This Effect*, although a Thousand other *Causes* should be supposed to intervene: For all Nature is as one Great *Engine*, made by, and held in His Hand.[1]

A general sketch of the horizon in biology about the year 1680 would show virile activity, a steady expansion of the sphere of interest, and the fruitful exploitation of new techniques. Admittedly Man was still the prime focus of attention, whether in the Royal Society's endeavour to introduce a scientific spirit into agriculture, or in the relics of the belief (still held by a plant-anatomist like Grew) that all vegetables have "virtues," or in the frequent backward glances of the zoologist at the human body. Nevertheless, as the peripheries rapidly became more remote they assumed, as it were, a territorial autonomy. The survival of anthropocentricity in the feverish concentration of Swammerdam was small. It is significant that naturalists no longer defended their preoccupations as useful, but rather as contributions to knowledge of the universe, of the organic part of the divinely created machine. And, though description and cataloguing of macroscopic flora and fauna remained their principal tasks, natural history showed clear signs of entering into partnerships in which the skills of the human anatomist and physiologist, the chemist, and the physicist, should be placed at its service. Gradually, through the seventeenth century, biology had returned to the philosophic attitude of an Aristotle; now it seemed likely that the borrowing of modern knowledge and techniques would permit the ancients to be as greatly excelled in these sciences as in physics and mechanics.

Briefly, there was promise—a promise of growth in depth and extent that was hardly fulfilled during the next century and a half. That it was not fulfilled may be attributed partly to the

---

[1] *Op. cit.*, p. 80.

fallaciousness of the early hopes, for neither microscopic technique, nor chemical experiment, were capable of changing the pattern of activity so permanently as the work done during the two decades 1660–80 would suggest. These crude tools were soon blunted. The close connection between biology and medicine, which had encouraged study of the former science in the seventeenth century, tended to hamper its later development, for as medical studies were permeated by the influence of Galen's ideas until the nineteenth century, it was impossible that animal and plant studies should escape the limitations of those ideas. Unable, as yet, to build freely upwards upon the half-finished foundations of their predecessors, eighteenth-century naturalists might well be discouraged by the splendour of their inheritance. Discouragement was all the more harsh because this inheritance included such a feeble element of hypothesis to serve as a scaffolding for their own researches. Thereafter it is not surprising that they felt strongly a positive attraction that was both old and new. Like the sixteenth-century encyclopædists, they were subjected to a vast incursus of new species, fruit of a renewed urge towards exploration that drove Linnæus into the sub-arctic tundra, and Joseph Banks to the Pacific and Australasia. Moreover, this invasion synchronized, not with a sense of confusion before the profligacy of nature, but with an increasingly dogmatic confidence in a System, the system of Linnæus. Quite suddenly, about the middle of the century, classification became one of the easiest, instead of one of the most difficult, biological exercises. Not for the first or the last time in science there was a rush to gather the harvest, while the unbroken fields were neglected.

Admittedly, there was no very abrupt transition. In the early part of the century Leeuwenhoek was still active, and a little later Réaumur, one of the most versatile experimentalists of any age, began to publish his monumental *Mémoires pour servir à l'histoire des Insectes* (1737–48) which continued the work of Malpighi and Swammerdam. Low-power microscopy was applied to aquatic subjects by Trembley, who experimented upon the capability for regeneration and asexual reproduction by budding that he discovered in fresh-water "polyps" (*Hydra* and *Plumatella*),[1] and by Ellis (*Natural History of the Corallines*, 1755). Plant physiology

[1] *Mémoires pour servir à l'histoire d'un genre de polypes d'eau douce* (1744); cf. J. R. Baker: *Abraham Trembley of Geneva* (London, 1952).

was studied by Hales with the aid of quantitative physical methods largely derived from Boyle (*Vegetable Staticks*, 1727). He measured the upward pressure of the sap in the roots of plants, the quantity of water absorbed by the root and transpired by the leaves, and the rate of growth in different structures. He proved that some atmospheric substance entered into the composition of plants. Such experiments he tried to elucidate mechanically by others on the capillary attraction of water in fine tubes and porous substances; somewhat similar ones were extended to the circulation of the blood in his *Hæmostaticks*. Further important contributions to physiology were made by the contemporary chemist-physicians G. E. Stahl, Friedrich Hoffman and Hermann Boerhaave, who also continued a seventeenth-century tradition.

The physiological processes involved in reproduction and the formation of the embryo, in particular, remained a matter for heated controversy, on much the same lines as in the seventeenth century. In J. T. Needham spontaneous generation had a new champion, who found microscopic animalculæ in boiled broth that was (as he thought) effectively sealed from the air (1748). He was answered in more precise experiments by Spallanzani (1767), but this question was not regarded as decisively settled up to the time of Pasteur. The great debate between Ovists and Animalculists was more widespread, and even less fruitful. Harvey had believed in epigenesis, that is, that the growth of the embryo proceeded both by the gradual differentiation of its parts, and by their increase in size: 'there is no part of the foetus actually in [the egg], yet all the parts of it are in [the egg] potentially.' The effect of microscopy, soon after Harvey's death, was to give immediate advantage to the alternative theory of preformation, according to which the embryo merely swelled from being an invisible speck which was from the first completely differentiated; as Henry Power said: ' So admirable is every organ of this machine of ours formed, that every part within us is intirely made, when the whole organ seems too little to have any parts at all.' Preformation was developed especially by Malpighi and Swammerdam. Since the embryo, among oviparous creatures, develops in the maternal egg, and microscopists believed that the first signs of its future form could be detected as soon as the egg appeared, it was naturally assumed by them that the embryo, or potential embryo, was solely derived from the female. This view conveniently opposed the unfashionable Aristotelean conception that the male, supplying the active "form," was the prime agent in generation, and the female responsible merely for the passive "substance" of the offspring. Aristotle seemed to be further confounded by the discovery of the mammalian ovum attributed to De Graaf (1672). This supposed discovery was premature—De Graaf saw the follicles since known by his name, and the true ovum was first described by von Baer a century and a half later. However, it brought about an essentially correct change of thought, to the view that both viviparous and oviparous reproduction begin with the fertilization of an egg formed in the female. According to the Ovists, the ovum contained the embryo not potentially but actually, and in the version of their theory known as *emboîtement* they supposed that this held within its own organs the ova of the next generation, and so on *ad infinitum* like a series of Chinese boxes: in the ovaries of Eve were confined the future forms of all the human race.

The discovery of spermatozoa opened up a contrasting but parallel theory. Leeuwenhoek, in one of his rare flights of hypo-thesis, suggested that these "little animals" were the living embryos, which were enabled to grow by transplantation into the egg: 'If your Harvey and our De Graaf had seen the hundredth part they would have stated, as I did, that it is exclusively the male semen that forms the fœtus, and that all that the woman may contribute only serves to receive the semen and feed it.'[1] He supported this doctrine by reference to well-known cases where the offspring was strongly marked with the characteristics of the male parent. Hartsoeker (1694) and Plantades (1699)—the last perhaps as a deliberate fraud—published illustrations of a "homunculus" enclosed in the head of a spermatozoon. *Emboîte-ment* was also taken up by the Animalculists in the eighteenth century. Rival interpretations of observations that were commonly very imperfect and carelessly recorded continued for over a hundred years. Some regarded the spermatozoa as products of corruption, like the eel-worms in vinegar, for it was only in 1824 that they were proved essential to fertilization by Dumas and Prévost; at about the same time the experiments of Geoffroy Saint-Hilaire on the production of monsters proved that the

---

[1] Letter to Nehemiah Grew, 18 March 1678. *Collected Letters*, vol. II, (Amsterdam, 1941), p. 335.

appearance of the embryo is not preformed or predestined. At an earlier date the scientist-mystic Swedenborg favoured epigenesis (*c.* 1740), and more powerful support came from the researches of Caspar Wolff at St. Petersburg (1768), who pointed out that, as in plants the rudiments of flowers develop from undifferentiated tissues and are at first indistinguishable from those of leaves, so in animals no miniature fœtus could be found in the earliest phase of development, after which the nervous system, the blood-vessels, and the alimentary canal were observed to arise in successive stages. Generally speaking, however, the reaction against pre-formation (and the consequent expiry of the Ovist-Animalculist controversy) did not occur until the close of the eighteenth century, although, as F. J. Cole has said, the admirable iconography of Malpighi carried its own refutation of its author's doctrines.[1]

In most other branches of biology the eighteenth century appears equally unproductive of truly creative investigation. In its more medical aspects, such as human and comparative anatomy, and human physiology, the successes of the seventeenth century were extended, particularly during the last two decades. Lavoisier, for example, was able as a result of his new theory of combustion to throw fresh light on animal respiration. But the greatest biological achievement of the period, that which won the greatest fame and attracted the greatest number of pupils, was certainly that of Carl Linnæus (1707–78). His mastery of the order of nature—'God,' as he complacently acknowledged, 'had suffered him to peep into His secret cabinet'—touched the imagi-nation of a generation already turning towards a romantic naturalism, which was soon to cherish Gilbert White and Thomas Bewick, to prefer landscapes to portraiture, and to talk contentedly of the noble savage. Linnæus was the prophet of Wordsworth. His arrogance, like Samuel Johnson's, enslaved admiration, while the confidence with which he wrote as though personally present at the Creation was the more acceptable (in that, in all respects, it reassured the somewhat conventional religious conscience of the age) because it counteracted the scientific agnosticism of Voltaire and the French *philosophes*. Most of the major advances of science have, in one way or another, imperilled the comfortable security of the popular understanding. One great merit of Linnæus' in-tellect was that, save in his great gift for classification, it was

---

[1] F. J. Cole: *Early Theories of Sexual Generation* (Oxford, 1930), p. 147.

remarkably undistinguished; he had neither the wish nor the power to *épater les bourgeois.*

That does not detract from the importance of his work. Linnæus was a sincere amateur of nature in all her moods; a competent teacher of many subjects in biology and medicine, able to inspire affection and devotion in his students; a voluminous and lucid writer. Like Newton, he possessed a strong, though not wholly attractive, character. Like Newton too, he saw the essentials of a problem and solved it. And he was only slightly less dominant than Newton in forcing future naturalists to follow the path that he had cleared.

The Linnean systems of classification embraced the animal, vegetable and mineral kingdoms, and even diseases. Of these the second was by far the most complex, and the most influential. It was derived by applying to the immense descriptive materials amassed by Ray, Tournefort and others of his predecessors the principle of plant sexuality firmly demonstrated by the experiments of Camerarius (1694). Thus plants were distributed into twenty-four Classes according to the number, proportions or situations of the stamens (male organs), and each Class further subdivided in accordance with the number of styles (female organs). The first three Classes in this system were Monandria (1 stamen), Diandria (2 stamens), and Triandria; the Class Polyadelphia had "stamens united by their filaments into three or more sets." In each Class plants with one style were assigned to an Order named Monogynia, those with two to the Order Digynia, those with three to the Order Trigynia, etc. Each Order, easily determinable by inspection of the flower, contained a number of genera, which groups Linnæus regarded as primary and as "naturally" distinct. Each genus had a brief description of the features common to all the species included in it, mainly derived again from the method of fructification. In further division into species the shape of the leaves and other characteristics became important. For example, Genus 696 in Linnæus' *System of Nature* is found in the Class Hexandria, Order Monogynia; it is called *Berberis*, and is distinguished by "Calyx 6-leaved, petals 6, with two glands at the base of each, style 0, berry superior, 2-seeded." In this genus Linnæus reported five species, one European, one Cretan, one Siberian, and two from Tierra del Fuego.

The main principles of this classification apparently became clear to Linnæus at an early age, and were worked out with the aid of the experience gained in the course of his journey through Lapland (1732) and during later European travel. The first version of the *System of Nature* (1735) was written in Holland when he was acting as botanical curator to Boerhaave. The illustrations of the principles were extended vastly in the subsequent editions which speedily issued from the press. (The tenth, of 1758, is now the standard work of reference.) The utility of the work depended very much upon the rigidly methodical, and extremely succinct, descriptions, which in turn were made possible by the use of a technical terminology that was largely of Linnæus' own creation. He also paid great attention to nomenclature:

> As I turn over the laborious works of the authorities, I observe them busied all day long with discovering plants, describing them, drawing them, bringing them under genera and classes: I find, however, among them few philosophers, and hardly any who have attempted to develop nomenclature, one of the two foundations of Botany, though that a name should remain unshaken is quite as essential as attention to genera.[1]

The rules which Linnæus drew up for coping with this problem anticipated in part those now accepted by international agreement, and within a few years his binomial system was universally accepted among naturalists.

The Linnean Order was determined by a purely mechanical procedure. To it might be assigned thousands of different plants. How were the further distinctions to be defined, and the lines drawn between mere differences of variety, those of species, and those of genus? Even post-Linnean taxonomy has not succeeded in drawing such lines firmly, and in practice Linnæus made much use of earlier experience. The concept of species is indeed something that can easily be grasped intuitively—Aristotle had it in perfect clarity—for in very primitive languages the *kinds* of animals and plants are named even though the general concept "plant" or "animal" is lacking. In surveying *kinds* of creatures, it is not difficult to see why dog, wolf and hyena are more like each other than any member of another group containing lion, tiger and panther. The difficulty arises as the analysis of what constitutes an effective likeness or unlikeness has to be made finer and

---

[1] *Critica Botanica*, translated by Sir Arthur Hort, Ray Society (London, 1938); Preface.

finer: should the wolf (which is interfertile with the dog) be placed in the same genus with the pekinese? Linnæus realized that though the species is the fundamental unit of taxonomy, the assignment of the generic boundaries is the classifier's most tricky task. Here he could offer no rigid rules, though he stated certain negative propositions, and therefore his success was empirical rather than theoretical. His system could be made to work, but because generic groupings depended upon one man's notion of significant similitude, it was far from infallible.

Although unable to define the characteristics of a species and a genus precisely, Linnæus did associate ideas with each of these concepts that were received as dogmatic truths up to the time of Darwin. Some had long been current without winning universal credence. The most important of them was the fixity of species, which Ray had not admitted. In Linnæus' view each species represented the descendants of an original entity, or pair of entities, individually created at the beginning of the world. Its flora and fauna had always been exactly as they are now, and hence the concept of species was justified (if not, in practice, defined) by common descent from a unique created form, just as humanity was defined by common descent from Adam. Equally, the disappearance of species was ruled out, despite the evidence of fossils, which were thus denied an organic origin.[1]

Until late in life Linnæus regarded the production of fertile, stable hybrids by crossings between members of the same genus as impossible; intergeneric hybrids were unthinkable. Shortly before 1760 new evidence led Linnæus to believe that new species could arise—have arisen—through differentiation by crossing. Perhaps after all only the ancestors of the Orders were created—but he never perfected this thought, of which he seemed ashamed, and though he withdrew from subsequent editions of the *System of Nature* confident statements that new species never occur, the recantation came too late. It is strange that the chief legacy of Linnean biology, the main scientific argument against Darwin, was a doctrine in which the mature Linnæus had himself lost faith.[2]

[1] Linnæus was more orthodox than Steno, Hooke, Ray and others who admitted that living species represented only by fossilized shells, teeth or bone had disappeared from the world, incompatible as this seemed with divine Providence, and the care taken to preserve all the land-animals at the time of the Flood.
[2] Cf. Knut Hagberg: *Carl Linnæus* (London, 1952), Chapter XII.

It is important to recollect that there was already, before the end of the century, significant opposition to the doctrine of the immutability of species. Evolutionary ideas were first applied to the formation of the earth's crust, not its peopling with creatures. Descartes and Leibniz, Burnet and Whiston, had each before 1700 devised an hypothesis to account for the separating out of rock and water from an incandescent mass or primitive chaos, and traced the depression of ocean-beds, the elevation of mountain ranges, to the action throughout long periods of purely mechanical forces, though the time-scale imagined was absurdly short. The first two of these ignored the story of the Flood completely. Fossils, if accepted as organic remains, enforced the conclusion that species had either changed, or disappeared. The great French naturalist, Georges Cuvier (1769–1832), who founded the science of palæontology, favoured the second hypothesis. Successive catastrophes— of which the Biblical Flood was the most recent—had swept the earth of life; successive creations had re-peopled it with new species. His influence, more than any other, reinforced that of Linnæus in deriding evolutionary ideas, and in creating the intolerant assurance in immutability which faced Darwin. But his compatriot Buffon (1707–88), a man of less pretentious scientific authority, whose main effort was directed towards descriptive biology in the forty-four volumes of his *Histoire Naturelle* (once immensely popular, now a dead weight in booksellers' basements) chose the alternative explanation. For Buffon, as for Aristotle and Ray, the living process became more complex by infinitesimal gradations: 'Le polype d'eau douce sera, si l'on veut, le dernier des animaux et la première des plantes.' The power to reproduce in kind, and to grow, he regarded as the prime characteristic of the living state. This power resided in "organic molecules," the basic units of both plants and animals; death was the dispersion of these molecules, nourishment their assimilation into the body of the creature. Denying preformation in all its aspects, Buffon asserted that the segregation of the "organic molecules" in the sexual organs was the cause of reproduction, just as, in a different way, it caused the generation of parasites. The organic molecules had been the same since the beginning of the world; but, as he stressed in his *Époques de la Nature*, the world itself had altered, evolved. In the Fifth Epoch elephants had roamed the far north; the earth was hotter in its youth, and of greater vitality. Animals were larger, witness the ammonites big as cartwheels, the huge tusks found in Siberia, perhaps even man was then a giant. In general, Buffon imagined that the change in nature was towards degeneration, but he was acute enough to see that where human purposefulness had intervened, species had been changed for the better. Bread-wheat, for example, was not a gift of nature, for it is unknown in the wild state; it is evidently a herb brought to perfection by man's care and industry. 'Our best fruits and nuts' he wrote, 'which are so different from those formerly cultivated, that they have no resemblance save in the name, must likewise be referred to a very modern date.' By selective breeding, man has 'in a manner created secondary species which he can multiply at pleasure.' Although, in such passages, Buffon seems to tread upon the heels of Darwin, and although the fixity of species (together with the taxonomical nicety attached to that doctrine) was alien to his mind, he made no great play with the idea of evolution. It was not, for him, an explanatory concept. He did not make it his task to account for the origin of specific differences.

Thus, in a variety of ways, the elements of modern biology grew out of the older natural history. Their growth was necessarily spasmodic since it was not autogenous, the stages of the transformation of the naturalist into the biologist being fixed by the availability of techniques and ideas borrowed from physics, chemistry, medicine and philosophy. In the sixteenth century natural history became a respectable branch of study; in the eighteenth it was moulded into a formal discipline by Linnæus; but the special quality of the seventeenth century, which made biologists of men like Harvey, Descartes, Hooke, Redi and Leeuwenhoek, who were not naturalists, was the rich opportunity for opening up new subjects, largely with the aid of imported methods. That the physical science, and general intellectual climate, of the eighteenth century had less in comparison to offer to the biologist was the chief reason for the failure of these new subjects to continue their startling initial progress.

The distinction between natural history and biology is not, of course, recent. Aristotle was a naturalist in the *History of Animals*, a biologist in the *Generation of Animals*. In the scientific renaissance, the writings of such an embryologist as Fabricius clearly fall into a different category from those of Gesner. But the distinction is one which the events of the nineteenth century finally made plain. Through this perspective, a man who studies the courtship of birds is certainly a naturalist; another who examines the differences between bird-blood and mammalian blood—who may not know a teal from a tern—is as certainly a biologist. The naturalist takes the whole rural scene for his province, and is primarily interested in creatures as individuals; the biologist is predominantly concerned to answer specific questions, and seeks general truths. The biological sciences are descriptive indeed, in the same sense, however, that the chemical sciences are descriptive for they use concepts like "genes," "evolution," "photosynthesis" which are as much the product of scientific inference as "atom" or "polymerization." Natural history, on the other hand, "describes" in the commonplace sense of the word; its concepts like "mating," "hunting," "feeding young" are those of ordinary thought, rarely the product of scientific inference. The naturalist is indeed a trained observer, but his observations differ from those of a gamekeeper only in degree, not in kind; his sole esoteric qualification is familiarity with systematic nomenclature.[1]

Natural history, as here differentiated from biology, seems at the present time to be of small and shrinking scientific significance. The naturalist has inevitably become dependent upon the deeper insight of the biologist so that (for example) questions of classification, after being his main concern since the sixteenth century, have been radically modified during the last hundred years by biological research into evolution and embryology. The introduction of new and more rigorous techniques, making large use of the experimental method, has carved from the naturalist's former province such subjects as ecology and animal psychology. In any case, the observations of a Gilbert White are no more repeatable than those of a Marcel Proust, and partly for the same reason that their enduring interest resides in individual qualities of imagination and literary skill applied to a particular social setting. The writing of natural history may continue, like that of the novel, ever different and ever the same, but the evolution from it of scientific biology could only happen once (or, more accurately, in successive

---

[1] These distinctions are, it is recognized, subject to innumerable qualifications. In practice, many biologists share the naturalist's attitude in part, and many naturalists the biologist's; more exact experimental methods are being applied to field-work, etc.

unique stages) just as the science of psychology could only emerge once from the physician's interest in mental disorders.

The emergence of biology was certainly the leading contribution of the scientific revolution to the study of living things. A totally new *kind* of knowledge was thereby made possible, supplementary to that already obtained through the renaissance of natural history in the sixteenth century. Necessarily the first steps in this emergence were mainly descriptive, in studies of comparative morphology, minute anatomy and physiology, and the like. The description of an animal or plant which permits its identification may, however, be very different from a description of the same creature in relation to an appropriate group of scientific inferences. The structure of the hoof and leg of the horse may be so depicted, or portrayed in words, that the member may be easily distinguished from that of a dog; but to describe the same leg of the horse in terms of its evolution by extension of the middle digit of the foot and reduction of the side-digits is to embark upon a very different procedure. Or, to take an example from mineralogy (since this was a part of natural history until recent times) when metallic ores have been classified and labelled under such names as malachite, chalcocite or chalcopyrite, a new kind of knowledge is brought in by assigning to them their appropriate chemical formulæ. Though these formulæ are merely descriptive of the composition of the minerals, they enable comparisons to be made which are otherwise hidden.

A mineralist who considers gypseous alabasters, plaster stone, lamellated gypsums, . . . and a great many other bodies as proper to be distinguished from one another, and who is able to ascribe any particular body to its proper species from considering its external appearance, is possessed of a particular kind and degree of knowledge: He who besides being acquainted with the external appearances, is able to prove that all these different bodies are composed of a calcareous earth, united to the vitriolic acid; and thus makes several species of things coalesce together, and unite, as it were, under one general conception, hath a knowledge of these bodies different in kind and superior in degree.[1]

In the former example, appeal was made to the concept of evolution, in the latter to the concepts of inorganic chemistry.

[1] Richard Watson: *Chemical Essays*, vol. V (London, 1787), p. 127. Cf. L. J. M. Coleby in *Annals of Science*, vol. IX (1953), p. 106.

But the naturalist's description is *sui generis*, and all its complexity is no deeper than that of an unfamiliar vocabulary.

At the close of the eighteenth century many of the procedures followed by the biologist in the course of his still mainly descriptive work—as one may see from the investigations of Cuvier, of Haller, of Lavoisier, of John Hunter—already anticipated those of the twentieth century. They were at least as "modern," by the same comparison, as those of the contemporary physicist and chemist. The great change has taken place in the intellectual framework within which such procedures are ordered. Biology had, indeed, successfully constructed an unco-ordinated group of scientific inferences, but it was still devoid of basic guiding principles: or rather, those which it possessed were derived from extra-scientific sources. Comparable in this state to mechanics before the seventeenth century, the extent and the accuracy of the observational material collected together was by 1800 far greater. The science had followed a rather haphazard Baconian course, for although information had been compiled piecemeal with great diligence, the elucidation from it of keys to the understanding of living nature had hardly begun. The "laws of nature" were as yet purely inorganic: all the great theoretical principles of biology have won their dominion during the last century. This being so, it may be recognized that the chief difficulties confronting biologists during the critical period 1750–1850—that in which the influence of the natural-historian Linnæus was at its height— were those of conceptualization; as, again, had been the case in mechanics long before. The possibilities for significant theoretical thinking open to men who accepted the futilities of *emboîtement* or the doctrine of successive catastrophes were as limited as those available to the Aristotelean opponents of Galileo.

The similarity ceases, however, when the intrinsic difficulty of establishing the "laws" of biology is compared with that of clarifying the "laws" of mechanics. More mental effort is required to grasp the necessity for the concept of evolution than to see the plausibility of that of inertia; more important, it is possible to justify the former only by elaborate discussion of varied and obscure evidence. Darwin's *Origin of Species* has a very different character from Galileo's *Discourses on Two New Sciences*. Darwin devoted more than twenty years to the filling of note-books with materials bearing on his problem, materials which were in part made accessible to him by a long tradition of descriptive natural history, and which owed much to the taxonomical precision of his authorities. He was himself an expert taxonomist.[1] Similarly, the elements of the cell-theory were pieced together as the result of an even greater mass of cumulative observation, extending to the seventeenth century. The principles of biology could not spring from the analysis of common and simple observations, as did those of mechanics; and although it may truly be said that much current research owes little directly to the tradition of classificatory refinement stretching through and beyond Linnæus, its origin lies in descriptive work to which that refinement was essential. A novelist does not need to be a lexicographer, but dictionaries are an essential foundation of good literature. Although biology has emerged from the natural history stage, its methods are not, and perhaps can never be, identical with those of physics. However great its development within laboratory walls, it must always have room for those techniques of description and discrimination with which the naturalists of the sixteenth century first strove to create a science.

[1] One reason for undertaking his monograph on the *Cirrepedia* (1851) was to prove to himself (and the world) that he was no mere philosophical biologist; though he afterwards doubted 'whether the work was worth the consumption of so much time.'

CHAPTER XI

THE ORIGINS OF CHEMISTRY

CHEMISTRY, as an integrated science with its own concepts, its own techniques, and its own area of applicability, is a product of the scientific revolution. The ancients, and their medieval successors, had no such distinct science, though they had much scattered empirical knowledge of this type, and some theoretical ideas that would now be described as of a chemical nature. Alchemy was not a primitive or pre-scientific chemistry, for it was both less (in the restricted range of its pretensions) and more (in its mystical affiliations) than a natural science. Chemistry, like biology, grew from a number of distinct roots, and not by the expansion of a single tradition, as did mechanics and astronomy. Hence the attempt to classify the sources of chemical knowledge and theory in the early modern period becomes complex, and shows that these had little relation to the subdivisions which now prevail. Researches of chemical significance occurred in mineralogy, in physiology, in physics, in pharmacology, and most obviously in the development of technology. Likewise, the concepts used by writers on chemistry before the time of Lavoisier were often common property among natural philosophers, rather than exclusive to their own science. Such is the case with the four-element theory of matter, and with the corpuscularian "mechanical" hypothesis; both were physical, or rather cosmological, in origin. Chemistry, again like biology, shows that a science gains stature as it acquires its own specialized concepts as instruments of thought. Stage by stage, from the theory of the three principles in the sixteenth century, through that of phlogiston to the Lavoisierian notion of the chemical elements and their combinations, the science developed coherence and independence.

Hence the idea that there is a particularly *chemical* way of studying matter and its properties, whether inorganic or organic matter, was almost totally absent before the late sixteenth century, and even then gained ground but slowly. Early speculations about the nature or composition of matter, and about the processes involved when one kind of stuff is turned into another kind of stuff, were a part of physics, as little empirical as the rest of physical theory. They were scarcely at all connected with the practical knowledge of certain groups of craftsmen. In a similar manner it was far from obvious that a distinct science was required to explain how the bread that man eats is transmuted into flesh and bone. A pre-scientific physiologist might speak of "concoction" in the stomach, but the term, though frequently used by chemists, had no specific meaning. It was an empty word that described nothing and explained nothing. When, however, certain experimenters adopted the belief that all metals are variously compounded of sulphur and mercury—using the names sulphur and mercury in a particular sense distinct from that of ordinary language—it does become possible to speak of a chemical attitude to substance. Robert Boyle frequently applied the word "chymist" in this way, as describing those who thought and worked in accordance with the three-principle theory. Otherwise the chemist was only distinguished from other men by the nature of his methods: 'What is accomplished by fire,' wrote Paracelsus, 'is alchemy—whether in the furnace or in the kitchen stove.'[1] The chemist was indeed primarily a pyrotechnician, who knew (or tried to discover) how to obtain certain results by long and gentle, or short and fierce, heating. To the end of the seventeenth century chemical analysis was practically confined to destructive distillation by fire, in which the substance to be analysed was forced to yield its waters, oils, sublimates, salts and *caput mortuum*. In this sense the metal-refiner, the soap boiler and the distiller were chemists; the practices of more learned men in the early modern period were hardly less haphazardly empirical than theirs, and owed little more to the guiding influence of a distinctive theory. And, as chemical ideas were but slowly differentiated from those generally current in natural philosophy, so chemical techniques were very gradually differentiated from those of the kitchen and workshop. Even in the time of Lavoisier they still bore strong marks of their craft origin.

Two possible approaches to the early history of chemistry are, therefore, bound to prove misleading, and to conceal the

[1] Alchemy to Paracelsus meant something much wider than the search for the secret of the transmutation of base metals into gold.

significance of the development of the science during the period of the scientific revolution. It is futile to attempt to trace the progressive evolution from primitive beginnings into a modern form of a texture of chemical theory, because the idea that it is useful to apply a group of characteristically chemical concepts to the study of nature is itself modern. It is equally futile to derive chemistry from the elaboration of certain techniques of investigation, firstly because it is doubtful whether these enabled useful empirical facts to be discovered before a late date, and secondly because the intellectual background to the refinement of technique is far more significant than the refinement itself. If modern chemistry is not, as mechanics is, the result of the progressive emendation of an autogenous conceptual structure, it is also not that of purely deductive reasoning applied to "natural history" information collected about the properties of minerals, acids, alkalis, etc.— though this view seems to contain more truth than the former. On the other hand, two analogous questions may usefully be asked. What attempts were made to account for changes in the properties of bodies effected by various manipulations (mainly with the aid of heat) in the light of existing scientific theory? How successfully were techniques adapted or invented with the double object of advancing technology and understanding? By asking questions of this form it is possible, without plunging into confusion, to recognize the indubitable facts that chemistry was always an eminently practical science, as well as a branch of natural philosophy, and that it lacked (before Lavoisier) a coherent conceptual scheme.

To state the problems in such terms is not to deny that theory and practice were interdependent, or that individual chemists like Boyle might both add to factual knowledge, and seek for an explanation of chemical phenomena in current scientific thought. It does, however, admit that these two strands in the history of chemistry are logically distinct. This is very clear in the late medieval period. Then, natural philosophers were engaged in trying to fit the known facts of chemical change into the pattern of Aristotelean ideas concerning the nature and properties of matter, exactly as, more assiduously, they tried to arrange the facts of physics and astronomy according to the same pattern. Meanwhile, the pattern itself preserved its character, and no essentially new ideas were brought out. By contrast, empirical knowledge of the phenomena of chemistry increased rapidly during the same period, owing progressively less to a remote Hellenistic ancestry. In glass-working, in the smelting and refining of metals, in dyeing and leather-dressing, in military pyrotechnics, in the distillation of alcohol and other "waters," in the glazing of pottery and the preparation of pigments generally, in the manufacture of medicaments of all kinds—briefly, in every aspect of chemical technology the European world of about 1500 enjoyed a consistent superiority over the Græco-Roman. In the millennium between the fall of Rome and that of Constantinople alcohol and the mineral acids were discovered, saltpetre was distinguished from soda (this made gunpowder possible), many new minerals were recognized, named, and their usefulness exploited, the known compounds of metals increased in number, chemical apparatus assumed a definite form, the control of furnaces was improved, and operations like reduction and oxidation were mastered (though their nature, of course, remained unknown). Some of this knowledge emanated from a Græco-Egyptian tradition, centred on Alexandria; important elements were derived from India and China, but practically nothing came from the academic scientific line of the ancient world, stemming from Plato and Aristotle through Pliny. The whole was synthesized and further advanced by the Islamic peoples, who were excellent chemical craftsmen, and further considerable progress was made in the Latin West from the twelfth century onwards. As some of the important sources, such as the writings of Geber and the *Book of Fires* of Marcus Græcus, are now accredited to Latin compilers who made use (to a degree which cannot be exactly estimated) of unknown originals, the detailed ascription of inventions in the chemical arts to Byzantium, Islam, or Latin Europe is often impossible. But it is quite certain that they are post-classical.

Little reflection of this growing empirical knowledge is to be found in the writings of medieval natural philosophers, with the rare exception of Roger Bacon. Only such discoveries as were possibly effected, or alternatively adopted, by the alchemists (like the preparation of mineral acids and the solution of metals in them) were of interest. More was done by physicians, with regard to pharmacologically useful discoveries. The third class of writing which gives an insight into medieval chemical technology consists of recipe books of many types, among them the different versions of the *Mappæ clavicula*, the *Note on various crafts* of the monk Theophilus (*c.* 1100), the *Book of Fires* already mentioned, and the *Book on colour making* of Peter of St. Omer (*c.* 1300). This class becomes much fuller from the fourteenth century. While, therefore, texts in this group may be drawn upon to form a picture of the level of factual knowledge concerning the preparation and properties of substances attained by the close of the middle ages, it is almost useless to look to them for the beginnings of a chemical attitude.

The literature of alchemy is copious, and many have searched in it for the beginnings of chemistry. There the grain of real knowledge is concealed in a vast deal of esoteric chaff. The view that alchemy represents the pre-history of chemistry (as developing chemical techniques, and factual knowledge of substances) is, after all, primarily based on the fact that such information is frequently set out, in the surviving texts, in an alchemical context. But there is no means of knowing that, because a discovery or an observation is first reported by an alchemist, it was made as a result of his inquisitive experimentation. The most remarkable feature of all alchemical writings is that their authors prove themselves utterly incapable of distinguishing true from false, a genuine observation (according to our modern knowledge) from the product of their own extravagant imaginations. It seems unnecessary to give them credit for making important truths known, when they were so obviously incapable of discrimination. It is certain that many practices and observations of alchemy were older than alchemy itself, just as observational astronomy preceded astrology. It is also certain that in the medieval period much knowledge was gained outside the alchemical context—which was restricted, almost exclusively, to metallic compounds. Taken together, these facts suggest that were the early history of practical, industrial chemistry more fully revealed, the inventiveness attributed to the alchemical dream would be found exaggerated. Some of the discoveries attributed to it may well have come from a differently directed experimentation; some alchemists were also physicians.

The theoretical contribution of alchemy to science was very small. Its own pretensions forbade the application of the usual notions of natural philosophy to the phenomena studied, and despite the interest of some philosophers in the art, there were always others who derided it. The theory of transmutation held by the later alchemists was originated by Jābir ibn Hayyān and al-Rāzi. It was made known to Europe in the twelfth century. They believed that all metals are composed of "philosophic" sulphur and "philosophic" mercury, which could be obtained by art from the base metals, and recompounded to form the precious metals:[1]

> Therefore if clean, fixed, red, and clear *Sulphur* fall upon the pure *Substance* of *Argentvive* (being itself not excelling, but of a small *Quantity*, and excelled) of it is created pure *Gold*. But if the *Sulphur* be clean, fixed, white and clear, which falls upon the *Substance of Argentvive*, pure *silver* is made . . . yet this hath a *Purity* short of the *Purity* of *Gold*, and a more gross *Inspissation* than *Gold* hath.[2]

It is enough to say that this theory was never developed (save in mystical embroidery), that it was never attached to any sound body of empirical knowledge, and that its persistence was the greatest obstacle to the development of a rational chemistry.

In any case, the pursuit of alchemical chimæras had long ceased to bring any useful information to light by the beginning of the sixteenth century. The significant event of this time was the emergence, from the obscure and laconic notes of recipe-books, of literate descriptions of the operations of chemical industry, especially those connected with metallurgy. These books have been discussed already (pp. 221–3). Here at last was metallurgical chemistry (and much more) free from the extravagances of alchemy. Here was a clear discrimination between fact and fiction with regard to the "transmutations" effected by chemical art. Theoretical speculation was almost entirely absent from these accounts, Agricola alone being notable for an attempt to explain his observations in terms of Aristotelean science, without, however, improving its texture. Thus was founded a serviceable and lasting tradition, which after a recession lasting through a couple of generations (those most subject to the influence of Paracelsus) was again revived by the scientific societies. From that time the academic study of industrial chemistry was never neglected. Eighteenth-century chemists like Black and Macquer were closely associated with it. Lavoisier did an enormous amount of work as a government consultant, reforming the administration and chem-

---

[1] These principles were ultimately compounded of the Aristotelean elements, but could be procured, as it were, as intermediate states.

[2] *Argentvive* = mercury. *Works of Geber, Englished by Richard Russell 1678.* Ed. by E. J. Holmyard (London, 1928), p. 132.

istry of the French explosives industry. And it may safely be said that the rapid extension of chemical research in the nineteenth century and later would have been impossible but for its close connection with manufacture. Much earlier the remarks of the Fellows of the Royal Society, especially Robert Boyle, on the necessity of close attention to trade methods give a hint of a source from which much was learnt—not least, in the way of suggesting the type of problems which might most usefully be attacked. For on this, it is obvious, modern chemistry largely depended for its early success. The only problems which the early chemist could hope to solve in a rational way were the simple ones—dealing with the oxidation of metals, the calcination of limestone, quantitative analysis of simple inorganic compounds—which were in fact posed by the basic chemical industries. However stimulating the fascinating questions suggested by more elaborate chemical changes might be, including those raised by the known connection between chemistry and medicine, answers to them remained far over the horizon until such time as organic chemistry slowly increased its scope through the mastery of inorganic reactions.

Though the endeavour to render physiology a branch of chemistry inevitably failed in the sixteenth and seventeenth centuries, nevertheless it did on the one hand promote the development of a distinctively chemical attitude to physiological and other problems, and on the other led to extensive exploration of chemical compounds with the object of using them as drugs. Inorganic medicaments were not new. Salt has always been collected as a necessity for life; antimony sulphide, copper salts, sodium carbonate, ochres, alum, and other minerals are prescribed for various purposes in the *Papyrus Ebers* (sixteenth century B.C.).[1] Possibly less, and certainly not greater, faith was attached to them in late medieval Europe. These were, however, natural substances, not the factitious products of art, few of which had yet been identified with naturally-occurring minerals.

Roger Bacon had taught that medicine should make use of remedies provided by chemistry, but it was not until the sixteenth century that his idea was fully developed, by Theophrastus Bombastus von Hohenheim (1493–1541), called Paracelsus, the founder of iatrochemistry (medical chemistry). His was not in any sense a modern mind. He believed in the philosopher's stone.

[1] H. E. Sigerist: *History of Medicine*, vol. I (Oxford, 1951), p. 343.

He believed in the alchemical theory of transmutation, and in others yet more wonderful:

> If the living bird be burned to dust and ashes in a sealed cucurbite with the third degree of fire, and then, still shut up, be putrified with the highest degree of putrefaction in a *venter equinus* so as to become a mucilaginous phlegm, then that phlegm can again be brought to maturity, and so, renovated and restored, can become a living bird. . . .[1]

He had in full measure the faculty for self-deception characteristic of the Hermetic tradition. For him, the physician and chemist were one, a magus whose operations influenced the natural and supernatural worlds together. In the words of Lynn Thorndike:

> for Paracelsus there is no such thing as natural law, and consequently no such thing as natural science. Even the force of the stars may be side-tracked, thwarted or qualified by the interference of a demon. Even the most hopeless disease may yield to a timely incantation or magic rite. Everywhere there is mystery, animism, invisible forces.[2]

But he was an iconoclast. He poured scorn upon the revered writings of Galen and other authorities. For their dietary rules and herbal preparations he wished to substitute new drugs purified by the action of fire. 'The work of bringing things to their perfection is called alchemy, and he is an alchemist who brings what nature grows for the use of man to its destined end.' Vulcan was to be his apothecary. In his medical practice he made much use of chemical preparations of herbs (to extract their virtue), laudanum, alcohol, mercury and metallic compounds, obtained by techniques familiar to alchemists. In theory, he added *salt* to the other alchemical principles, sulphur and mercury; otherwise his theoretical notions were fully as chaotic as those of other alchemists.

No great reformation was to be expected from Paracelsus' incoherent, obscure, megalomaniac writings. Yet the iatrochemical school flourished; the greater part of what was done up to Boyle's time may be attributed to it, and during two generations chemists were to a greater or less degree Paracelsians, known by their attachment to the three principles. Indeed Paracelsus' theses with

[1] A. E. Waite: *Hermetical and Alchemical Writings of Paracelsus* (London, 1894), vol. I, p. 121.
[2] *History of Magic and Experimental Science*, vol. V (New York, 1941), p. 628.

regard to medicine, that useful drugs could be made in the laboratory, and that there was room for bolder experiment in the treatment of disease, were obviously correct when purged of the fantasies and occultism with which, in him, they were always enmeshed. No one could doubt the efficacy of Glauber's *sal mirabile* (sodium sulphate), and long before Paracelsus' time the administration of mercury had been proved a specific against the common and dangerous disease, syphilis.[1] The ambitions of the iatrochemists were to reveal the chemical nature of physiological processes, by discovering the secret laws by which the combinations and recombinations of matter are governed, and to enlarge the list of compounds known to be effective against disease. These were rational objects, though the manner in which they were pursued was haphazard. Paracelsus had done little in chemistry himself. His successors devoted themselves more fully to mastering the techniques which would yield hitherto unknown substances. They did not hesitate to draw, where they could, upon the experience of industrial chemistry.

These men were the "chymists" or "spagyrists" known to Boyle and his contemporary philosophical chemists. They were men of learning and wrote Latin. Their view was narrow, being limited by the doctrine of the three principles, and they were often subject to the delusions of alchemy, but their books were meant to be understood. They created no secret language. Instead they began to describe, as plainly as their knowledge and terminology allowed, how the operations of chemistry are performed, from what materials and by what methods a large number of compounds are prepared, and for what purposes they might be employed. They began to compare the method used in one case with that used in another, to detect analogies between different compounds, and to try to explain what happened when a chemical reaction occurred by means of concepts which they invented or adapted. Here was the beginning both of a "natural history" of chemistry, and of a chemical theory.

Andreas Libavius (d. 1616), the first important iatrochemist

---

[1] Salivation was condemned by many non-iatrochemical physicians, because death was so often caused by mercurial poisoning. Medical faculties forbade the use of iatrochemical methods for well over a century, Sir Theodore de Mayerne, James I's physician, being expelled from Paris on this account. They were gradually admitted to official pharmacopœias during the seventeenth century.

after Paracelsus, refused to number himself among the latter's followers. From his *Alchemia* (1597) he claimed, justifiably, to have omitted the magical and superstitious elements introduced by Paracelsus, and to have purified the art of his figments and phantasms. 'Unhappy would chemistry be, if it had been founded by Paracelsus. . . . Filthy are the Paracelsian lies and blasphemies.' Nevertheless, he still believed in transmutation. The *Alchemia* was Libavius' most important work, though he admitted that it was largely compiled from other writers.[1] It is a methodical account of the chemical knowledge and laboratory technique of his time. He described many antimonial and arsenical compounds,[2] sulphurous acid, the stannic chloride ($SnCl_4$) known as "Libavius' fuming liquor," the extraction of many oils, waters and essences, and the analysis of mineral waters. In the second edition there is a long discussion, with wood-cuts, of the design of a chemical laboratory. From such a work as this it was but a step to the treatment of chemistry simply as an auxiliary of medicine. This is the attitude of Jean Beguin in his *Tyrocinium Chymicum* (1610) who thus defined his subject: 'Chymistry is the Art of dissolving natural bodies, and of coagulating the same when dissolved, and of reducing them into salubrious, safe, and grateful medicaments.'[3] The *Tyrocinium* is a very straightforward recipe book, forerunner of many others of the same kind.

The greatest of the iatrochemists, Johann Baptista van Helmont (1577 or 1580–1644) was a figure in some ways only less flamboyant than Paracelsus, for he also gave his imagination free licence. His criticism of academic medicine was more reasoned than Paracelsus', but not less severe. 'The art of healing,' he thought, "was a mere imposture, brought in by the Greeks. . . . To this day the schools do scarcely acknowledge any other remedies than blood-letting and their stock of laxatives; their whole endeavour is with bleeding, evacuations, baths, cauteries, sweats, so that they presume to cure all ills of the flesh by weakening the

---

[1] There are two editions, *Alchemia Andreæ Libavii Med. D. Poet. Physici Rotemburg*, etc. (Frankfort, 1597), p. 424; and the large folio with wood-cuts *Alchymia Andreæ Libavii, recognita, emendata, et aucta etc.*, (Frankfort, 1606), pp. 196 + 402 + 192.

[2] This, of course, was before the *Triumphal Chariot of Antimony* of the pseudo-Basil Valentine.

[3] *Tyrocinium Chymicum, or Chymical Essayes Acquired from the Fountaine of Nature and Manuall Experience* [trans. by Richard Russell] (London, 1669), p. 1. This primer was deservedly popular.

body and its strength, and by corrupting the blood.'[1] Despairing
of such methods, van Helmont turned to chemistry because
insight into nature generally, and the workings of the human body
in particular, could only be gained through a sound knowledge
of substance. To this end he developed a totally novel theory by
which he proved himself far more than a mere chemist like
Libavius. In the first thirty chapters of the *Ortus Medicinæ* he
sketched out a new natural philosophy, often obscure and
muddled, and more than tinged with a naive credulity, which
was as much directed against Aristotle as against Galen. His
speculations ranged far beyond the bounds of chemistry and
medicine, to meteorology and the causes of earthquakes. The
most important of them was his contention that there are only
two elements, Air and Water. Fire he did not regard as a body,
and therefore it could not be an element. All solid bodies, including the earths, were generated from water by the action of seeds
or ferments: 'The first beginnings of bodies, and of corporeal
causes, are two and no more. They are surely the element Water,
from which bodies are fashioned, and the ferment by which they
are fashioned.'[2] These divinely created ferments were the specific
organizers of water, the *prima materia*, into minerals as well as
living things; they were immaterial, though the seeds to which
they gave rise were not. Van Helmont referred in support of this
doctrine to the success of chemists in reducing solid bodies to
an "unsavoury water" (e.g. by solution in acids, followed by
neutralization); to the fact that fishes are nourished solely on
water (!); and to his famous experiment on the growth of a
willow tree in a tub of earth. Although nothing but water was
added, the tree gained 164 lb. weight in five years without any
diminution of the earth in which it was planted; water had
clearly been transmuted into solid matter by the action of the
"seed" in the tree.[3] Like Boyle later, van Helmont attacked the
three principles of orthodox chemists on the ground that some
bodies could not be resolved into them. He accepted the existence
of vacua in solid matter: for this explained how metals could be

[1] *Ortus Medicinæ*, Col. I, § 6; III, § 7. This collection of all van Helmont's
works was published by his son at Amsterdam in 1648. An English translation
by John Chandler appeared in 1662.
[2] *Ortus Medicinæ* (Lyons, 1667), p. 23.
[3] *Ibid.*, p. 68. The experiment had been suggested by Nicholas of Cusa in the
fifteenth century.

more dense than water. Air, however, could not be turned into water, even by great compression, and was therefore a distinct element.

False thinking on these matters was, in van Helmont's view, the main cause of error in science down to his time. He also attached great importance to two new conceptions, for which he coined new names. One was "Blas"—a sort of intrinsic motion in the stars and planets (which thereby affected the earth beneath), in the air, and in man. The other was "Gas," which name had a significance quite other than that now attached to it. Van Helmont's gas was simply a form of water, and he used the idea to confirm further his theory that water was the *prima materia*. Any matter, such as water vapour, carried into the extreme upper air, was turned into gas (i.e. was finely divided, since 'gas is far more subtle than vapour, steams, or distilled oils, although much denser than air') by the sharp cold and the "death" of the ferments. Then this gas might itself condense into vapour and fall as rain: at any rate, it was the chief cause of meteorological phenomena. Again, when substances burnt, the greater part of them disappeared as gas, which was also water, or 'water disguised by the ferments of solid bodies.' The point of van Helmont's observation, that there is only 1 lb. of ash obtained from 62 of charcoal, the rest disappearing as *spiritus sylvester* (wild spirit), seems to be that the charcoal itself is gas, i.e. water.[1] He knew that if the escape of the spirit was prevented, by enclosing the charcoal in a sealed vessel, combustion would not occur. This led him into his celebrated definition: 'This spirit, hitherto unknown, which can neither be retained in vessels nor reduced to a visible body (unless the seed is first extinguished) I call by the new name *Gas*.'[2] Thus van Helmont's element-theory might be symbolized

water + seed (ferment) —→ substance

and with his concept of gas this became

water —→ vapour —→ gas --→ water,

and   (water + seed) —→ substance —→ gas —→ water

[1] He had already "proved" that ash could be turned into water.
[2] Cf. *Ortus Medicinæ* "Progymnasma meteori," "Gas aquæ," "Complexionum atque mistionum elementalium figmentum."

In other words, the concept of gas was simply a complexity added to the doctrine which van Helmont most wanted to drive home, that is: water is all. Its main function was to explain the otherwise rather awkward fact that the products of combustion do not normally include much water-vapour. Just as he had to deny corporeal status to flame and fire (which could not be supposed to come from, or turn into, water) so van Helmont had to invent gas—as a form of water.

He noticed that gas might be evolved from a combination of substances, or a substance acted upon by a ferment, though not if the substance was taken by itself. Thus he found that saltpetre alone yielded no gas, but that gunpowder does—hence the violence of the explosion. Grapes yielded a gas when bruised and fermented. He also applied the name gas to the red fumes given off when nitric acid reacts with silver, to the product of the reaction between sulphuric acid and salt of tartar, to the fumes given off by burning sulphur, to flatus, to the poisonous substance which collects in mines and extinguishes candles, etc. While unable to recognize and classify these different gases systematically, he was of course able to make qualitative distinctions between nitric oxide, sulphur dioxide, carbon dioxide, mixtures of hydrogen and methane, and this particular attention to "fumes" (which must have been very often observed in the past) was perhaps van Helmont's best service to chemistry. But there is no evidence that he ever advanced beyond the notion that gases were substantially water, modified by the characteristic of immaterial ferments: indeed, he does not seem to speak of "gases" at all, as a plurality.

It has often been said, and rightly, that the long failure to understand the rôle of the common gases was a grave obstacle to the development of chemistry. It seems, however, that van Helmont, despite the praise frequently meted out to him as the first student of gases, really did little to lessen this obstacle. The modern notion of gas involves two propositions, the second being logically unnecessary, but required by experience: (1) There is a state of matter in which the particles are separated and free to move, rendering the matter tenuous and elastic; (2) Some forms of matter are normally encountered only in this state. Now the first of these propositions, though it was little developed theoretically before the nineteenth century (having been disregarded, for the most part, by the practical chemists who really worked out something of the importance of gaseous elements in chemical combinations) would not have seemed at all unfamiliar to seventeenth-century physicists, to whom the particulate nature and elasticity of "airs" were commonplaces. Their notions of "factitious airs" and "elastic fluids" clearly foreshadowed the modern physical idea of a gas. But they had no corresponding *chemical* conceptions. The pivotal discovery, the apprehension that made Lavoisier's chemistry possible, was the enunciation of the second proposition in a chemical setting. It was the fact that some participants in chemical reactions—both elementary and compound—could normally be isolated in the gaseous state only, though capable of entering into solid and liquid combinations, which was of strategic significance.

Therefore van Helmont's view of gas as a state of matter intermediate in fineness between a vapour and the element air, and as the fume, smoke or spirit evolved from a chemical process, was of very little profit in itself. While his ideas helped to encourage interest in these curious products, there was little real merit in the attachment of a special name to something so vaguely conceived, and so deeply involved in an unacceptable doctrine of watery transmutations. Chemists—especially the English—continued to speak of "airs" and "elastic fluids" with good reason; they chose rather to believe that gases were modified air than (with van Helmont) that they were modified water, and air an inert element. It cannot be said that they preferred the less plausible of the two hypotheses. To have converted the Helmontian view "gas is a state of water" into the statement "gas is a form of material substances, distinct from air, which participates in chemical reactions" was impossible. If van Helmont was right (as we now know) in suggesting that gas is not air, it was far more obvious to contemporaries that he was wrong in maintaining that gas was a phase in water $\longrightarrow$ substance $\longrightarrow$ water transmutations. Chemists like Boyle could make nothing of his Gas and Blas. They could not see that of these two whimsical notions, the former was far more worthy, since to them the single-element doctrine in which van Helmont's gas is involved was incomprehensible. Even to hint that the subsequent neglect of van Helmont's work on gas retarded the development of chemistry is to misunderstand the content of his thought, and to convey a thoroughly misleading impression.

About the middle of the seventeenth century the position in chemical theory (to use what is still an anachronism) was that the Aristotelean doctrine, now moribund and opposed to the broad trend of the scientific revolution, was still respectable; practising chemists, as a class, stood by their three principles and the general tenets of iatrochemistry; van Helmont's single-element theory aroused much interest, but won few adherents. Meanwhile, a fourth approach to the problems of chemical combination, based on the mechanical, particulate view of matter was taking shape, and was soon to be developed elaborately by Robert Boyle. Factual knowledge of chemical reactions and processes had also ramified considerably since the time of Libavius. Van Helmont himself, for all his *natur-philosophische* outlook, was a skilled practical chemist, who reported a good number of new preparations. He taught that matter was indestructible, illustrating this belief by the recovery of the original weight of a metal from the compounds in which it was apparently disguised.[1] It is said that he made much use of the balance: his famous tree-experiment shows how deceptive quantitative methods may be when applied within an inadequate conceptual scheme. A younger iatrochemist (who also pursued the philosopher's stone) was Johann Rudolph Glauber (1604–70). He described for the first time the preparation of spirit of salt (HCl), sodium sulphate, and perhaps chlorine. Glauber had a sound insight into certain types of chemical reaction, such as double decomposition; for example, he explained the formation of "butter of antimony" with sublimation of cinnabar from stibnite heated with corrosive sublimate by saying that the spirit in the latter, leaving the mercury, preferred to attach itself to the antimony in the stibnite; the mercury then united with the sulphur in the stibnite to form cinnabar.[2] From this example it is also clear that Glauber employed the concept of chemical affinity—he understood that one unit in a reaction might attract another more than a third.

[1] 'Since nothing is made from nothing, the weight [of one body] is made of the equal weight of another body, in which there is a transmutation of matter, as well as of the whole essence' (*Prog. Meteori*, § 18). Van Helmont's argument seems to be, that the same weight of *prima materia* (water) is represented by equal weights of any substance, irrespective of their densities.

[2] i.e., $Sb_2S_3 + 3HgCl_2 = 2SbCl_3 + 3HgS$. Double decomposition in this same reaction was indicated rather less clearly by Beguin in the 1615 edition of his *Tyrocinium Chymicum*. Cf. T. S. Patterson in *Annals of Science*, vol. II (London, 1937), p. 278.

In 1675 there was printed for the first time a new textbook on chemistry by a practical man, Nicholas Lemery (1645–1715), which remained popular for upwards of half a century. The title of the *Cours de Chymie contenant la manière de faire les operations qui sont en usage dans la Medécine* sufficiently indicates its purpose. It was a straightforward recipe-book, dealing first with the metals, then with salts, sulphur and other minerals, and finally with preparations obtained from vegetables and animals. Lemery was not given to theorization, but he taught that there were, in addition to the three active principles mercury (spirit), sulphur (oil) and salt, two passive principles, water and earth. He also accepted the theory due to Otto Tachenius that salt = acid + alkali, and occasionally followed the particulate ideas of Descartes. The *Cours de Chymie* was the most successful of many similar practical books published at about the same time: an English example is George Wilson's *Compleat Course of Chymistry* (? London, 1699). By the end of the seventeenth century the best accounts of experimental chemistry were those written with medical applications in mind, and though the old, esoteric iatrochemistry deriving from Paracelsus had perished, future progress was to owe much to physicians and apothecaries, among them Boerhaave, Cullen, Scheele and Black. It is of significance also that the teaching of chemistry in universities, beginning *c.* 1700, everywhere placed it as an auxiliary to medicine. Even at the close of the eighteenth century most of Black's pupils at Edinburgh were medical students.[1] From the time of Boyle to that of Priestley and Cavendish the rôle of the "amateurs" in chemistry was relatively insignificant, and the reason is not far to seek. This was a period of rapid practical development in the subject, but of minor theoretical expansion.

It is therefore appropriate now to turn to this last of the great philosophical chemists, Robert Boyle, who is one of the most enigmatic figures of the scientific revolution. There can be no doubt that he was one of its leading personalities, nor that he made wide, and incisive, use of the new ideas maturing in his time. He was one of the outstanding theorists of the "mechanical or corpuscularian" philosophy (though he never expounded a coherent system) and a consummate experimenter in both physics, in his work on pneumatics with the air-pump, and chemistry;

[1] Cf. the interesting paper by Dr. James Kendall in *Proc. Roy. Soc. Edin.*, Section A, vol. LXIII (Edinburgh, 1952), p. 346 *et seq.*

Boyle's works are perhaps the first to be filled with descriptions of experimental research. These books, on which his fame was established, are hugely prolix and sometimes tedious, containing masses of observations haphazardly selected and arranged, through which the reader has to pick his way in search of the point of the discussion, often to find that Boyle has been unable to make up his own mind. Yet they are always redeemed by the sense of a sharp and continuous purpose, a far-reaching breadth of learning, and a humility in face of the facts of nature that forbade dogmatism.

The task Boyle set himself was to examine philosophically the natural phenomena made known by chemical art; to determine the underlying nature of the material transformations of which a cumulative description had been built up since the time of Libavius, and of the processes by which they were brought about. He undertook to prove to philosophers that chemistry could be more than a collection of recipes, and to the "chymists" that in revealing nature's secrets they would have a more noble aim than the concoction of medicines. Not that Boyle despised the application of chemistry to medicine. He regularly dosed himself and his friends, and welcomed any new compound that promised therapeutic value, but he was aware that science as a whole is greater than the cure of disease. Boyle wrote no textbook of chemistry, like Lemery's, rather he usually assumed such a level of knowledge in his readers; nor was he greatly interested in the piling up of more and more empirical knowledge for its own sake. Almost always he wrote with a definite problem in mind, some aspect of his ambition to restore natural philosophy as a unified whole, in which chemical knowledge should play its due part. He sought to build a bridge between chemistry and physics, the two sciences concerned with the properties of matter, which ought to start from common ground and be explicable one in terms of the other. In this respect Boyle had much in common with van Helmont, whom he greatly admired (while admitting his frequent inability to comprehend him); but whereas van Helmont had criticized natural philosophers for their ignorance of ideas to which he himself had been brought by his own quasi-Paracelsian development, Boyle proceeded quite differently. Essentially a physicist looking at chemistry, Boyle sought to demonstrate that the facts and processes of chemistry were explicable in terms of the corpuscularian, mechanical hypothesis of matter which he had adopted.

In the first of his major works to be printed, which is also the most widely read, this is certainly not very clear. The purpose of the *Sceptical Chymist* (1661), as its title suggests, is negative, not positive. It was to clear the ground of the three theoretical attitudes to chemistry which were then in vogue. Boyle disposed rapidly of the four Aristotelean elements, for they were no longer plausible entities (at least to the progressive experimental scientist).[1] His main attack was upon the three principles of the orthodox chemists, his real target. In this connection there was nothing novel in his own definition of a chemical element, nor in his insistence on the importance of discovering what the ultimate constituents of compound bodies are.[2] He argued that none of the chemists' principles could be extracted from metals like gold or mercury; that their criterion of analysis by fire was in any case faulty, since it could not divide glass into its known constituents, sand and alkali. He pointed out (like van Helmont) that the natures of bodies are not changed by entering into combinations, since the same bodies could sometimes be recovered separately in their original state. He emphasized acutely the illogicalities and contradictions in which the ideas commonly entertained by chemists were involved. This criticism was warrantable, but in the *Sceptical Chymist* Boyle equally proved that he could create nothing to take the place of that which he would destroy. No more than any other contemporary did he believe that ordinary substances— gold, mercury, sulphur—were elements, though they resisted analysis. He used van Helmont's tree-experiment (which he repeated) to show that vegetable matter could be formed from water alone, without the participation of earth and fire, or salt and sulphur, but he did not believe that all things were made of water. In the end Boyle failed not only to draw up his own list of chemical elements, but even to decide definitely whether such simple substances do exist at all.

[1] However, they were by no means dead. The four Aristotelean elements were still to recur in the eighteenth century, whether in *Chambers's Dictionary* or in P. J. Macquer's *Dictionnaire de Chymie*, where it is said (s.v. *Élémens*) 'on doit regarder en chymie le feu, l'air, l'eau et la terre, comme des corps simples.'

[2] Cf. The definition of van Helmont (*Prog. Meteori*, § 7): 'An element would cease to be a simple body, if it were divisible into anything prior, or more simple. But nothing among corporeal things is granted to be prior to, or more simple than, an element.' Boyle did not claim any originality for himself, rather admitting that his definition was a common one.

There was, indeed, sufficient reason in his corpuscular physics why he should have been doubtful, why he should have thought that anything might be transmuted into anything else, by nature if not by art.[1] Some of the evidence is set forth in the *Sceptical Chymist*: Boyle believed, with most of his generation, that metals and minerals like saltpetre "grew" in the earth. These substances were not themselves elements; neither were they formed from pre-existing elements, for such elements could not be traced in the earth where the growth occurred. 'From thence' wrote Boyle 'we may deduce that earth, by a metalline plastick principle latent in it ["seed"], may be in process of time changed into a metal'—a common enough opinion.[2] From a survey of phenomena of this kind he came to the conclusion that transmutations in the chemical sense were possible by means of this "plastic power" in the earth, as also by the "seminal virtue" in seeds (for the salts etc. in wood were certainly not present as such in the water with which the tree was nourished). However puzzling this might be from the chemist's point of view, it was not at all inexplicable to the physicist in Boyle, for his physical theory of matter taught him that all substances are made up of the same fundamental particles.[3] Since, therefore, substances differ from one another only in the 'various textures resulting from the bigness, shape, motion and contrivance of their small parts, it will not be irrational to conceive that one and the same parcel of the universal matter may, by various alterations and contextures, be brought to deserve the name, sometimes of a sulphureous, and sometimes of a terrene or aqueous body.'[4]

The theory of matter which Boyle favoured led him to believe (as he acknowledged in the early pages of the *Sceptical Chymist*) that in its basic and primitive form matter existed as 'little

[1] One transformation in which Boyle was naturally very interested was that of base metals into gold. Like Newton he seems never to have been altogether confident that this transmutation, though theoretically attainable, was actually practicable.

[2] *Works* (London, 1772), vol. I, p. 564. Boyle cited Cesalpino and Agricola for the statement that metals reappear in previously worked-out veins. For Agricola's belief that metals are generated from the Aristotelean elements earth and water, cf. *De re metallica* (New York, 1950), p. 51, note.

[3] A modern analogy would be the proposition that the chemical element is an unsound and illogical concept, since all are composed of the same sub-atomic particles.

[4] *Works* (London, 1772), vol. I, p. 494.

particles, of several sizes and shapes, variously moved.' These
were further organized into 'minute masses or clusters,' being
the 'primary concretions' of matter, some of which were in
practice indivisible. Hence the clusters that compose gold, being
inviolable by the ordinary chemist's art, could always be recovered
from any compound of the metal. But a mass of such corpuscles
was not an element, for as Boyle stated, the particles of two sets of
corpuscles might so regroup themselves that 'from the coalition
there may emerge a new body, as really one, as either of the
corpuscles was before they were mingled.'¹ Thus vinegar acting
upon lead formed the "sugar" of lead (lead acetate) but by no
means could the acid spirit of vinegar be recovered from the new
compound; Boyle thought its corpuscles were destroyed. The
indivisible corpuscles of glass were formed by a "coalition" of
those of sand and ashes; so that resistance to chemical analysis by
fire or acids was not a test of elementariness: for such indestructible
corpuscles could as well be found in the bodies due to art, as in
those due to nature. The same experience that taught Boyle that
glass was not to be numbered among the elements, prevented him
from knowing whether gold was one or not: he thought it was
probably not.

It is plain that Boyle's corpuscularian philosophy, in many ways
so fertile, prevented him from taking any step towards the modern
conception of the chemical element (an achievement usually
credited to him). In fact it led in the opposite direction, for so
long as Boyle, the physical chemist, concentrated on explanations
of chemical reactions in terms of corpuscles and particles, he would
be the less likely to believe that if elements existed at all, their
existence was very significant to his new outlook. Lavoisier determined that if a body resisted chemical analysis it should be
accepted, pragmatically, as an element. Boyle could not have
done this, because he thought in corpuscular terms. No body made
up of corpuscles could be simple, homogeneous or elementary in
the strict sense, because the corpuscles themselves were compounded of a variety of particles. The pragmatic test of resistance
to analysis proved only that some concretions of these particles
into corpuscles were more coherent than others, whether brought
about by nature (gold) or by art (glass). But coherence and
elementariness could not be equivalent, since Boyle knew that

¹ *Works* (London, 1772), vol. I, pp. 474–5, 506–7.

some corpuscles whose coherence resisted analysis were factitious, i.e. non-elementary.

The *impasse* was complete, and Boyle could not avoid it. Nor was it really important for him to do so: the truth that chemical changes occur by modifications of corpuscular structure, and by re-shuffling of the particles within corpuscles, was far more significant to him than a dubious search for primary elements, whose existence he was quite content to regard as hypothetical. In this he may be said to have anticipated the attitude of a modern physical chemist who (of course within a far more rich and exact conceptual framework) thinks in terms of molecules, atoms, ions and kinetic theory, without paying much attention to the elementariness of hydrogen or carbon.

That Boyle's chemical theory was predominantly shaped by his corpuscularian physics is apparent from many works beside the *Sceptical Chymist*. He was no dogmatist, he was no follower of systems, he believed that hypotheses should be framed to fit the facts, but it was quite clear to him that complete scepticism and abhorrence of all theory were antithetical to true philosophy.[1] Indeed, his ambition to bring chemistry into natural philosophy would have been meaningless had he not had a theory of natural philosophy, essentially a theory of matter, to hand for the purpose:

> I hoped I might at least do no unseasonable piece of service to the corpuscular philosophers, by illustrating some of their notions with sensible experiments, and manifesting, that the things by me treated of may be at least plausibly explicated without having recourse to inexplicable forms, real qualities, the four peripatetick elements, or so much as the three chymical principles.[2]

Thus for Boyle solution always represented the interspersion of the corpuscles of the solvent among those of the dissolved body, heat consisted of material particles, and "air" of all kinds mainly of elastic corpuscles of a particular type, among which the mingling of other corpuscles gave each "air" its own character. He constantly attributed the properties of acids, oils, salts, etc. to the nature of their component corpuscles. In fact the theory was brought into play whenever Boyle commented on a chemical experiment. Instances are particularly numerous in the *Origin of Forms and Qualities*, where (as examples chosen at random) he

[1] Cf. *Ibid.*, p. 591.
[2] *Certain Physiological Essays* (1661), *Works*, vol. I, p. 356.

spoke of 'the body of silver, by the convenient interposition of some saline particles [being] reduced into crystals' of *Luna cornea* (silver chloride); or, of another experiment, 'considered that the nitrous corpuscles [of nitric acid], lodging themselves in the little spaces deserted by the saline corpuscles of the sea-salt, that passed over into the receiver, had afforded this alkali';[1] or again declared that

> the more noble corpuscles which qualify gold to look yellow, [and to resist nitric acid] . . . may either have their texture destroyed by a very piercing menstruum, or, by a greater congruity with its corpuscles, than [with] those of the remaining part of the gold, may stick closer to the former and . . . be extricated.[2]

Corpuscularian ideas were especially developed by English scientists in relation to the allied reactions of combustion and calcination. Both were anciently regarded as separations: the ash or calx (oxide) remaining was an earthy residue left after the more volatile parts of the combustible or metal had been driven off by fire. It was well known, however, that the calx (oxide) exceeded the original metal in weight: whence Jean Rey was led to believe (in 1630) that though the calx was lighter than the metal in nature, it became heavier by the attachment to it of air which had become thickened in the furnace. Boyle later thought that the increased weight came from fire-particles penetrating the walls of the crucible and impregnating the calx. This explanation was widely accepted until it was decisively confuted by Lavoisier. With regard to combustion it was well known to Boyle and others that bodies would not burn without air, unless (like gunpowder) they contained some nitrous material. A familiar experiment (originally described by van Helmont) showed that when a candle was burnt in a closed vessel over water, the air within diminished and the water rose up inside. From these and other observations Robert Hooke sketched a theory in *Micrographia* according to which combustible bodies were dissolved by a certain substance present in the atmosphere, this solution (like others) evolving great heat, and thus flames, which Hooke took to be 'nothing else but a mixture of Air, and volatil sulphureous parts of dissoluble or combustible bodies, which are acting upon each

---

[1] i.e., $KCl + HNO_3 = KNO_3 + HCl$.
[2] *Origin of Forms and Qualities*, Section VII. Boyle is obviously far from thinking that gold might consist of a *single* kind of corpuscle.

other.' This aerial substance he identified with 'that which is fixt in *Salt-peter*.' This same theory was elaborated in much greater detail by the physician John Mayow in 1674, who also set it uncompromisingly in the corpuscularian framework. In his view the atmosphere consisted of a mass of air-corpuscles with which were mingled others that he called "nitro-aerial particles," because they were fixed in nitre and nitric acid. These nitro-aerial particles were highly elastic, and responsible for the hardness of quenched steel. They also corroded metals exposed to the air. The same particles, reacting violently with the sulphureous corpuscles of combustible bodies, produced heat and flame, as Hooke had said; therefore combustion could not occur without them. Air in which combustion had taken place was (in some way which Mayow did not explain) deprived of its nitro-aerial particles, thereby decreasing in elasticity and causing the water to rise in van Helmont's experiment. Similarly air breathed by animals lost its nitro-aerial particles, so that the expired air was less elastic (i.e. occupied a smaller volume) than that inspired; the same particles caused the blood to ferment and provoked animal-heat. They caused fermentation generally; they formed the fiery body of the sun, and the flash of lightning; in short, the nitro-aerial particle became in the hands of Mayow a *deus ex machina* for explaining many totally unrelated phenomena. He had no idea that these particles made an "air," or that the atmosphere was a mixture of "airs" of which that was one. The notion (first put about by Thomas Beddoes, Davy's patron) that Mayow had closely anticipated the oxygen-theory of Lavoisier cannot support serious examination: his was a typical corpuscularian philosophy.

During the next hundred years studies on combustion and calcination were to prove vital to the progress of chemical theory, while the accumulation of empirical fact was proceeding in a way which rendered the dubiety of theory almost irrelevant. The attempt to explain chemical reactions in terms of the mechanistic theory of matter, begun by Descartes and continued by Boyle, Mayow and others, was abandoned in the early years of the eighteenth century. Its claims were too wide, its achievements too lacking in definition. Not that chemists doubted the particulate structure of matter, but since nothing was known about the size, hardness, shape, etc. of corpuscles, statements in which these properties of the fundamental constituents of matter were involved were recognized as meaningless. This downfall of Boyle's hopes— the cause of the subsequent neglect of the theoretical import in his writings—was accelerated by the ascendancy of Newtonian mechanics. The laws which governed the stars must be applicable also to the smallest particles of matter, and the laws of chemical affinity be subject to the supreme law of gravitational attraction. The elements of this theory, already sketched by Newton in the *Queries* to his *Opticks* (1704), were soon after worked out more completely by Keill and Freind. The later empirical study of affinity (e.g. by Geoffroy, who published tables of affinity in 1718) was certainly encouraged thereby, but after the middle of the century (when attraction had long been abandoned by practical chemists) it became clear that the property of attraction in corpuscles was as intangible as their other properties: nothing could be said about it, other than that it existed. And so the ambition to render chemistry a branch of physics was, for the time, frustrated. Dalton's atomic theory was conceived entirely in a chemical context, and it was under the ægis of chemistry that the second penetration of atomism into modern scientific thought was to take place.

If to many chemists of the early eighteenth century it seemed hopeless to base their theories directly on the properties of atoms or corpuscles, they were not the less convinced of the existence of elements or principles of an ultimately corpuscular nature. The principles adopted by the German chemist G. E. Stahl (1660– 1734) from the writings of J. J. Becher (whose *Physicæ Subterranæ* of 1669 Stahl republished in 1703) were generally accepted from about 1730 (when phlogiston appeared to be a great unifying chemical concept, making sense of many reactions) until near the close of the century. Stahl's principles were water and three kinds of earth: air was not chemically active. The three earths corresponded to the three principles of the iatrochemists, but Stahl preferred to call the second (sulphur) *phlogiston*. The followers of Stahl made much use of affinity in explaining reactions (meaning, in this case, the tendency of like to react with like)[1]

---

[1] Thus Junker on the formation of corrosive sublimate (Helène Metzger, *Newton, Stahl, Boerhaave et la Doctrine Chimique* (Paris, 1930), p. 142): 'Common salt consists of an acid and an alkaline base. Vitriolic acid, being more powerful, seizes the base and drives out the acid from the salt. This acid would escape

and of a novel theory of salts (as earth + water). The theory of phlogiston was far from being the whole of their chemical doctrine, though it was the dominant feature of that doctrine because it permitted the co-ordination of many previously isolated facts.[1] Phlogiston was the substance emitted during combustion and the calcination of metals, the "food of fire" or "inflammable principle." The complete, or almost complete, combustion of charcoal, sulphur, phosphorus, etc. demonstrated that these bodies were very rich in phlogiston: while the formation of sulphuric acid, phosphoric acid, etc. from the solution of the fumes produced by combustion demonstrated that the substances themselves actually consisted of nothing but the acid joined to phlogiston (i.e. sulphur *minus* phlogiston $\longrightarrow$ sulphuric acid: therefore sulphuric acid + phlogiston = sulphur). When a metal was heated, the phlogiston driven off left a calx behind (therefore calx + phlogiston=metal). Conversely, by heating the calx with charcoal, phlogiston was exchanged and the metal restored. Many reactions became comprehensible when interpreted in terms of an exchange of phlogiston, so that often where a modern chemist sees a gain or loss of oxygen, Stahl saw an inverse loss or gain of phlogiston.

Oxygen has weight: and a modern chemist would insist that phlogiston ought to have negative weight, a suggestion actually mooted when Stahlian chemistry was already moribund. The matter was not serious at first. Boyle himself had explained the increased weight of a calx in a way that nowhere conflicted with the idea of phlogiston. Until such time as gases were collected, and the gain or loss of weight due to the participation of a gaseous element in a reaction could be correctly estimated by means of the balance, it was impossible to achieve a balance of masses in a chemical equation. Chemical mathematics, before 1775, was such that there was no palpable absurdity in the conception of phlogiston as a material fluid: as Watson said,

> You do not surely expect that chemistry should be able to present you with a handful of phlogiston, separated from an inflammable body; you may just as reasonably demand a handful of magnetism, gravity,

freely, but meeting with another substance (mercury) with which it has some analogy (though not so great as with its own base), it forms a saline substance with it. This substance (corrosive sublimate) is volatile because both mercury and the acid from salt are volatile: they have an identity of "mercurial principle".'

[1] This is particularly emphasized by Mme. Metzger, *op. cit.*

or electricity. . . . There are powers in nature which cannot otherwise become the objects of sense, than by the effects they produce; and of this kind is phlogiston.[1]

The purely logical objection which has often been raised against the phlogiston theory is therefore of small value. Eighteenth-century scientists readily admitted the existence of weightless, impalpable fluids such as electricity and caloric. Caloric was indeed phlogiston revived, stripped of its chemical attributes to become Lavoisier's "pure matter of heat."[2] Nor did Stahl's influence check the progress of chemistry as an empirical science; rather his views provided a useful provisional scheme for the explanation of many experiments. As such many chemists accepted them, without sacrificing their liberty to interpret experiments in the light of the evidence alone. Lavoisier himself was at first far more keenly aware of the need to scrutinize experimental data, than of any implausibility inherent in the phlogistic doctrine itself. The practical chemists were always far more interested in particular problems than in universal theories. Hence it seems unnecessary to suppose that the fortunes of chemistry were inextricably bound up with those of the phlogiston theory, whose scope was largely restricted to the phenomena of combustion and calcination; but it is certain that the age of phlogiston witnessed great progress in chemical experimentation, though the hypothesis has so often been characterized as futile and retrogressive. In the phlogiston period the chemistry of gases was begun by Black, Cavendish, Scheele and Priestley; there were striking improvements in analytical technique; and there was a wise concentration on the understanding of the simpler reactions and compounds to which this concept afforded a useful key. There was also considerable advance in industrial chemistry: in France, where Réaumur, Macquer, Duhamel du Monceau and others explored the theory of manufacturing processes with profit, and in England where new industries developed with the aid of chemical science. Most important of all, through the phlogiston concept, chemistry at last gained coherence as a theoretical science in its own right.

Irrespective of its ultimate redundancy, phlogiston was undoubtedly a useful concept until about 1765, when the systematic

[1] Richard Watson: *Chemical Essays*, vol. 1 (London, 1782), p. 167.
[2] For the view that phlogiston is "fixed" heat, cf. *ibid.*, p. 165.

study of gases was begun by Henry Cavendish (1731–1810). It enabled a consistent interpretation to be given to experiments on combustion, and many others involving oxidation and reduction.[1] Chemists gained a valuable insight into a number of reactions in phlogistic terms, and so learnt to treat natural substances— sulphur, carbon, salts, alkalis, acids, metals, earths, and so forth— as the really active participants in their processes. Lavoisier's chemical revolution was based on a level of factual knowledge, and a pragmatism with regard to chemical combination, un- equalled at any earlier time. The fact that it often happened that Lavoisier could simply invert the phlogistic doctrine is evidence of the service which the phlogistonists had performed. Only with the discovery of the common gases—when hydrogen was taken to be pure phlogiston, nitrogen to be phlogisticated air, oxygen to be dephlogisticated air, etc.—did phlogiston become a serious impediment to the interpretation of experimental work; only then did the question of its usefulness as a hypothesis become really significant.

The technique of collecting gases over water was invented by Stephen Hales before 1720; Joseph Priestley (1733–1804) substi- tuted mercury when working on water-soluble gases. Hales examined "airs" produced in a variety of chemical processes, particularly in order to discover the quantity evolved from a given weight of materials, but he drew no new qualitative distinctions between them. From his experiments he concluded that "air" was capable of being fixed in substances as a solid. The first chemist to label such a "fixed air" decisively, at the same time discovering its function in a number of reactions, was Joseph Black (1728–99). The linkage is important, for many "airs" had been

[1] It is not difficult to find examples of reasoning in phlogistic terms which led to satisfactory experimental procedures. Thus Scheele (1779) wished to establish the ratio by volume of "pure air" (oxygen) to "foul air" (nitrogen) in the atmosphere. He argued: 'When this pure air meets phlogiston uncom- bined, it unites with it, leaves the foul air, and disappears, if I may say so, before our eyes. If, therefore, a given quantity of atmospheric air be included in a vessel, and meet there with some loosely adhering phlogiston, it will at once appear, from the quantity of foul air remaining, how much pure air was contained in it before.' He therefore took a small amount of iron filings, mixed with sulphur and a little water, which he knew to be a preparation very rich in phlogiston, and used it to effect the "disappearance" of the "pure air" in a known volume of common air. Scheele's mixture does in fact take up oxygen very readily. The result he obtained (9 : 24) was not good, but his procedure was perfectly correct. (Cf. *Chemical Essays* (London, 1901), pp. 190–4.)

roughly identified in the past (e.g. as inflammable, or extinguishing flame), but none had been clearly described as being distinct in species from common air, nor had any function been ascribed to them as participants in chemical processes. The most striking part of Black's work (*Experiments upon Magnesia alba*, etc., 1756) was his proof that quicklime was lime deprived of "fixed air": the quantitative relation was completely established. Black found that a mild alkali became caustic through loss of its "fixed air," and that "fixed air" was emitted with effervescence when lime was dissolved in acids. He made constant use of the concept of affinity in explaining his experiments: thus, discovering that when quicklime was added to a mild alkali, the original weight of lime[1] was precipitated and the alkali rendered caustic, he said that "fixed air" had been exchanged because of its greater affinity for quicklime than for the alkali. He carefully pointed out that though both water and the atmosphere contained "fixed air," it was not identical with "common air." Later he discovered it in exhaled breath, and in common air passed over burning charcoal: the precipitation of lime from lime-water proved to be a specific test. The whole essay was a neat model of scientific method: at one stage in his work Black drew up a list of predictions, each of which he was able to verify by experiment, thus proving that "fixed air" played the part he had assigned to it.

In all this there was no mention of phlogiston. Though Black was very far from rejecting the phlogiston theory in its entirety until long after this time, he was convinced that his experiments could be explained without phlogiston, and he resisted all attempts to argue that phlogiston was involved in them. Black had not, in 1756, actually collected his "fixed air," for the technique of imprisoning fugitive "elastic fluids" in vessels was still almost unknown to chemists. It was described by Cavendish in his paper *On factitious airs* (1766), and thereafter widely used. He distinguished between Black's "fixed air" and another (hydrogen) deriving (as he thought) from metals, which he found to be lighter than common air. At about the same time the Swedish apothecary Carl Wilhclm Schcele (1742-86) separated the two major constituents of the atmosphere "Feuerluft" (oxygen) and "verdorbene Luft" (nitrogen): he also prepared oxygen in various ways in the laboratory, as did Joseph Priestley, quite independently, in

[1] Calcined by Black to make the quicklime.

1774. Most of the characteristic qualities of these and other gases were noted.

Such were the strategic few, among a great number of other major discoveries made by chemists who thought without reservations in the framework of which phlogiston was an essential part. Two of the pioneers of gas chemistry, Priestley and Cavendish, were never reconciled to Lavoisier's doctrines. Their refusal was no doubt due to rigidity of mind, but it points also to the fact that the phlogistic theory had imposed no barrier upon the activities of these skilful experimenters. It also emphasizes Lavoisier's own originality in devising new interpretations of their experiments. While the adherents of phlogiston were by no means agreed in the details of their exposition—Priestley, for example, thinking of oxygen as dephlogisticated air, and Cavendish preferring to treat the gas as dephlogisticated water—a situation by no means unusual on the frontier of research, these hesitations did not inhibit inquiry; on the contrary, it is quite clear that Scheele, Priestley and Cavendish were each at times induced to make certain fertile experiments by reasoning in the phlogistic manner.

The situation in which the further progress of a branch of science is directly dependent upon an adequate matching of theoretical concepts and experimental facts is by no means uncommon. This was certainly the case when Galileo and Newton, respectively, revised the concepts of mechanics, and again with physics in the nineteenth century. But though such a matching of fact and theory is always useful it is far from being invariably essential. Did such a situation exist in chemistry at the end of the eighteenth century? The evidence would seem to suggest that it did not. The empirical attitude of the great experimenters was in reality far more important than their theorization: it is therefore the less likely that any plausible modification of the doctrines prevailing through the first three-quarters of the eighteenth century would have had much influence on the course of events. No one would deny that Lavoisier was the first chemical theorist of genius. No one would deny that his interpretation of the phenomena was far superior to that of the phlogiston theory: it was one upon which the ultimate advancement of chemical knowledge depended. Yet it is also perfectly clear that the inventive empiricism of his contemporaries was just as necessary for this as his own logical, interpretative intellect, and that, moreover, the rapid progress of chemistry in the nineteenth century owed a great deal to developments, such as electro-chemistry and the atomic theory, to both of which Lavoisier's own insight into the nature of chemical reactions contributed nothing.

Antoine Laurent Lavoisier (1743–94) was, like Newton, less the author of new experiments than the first to realize their full significance. Experimentation for its own sake, which delighted Priestley, had little appeal for him. In his laboratory work he proved himself a skilful analyst, and an able exponent of the quantitative methods of Black and Cavendish, but not greatly imaginative in the way that Cavendish (or, later, Faraday) was a supremely imaginative experimenter. None of his most famous experiments was new: the element of originality in them was limited to Lavoisier's insistence upon paying heed to the teachings of the balance:

> We may lay it down as an incontestable axiom, that, in all the opera-tions of art and nature, nothing is created; an equal quantity of matter exists both before and after the experiment; the quality and quantity of the elements remain precisely the same; and nothing takes place beyond changes and modifications in the combination of these elements.[1]

His first investigation was into the composition of the water in various localities about Paris. In his second he refuted the ancient fallacy that water was converted into earth by long heating and evaporation: he found that the "earth" obtained was dissolved glass. Thirdly, he discovered that air was "fixed" in phosphorus pentoxide and sulphur dioxide (made by burning phosphorus and sulphur), bringing about an increase of weight. He predicted that the same would be found true of all combustibles, and that the increased weight of metallic calces was due to a similar fixation of air. This last prediction he confirmed (1772), by reducing lead oxide to lead with charcoal: a large quantity of air was evolved.

Lavoisier had set out to produce facts, and by quantitative methods he had succeeded. Some of them were not exactly new, though they had formerly been less precisely stated, but he per-ceived that there was enough to make the phenomena of combustion

[1] *Elements of Chemistry* (trans. by Robert Kerr), (Edinburgh, 1790), p. 130. But Lavoisier was in no sense the first chemist to subject notions of chemical composition to quantitative tests. There are many earlier examples of this procedure.

and calcination worthy of detailed examination. All his future success followed from his apprehension of this significance. He embarked upon 'an immense series of experiments' intended to reveal the properties of the different "airs" involved in chemical reactions, which seemed 'destined to bring about a revolution in physics and chemistry.' In fact, however, Lavoisier's qualitative study of gases did not advance very far between February 1773 (when he wrote the note just quoted) and October 1774, though he did satisfy himself that the "air" given off in the reduction of lead oxide with charcoal was Black's fixed air, and that Boyle's explanation of the origin of the increased weight of metallic calces was false.[1] At this stage he was still uncertain whether the "air" combining with metals to form their calces was Black's fixed air, or common air, or something occurring in the atmosphere. He had, apparently, missed the importance of Black's own observation that burning charcoal was a source of his fixed air. He could hardly have claimed, at this point, to have done more than indicate a series of reactions in which some kind of air was involved, parallel to those already described, more completely, by Black.

In October 1774 Lavoisier was informed by Priestley himself of the evolution of an "air" (oxygen) from the red calx of mercury (mercuric oxide) when reduced by itself to the metal with the aid of a strong heat.[2] Some months later Priestley had established that this "dephlogisticated air" (as he now called it) was respirable as well as capable of supporting combustion. Soon after Lavoisier was able to report new experiments to the Académie des Sciences: the calx of mercury reduced with charcoal gave fixed air, reduced alone it gave off Priestley's new air. Therefore it was, said Lavoisier, common air in a highly active state which combined with metals to form their calces. Evidently he did not yet regard it as a particular constituent of the atmosphere, a view which he adopted in the course of the next two years, when he realized that one fraction of the air was inert during combustion and calcination. This he called *mofette*. The atmosphere, therefore, contained two "elastic fluids," as Scheele had discovered some time before.

---

[1] In this he had been anticipated by the Italian chemist, Beccaria. The equivalence of the increased weight of the calx with the weight of the air admitted when the retort was opened was not as exact as Lavoisier wished.

[2] Lavoisier's own claim to the discovery of oxygen is generally rejected.

By August 1778 Lavoisier could announce that "eminently respirable air" (the "dephlogisticated air" of Priestley) combining with a metal, formed a calx, and that the same air combining with charcoal gave Black's fixed air (carbon dioxide). The problem of calcination was thus solved, and other discoveries followed rapidly. Lavoisier found that sulphur and phosphorus, in burning, also combined with "eminently respirable air" only, and that it was this combination that yielded acids by solution in water. He found the same air in nitric acid, and was led to conclude that it was present in all acids. In respiration, the *mofette* part was exhaled unchanged, but the respirable part was exhaled as the fixed air which Black had found in the breath. From experiments carried out in collaboration between 1782 and 1784 Laplace and Lavoisier judged that respiration was a process of slow combustion, for the amount of heat produced by an animal in breathing out a certain volume of fixed air, was almost equal to the amount produced in burning enough charcoal to give the same volume.

In his memoirs before 1778 Lavoisier had not eschewed the word "phlogiston" completely, though he was uncertain of its rôle in his experiments, if it existed at all. By 1778 the elements of his new theory of oxidation were quite firm, and in it phlogiston had no part for he had proved that the phlogiston-concept was the inverse of the truth. The processes of oxidation (combustion, calcination) were not analyses but syntheses, in which phlogiston was as redundant as it was in Black's fixed-air reactions. Once the pattern of the substitution of the oxygen-concept for the phlogiston-concept became clear, it could be extended to whole groups of reactions, over almost the whole area of chemistry, by the development of suitable analogies. Indeed it was possible to be deceived by such analogies.[1] Proceeding in this way, it became ever more plausible to propose the exclusion of phlogiston from

[1] e.g. in Lavoisier's account of the relation of "oxygenated muriatic acid" (chlorine) to "muriatic acid" (hydrochloric). Lavoisier was convinced that the latter was a compound of an unknown base with oxygen (as were all acids, in his view), and the former a compound of the same base with a higher proportion of oxygen (by analogy with sulphurous, sulphuric acids, etc.). Therefore he postulated $2XO + \ldots O_2 \rightarrow 2XO_2 + \ldots$ for $4HCl + \ldots O_2 \rightarrow Cl_2 + 2H_2O + \ldots$ His name for it was simply the inverse of "dephlogisticated marine acid gas" (Scheele) and really considerably less appropriate, since if hydrogen = phlogiston (Cavendish), $HCl - H \rightarrow Cl$. (Cf. *Elements of Chemistry*, pp. 69–73, 233–5). Davy discovered the true composition of HCl, and established that chlorine is an element.

the doctrine of chemistry altogether. In 1783 Lavoisier was impelled to make a direct attack upon the phlogiston theory, which he seems to have disliked even before he had any very solid grounds for doing so. In this comparison of his own theory with that of the phlogistonists, he insisted upon their inconsistencies and duplications of hypothesis much as Galileo had analysed the weaknesses of his opponents long before. Phlogiston was not merely superfluous; it had come to mean quite different things in different contexts, and had become a mere drudge in chemical theory.

From this time the number of adherents to Lavoisier's views increased steadily, first in France, and then abroad. One serious difficulty remained. Lavoisier had not been able to discover the product of burning "inflammable air" (hydrogen) with common air, or with "respirable air" (oxygen). Nor could he account for the evolution of hydrogen when metals were dissolved in weak acids, and its non-appearance when the calces of metals were so dissolved. All this could easily be explained in terms of phlogiston. Once again enlightenment came from the English experimenters. In 1782 Cavendish continued experiments begun by Priestley and others in which an electric spark was passed through mixtures of hydrogen and common air. The work was done with wonderful neatness, and he was able to identify the condensate inside his glass vessel as common water, the gases disappearing in a ratio of 2:1 by volume. By June 1783 these experiments were known to Lavoisier, who, paying Cavendish scant tribute, hastily repeated them. Initial scepticism was replaced by a rapid reinterpretation in terms of his theory: water was the oxide of hydrogen, and hydrogen was evolved by the action of metals from the water present in the weak acids.

The last act in the "chemical revolution" was the establishment of a new terminology. The existing names of compounds were either misleadingly descriptive ("butter of antimony"), or redolent of the displaced theory ("dephlogisticated air") and even older notions ("spirit of salt"), or based on common words like "salt" and "earth." In 1787 Lavoisier and three of his French adherents devised a new nomenclature, still substantially preserved, in which the names were intended to identify the nature of the substances. Thus *hydrogen* (water-former) replaced "inflammable air" and *oxygen* (acid-former) "eminently respirable air." *Mofette* became *azote* (inert) or to non-French chemists *nitrogen* (nitre-forming). Compounds of an element with oxygen were all *oxides*, with sulphur *sulphides*, etc. This aim to establish regular patterns of nomenclature has been consistently followed. At the same time Lavoisier abandoned the attempt to seek (in chemistry) any reality more fundamental than that revealed by ordinary analysis. Substances which resisted analysis were, to him, elements; for example the common gases, metals, sulphur, carbon, lime, and a number of earths and other "radicals" some of which were soon analysed into other constituent elements. Lavoisier already used the word "radical" in the sense of a unit in chemical reactions, and an element was simply an indestructible radical. All this was fully explained in the first textbook of the new chemistry, the *Traité Élémentaire de Chimie* of 1789.

Its publication draws a convenient line of demarcation. Yet it is important to recognize how little, as well as how much, was really new in it. The theory, and the arrangement based upon the theory of chemical combination, was all Lavoisier's. The inorganic experiments and substances described had almost all been known for twenty years, and the majority for much longer. Lavoisier and his friends, however, had created the organic section (dealing with the composition of alcohol and sugar, etc.) almost unaided. If the table of elements was new, it derived (by a simple modification) from the older chemists' practical habit of treating phosphorus, sulphur, metals, etc. as virtually compounds of element + phlogiston. Lavoisier still spoke of earths and alkalis much as they had done, and knew little more about them. On the mystery of chemical affinity he said nothing, on the plea that it was unsatisfactory to discuss it in an elementary treatise. Perhaps it is most surprising, in the beginnings of modern chemistry, to discover Light and Heat listed as elements. 'Light,' wrote Lavoisier, 'appears to have a great affinity with oxygen . . . and contributes along with caloric to change it into the state of gas.' It also combined with the parts of vegetables to produce their green pigment.[1] Lavoisier also regarded light as a modified caloric, or *vice versa*; this caloric, the invisible, weightless matter of heat was essential to his view of matter. Caloric was interspersed between the particles of bodies: in small measure in the solid, to a great extent in the liquid, and most of all in the gaseous state when (so to

[1] *Elements of Chemistry* (trans. by Robert Kerr), (Edinburgh, 1790), p. 183.

speak) the particles of matter floated freely in the fluid. Oxygen gas was really oxygen matter plus caloric, which was liberated as free heat when the oxygen became "fixed" in combination with a combustible. Caloric also caused the physical expansion of heated bodies, by pushing their particles farther apart. Here are undoubted vestiges of corpuscular mechanism in Lavoisier's thought, when he handled (in a method which would have been familiar to Boyle, but unacceptable to him) the problem of the origin of heat and flame, which had so much puzzled Boyle himself. Here, on the boundary of physics and chemistry, the ideas of the seventeenth-century physico-chemical corpuscularians still survived.

What Lavoisier did is clear enough: how was he able to do it? Disregarding the noisy disputes of theorists and rival interpretations of this or that experiment, factors can be discerned beneath the surface, not always conspicuous to the great theorist himself, leading towards the chemical revolution. Some are fairly obvious: for example, increased reliance on quantitative procedures. What was important here was not the mere tabulation of weights and measures, still less the making of a serious mistake like Boyle's (p. 324)—for numerous exposures of the error failed to make plain its momentous importance—but rather use of measure for constructive purposes, to arouse or to answer questions. "Let us regard the facts without prejudice" said Lavoisier; but a table of facts concerning quantitative reactions is not, and can never be, a chemical theory. He might rather have said "Let us be astonished when we contrast the facts with the expectations created by our theory." This was the spirit of Galileo's, of Newton's, of Black's researches. Rigorous quantitative methods were only useful in proportion as they brought about a sharper juxtaposition of fact and theory, of the flint and steel to which Lavoisier supplied the tinder.

In the second place, the work of the great theorist co-ordinated that of the great experimenters in more senses than one. They supplied him with the pieces whose interlocking provided a perfect picture, and they also built the framework of chemical knowledge within which alone his theory could carry conviction. If Lavoisier's prime experiments were rarely of his own invention, they were the more telling in that they were familiar, in the same way that Galileo's reasoning was the more forceful because it was focused on commonplace experiences. As a third example, it may be observed that Lavoisier extended to its logical conclusion a practical, pragmatic way of thinking about the matter in a chemist's retort which went back at least to the mid-seventeenth century. Earlier chemists had thought of a salt as a metal-stuff and an acid-stuff in combination, had carried out analyses in terms of the sulphur-content, metal-content, alkali-content and so forth of the original material, and having come to see chemical reactions as a process of subtraction and addition, they had invented the theory of affinity to account for it. To this chemist's attitude to matter and its transformations—ignoring the question of the ultimate physical reality of what took place—Lavoisier brought rigour and precision, but it can hardly be said that he created it. Perhaps in this was the core of the chemical revolution, that it severed chemistry from physics, albeit temporarily. Boyle had been frustrated because he tried to explain chemical phenomena rigidly in terms of a physical theory, consciously denying himself the use of any other concepts; Lavoisier succeeded most where he was the pragmatic chemist, least when (in his theory of caloric) he in turn sought to bind physics and chemistry prematurely in one. Despite electro-chemistry and other developments of the first half of the nineteenth century, this hiatus continued, lasting throughout the formative period of modern chemistry. Only in the third stage of chemical atomism, with the acceptance of Avogadro's hypothesis after 1869 and the publication of the periodic table did the reconciliation of chemistry and physics, foreseen by Boyle, and partially abandoned by Lavoisier, become an attainable objective.

CHAPTER XII

EXPERIMENTAL PHYSICS IN THE
EIGHTEENTH CENTURY

THE dual character of eighteenth-century science has already
been remarked upon. On the one hand, those threads which
the historian of science has carefully followed through the
preceding hundred years tend to become dry and brittle; the ex-
citement had passed and there remained but its sequelæ, a rather
timid experimentation and a cool, logical extension of what were
now commonplace ideas. The mood changed too. The rebellious
spirit of a Galileo was no longer necessary, for the works of Aristotle had passed out of serious science into the care of classical
scholars, and the scientific revolutionaries were now respectable
citizens of the republic of learning. As the scientific movement
became more *comme il faut*, so it acquired an element of dullness:
no one can be very interested in the sort of chemical experiments
that Dr. Johnson performed to soothe his mind, in the lessons on
mechanics that were arranged for the royal children, or in the
scientific pot-boilers of John Wesley. As science became fashionable, patronage demanded that a major discovery be presented as
a humble tribute rather than as a challenge to established philosophy. But these idiosyncrasies of a society which left Priestley
to the mercy of the Birmingham rabble and made Banks President
of the Royal Society, and in which the *Ladies' Diary* was filled with
mathematical conundrums, are not after all of the first importance.
Nor was the England of Hume and Gibbon or the France of
Voltaire and Diderot immune to the challenge of intellect; and it
is obvious from what has gone before that in its science the
eighteenth century was by no means an age of mere continuations,
for the creative drive was not so much weakened as altered in
direction.

On the other hand, therefore, new threads of rich interest
strengthened the texture of science in the eighteenth century.
Strategic advances were not made along the broadest paths, but

341

where the giants of the seventeenth century had been least successful. This is strikingly apparent in physics. The natural heirs of Newton are found among the continental mathematicians—Clairaut, D'Alembert, Lagrange and Laplace among the French, and the German Bernoullis and Euler. Britain produced no one to carry on Newtonian physics, with the possible exception of Colin Maclaurin, partly no doubt because the Newtonian fluxions adopted there proved less fertile in mathematical ideas than the Leibnizian calculus adopted on the continent. Varignon (an early French supporter of Newton) had begun to translate mechanics into the language of the calculus in the early 1700s, while Johann Bernoulli used it to demonstrate errors in the first edition of the *Principia* and to make fresh discoveries. After 1740 the correctness in principle of Newtonian physics was generally admitted by the continental mathematicians—one of the first-fruits of this change in opinion was Clairaut's rectification of Newton's theory of the moon—but they always retained an independent, critical attitude. The fundamental concepts of mechanics were defined more neatly and exactly by them, while at the same time they moulded the mathematical structure of the science into a beautifully complete and harmonious series of equations. They achieved an elegance and precision in comparison with which the work of the seventeenth century seems involved and fumbling: even Newton's *Principia*, beside the *Mécanique Céleste* of Laplace, reveals a clumsy mathematical treatment. And Laplace, unlike Newton, saw no reason to bring God into his hypotheses. But the refinements of the continental mathematicians in no way modified the essential principles of the Newtonian mathematical, mechanical method in physics, which were already fixed, although their more penetrating analysis enabled some new problems to be solved, and some old errors to be corrected. The many new and useful ideas that they put forward must therefore be ascribed to the second order of discovery, as being derivative rather than fundamental.

At the same time, at a lower level, experimentation in mechanics, pneumatics, hydraulics and optics—the organized departments of physical science—was taken up extensively. In Leiden Musschenbroek, in London the Hawksbees, were already creating about 1710 a tradition of demonstrative teaching. It is perhaps invidious to call their lectures semi-popular, for the experiments they devised were often far more ingenious and conclusive than those suggested by the pioneer physicists. Such a work as Desa-gulier's *Course of Experimental Philosophy* (1734 etc.) was a valuable manual of laboratory practice and a sound introduction to physics without mathematics. Concurrently there was a steady improve-ment in the design of familiar instruments, such as the barometer, thermometer, hygrometer, air-pump, balance and so on; with these multitudes of observations were made, without, however, eliciting any major discovery. The situation in physics was clearly, so far as its more highly organized departments were concerned, very different from that in astronomy (for example), where Bradley's discovery of the aberration of light was the direct result of refinement in angular measure. Yet even so, problems of values arise, and it cannot be taken for granted that the less dramatic work of the eighteenth century had the less significance. It could be argued, for instance, that the graduated scales of Fahrenheit and Réaumur alone made thermometry effective in science, with important results in both physics and chemistry.

It is indeed very easy to undervalue second-order discoveries in a period of consolidation. Creative ability of the first order is extremely rare, and though its works must figure largely in any short account of the development of science, it would be absurd to suppose that it could flourish apart from the context created by men with smaller endowments. Between the peaks of scientific achievement there is a time when activities are re-phased, when perspectives alter, when with an enlarged range of facts the major problems gradually modify their shape. Without this Newton could not have succeeded Galileo, nor Darwin Linnæus. More-over, it very rarely happens that the statement or the demon-stration of a first-order discovery are so perfect that they win complete conviction, or that its potentialities are fully exploited. The task of welding the fabric of science together, of preserving the logic and homogeneity which originality in its highest degree often seems to imperil, usually falls to the derivative investigator, and it is by this consolidation, as well as by its most splendid feats of conceptualization and experiment, that science grows.

Two aspects of such consolidation are well illustrated in the experimental physics of the eighteenth century. The expedition to the head of the Gulf of Bothnia led by Maupertius and Clairaut in 1736–7 was a bold undertaking for the time. Forests tangled with fallen trees had to be penetrated, rapids navigated, and quadrants handled when the mercury had sunk far below zero, by these gentlemen fresh from the *salons* of Paris. Their purpose was to measure the length of a degree of latitude, in order to compare their result with that already obtained in France by Picard. Upon this comparison rested the verification of an important theory. For if the degree proved to be longer in northern than in southern latitudes, then the earth was flattened in the polar regions, as Newton had reasoned from dynamical considerations. This proved to be the case, and the ratio of the polar to the equatorial diameter of the earth was found to be $177/178$. Some years later a similar ratio was worked out from the results of another geodetic expedition to Peru (1735-43). The critics of the Newtonian theory of the earth's shape, including the astronomer Jacques Cassini, were thus decisively confuted.[1] Newton's theory, however, gave the ratio of flattening as $229/230$. A much better agreement was obtained in Clairaut's own reinvestigation of the theory (1743), which yielded a formula relating the earth's ellipticity to the gravitational attraction at any latitude.[2] By these means, not only was the application of the principles of mechanics to the great mass of the earth itself validated, but the actual method of applying them was improved to give a better coherence between theory and experiment. At the same time, new light was thrown on the pendulum experiments made in different parts of the globe. The impregnability of Newton's theory of attraction was further increased, at a time when it had still to win the confidence of many continental scientists.

The course of these events was predictable. It was certain that if there was a controversy over the shape of the earth, it would be settled by making measurements—that was the established spirit of the scientific revolution. It was also certain that if theory and measurement did not wholly coincide attempts would be made to re-examine the theory in search of some neglected factor. Exactly this happened—and the theory was improved. In contrast, other second-order discoveries in eighteenth-century physics, though continuations of what had gone before, were altogether unpre-

[1] Cf. Pierre Brunet: *La vie et l'œuvre de Clairaut*, in *Revue d'Histoire des Sciences*, vol. IV (Paris, 1951), pp. 105-32.
[2] The theory was also revised a little earlier by the Scottish mathematician Maclaurin using methods much less comprehensive than those of Clairaut.

dictable. Perhaps the most striking of them was made by the London optician, John Dollond, in 1759. Newton had supposed that when a beam of light passed through a prism, the dispersion of the colours in the spectrum cast was in an invariable proportion to the degree of refraction, and quite independent of the material of the prism. Consequently he had held that chromatic aberration (caused by the failure of lenses to bring all colours to the same focus, owing to dispersion) was incapable of correction. The German mathematician Euler, thinking of the human eye as a perfect lens-system made up of several media, suggested that a triplet "lens" formed by enclosing water between the interior concave surfaces of two juxtaposed glass lenses would likewise be free from aberration. He was able to work out (1747) the required curvatures, but attempts to put his idea into practice failed. Following Euler's suggestion, and the earlier empirical success of Chester Moor Hall who had made an achromatic lens in 1733, Dollond experimented on the relative dispersion and refractive power of water and glass, and later of the two kinds of glass, crown and flint.[1] He found that in a "doublet", consisting of a convex lens of flint-glass and a concave lens of crown-glass, the curvatures could be so adjusted that the dispersions were nearly equal and opposite, while the greater refraction of the flint-glass enabled the combination to behave as a convex lens. An almost achromatic object-glass for telescopes was possible: astronomers, who had been increasingly turning to the reflecting telescope, were once more able to make use of the more reliable refractor. Attempts were made almost at once to apply Dollond's discovery to the manufacture of achromatic objectives for microscopes, a task of which the solution occupied more than fifty years. Perhaps most important of all, Newton's authority in optics was seriously checked, almost for the first time. His confident belief, unfounded on experiment, was exposed as an error.

The principal innovations in eighteenth-century physics were, however, in totally new directions, showing that the conceptual fertility and experimental ingenuity so marked in the preceding hundred years were far from exhausted. Indeed, in England especially they received renewed inspiration from the example of Newton. Today he is (rightly) esteemed above all as a mathematical physicist; but the non-mathematical scientists of the

---

[1] Flint-glass contains a large proportion of lead, crown none.

eighteenth century who admired rather than understood the *Principia* found in *Opticks* a model of empirical enquiry, and in its *Quaeries* justification for developing speculative physical theories. While the Newtonian kind of mathematical physics was developed on the continent, British experimental scientists turned from mechanics to the investigation of heat, electricity, and chemistry, following the hints that Newton had left in the *Quaeries* and other non-mathematical writings. Men like Stephen Hales, Desaguliers and Franklin considered themselves to be Newtonians, but their ideas drew more upon his aetherial speculations than upon his solid mathematical reasoning, though most of them believed that Newton would not have printed those speculations had he not been sure of their truth.

The two major lines of advance in experimental physics during the eighteenth century were intrinsically far more difficult than those hitherto followed. The seventeenth century had failed, for example, both to place the science of heat on an exact quantitative foundation, and to make the conceptual distinctions which were essential to a true understanding of the phenomena. For these the vague notions of corpuscularian philosophers that heat was "nothing else but a very brisk and vehement agitation of the parts of a body" (Hooke) were a very unsatisfactory substitute. Even Boyle had not been able altogether to renounce the idea of "fire particles," or to distinguish firmly between combustion and other manifestations of heat. Eighteenth-century physicists found it more useful to think of heat as a fluid (called *caloric* in the new chemical nomenclature), which flowed from hot bodies to cold; an hypothesis which (like that of phlogiston) proved to be radically mistaken but which served well at a certain stage.

In particular, this view of heat was extraordinarily appropriate since most of the eighteenth-century experiments on heat were concerned with thermal capacity and calorimetry. Measurements of temperature by means of thermometers (though without regard to any widely accepted scale) had become fairly commonplace after 1660, but for a long time no distinction was made between the temperature of a body, and the amount or quantity of heat present in it. This was natural enough, for it was believed that equal weights of all substances had the same thermal capacity— that the same amount of heat was always required to raise (say) one pound of water, iron or anything else through the same number of degrees on the thermometer scale. Fahrenheit seems to have been the first to note a contrary observation. Comparing the heating and cooling effects of equal volumes of water and mercury, he found that the latter was not thirteen times as effective as the former, as the density-ratio would suggest, but only 60 per cent. as effective! From this Black judged (about 1760) that the amounts of heat required to raise two different bodies through the same number of degrees of temperature were in a very different proportion from that of their densities (assuming the volumes to be equal). Black went on to compare the amount of heat required to raise 1 lb. of water through $t°$ with that required to raise 1 lb. of any other substance through $T°$, which could easily be done by mixing the two together at different temperatures so that they attained a common temperature. From this experiment the relative heat-capacities (or specific heats) of various substances could be ascertained, taking that of water as a standard.[1] It was easy to imagine that different substances had different capabilities for absorbing the matter of heat (caloric), the association of the familiar notions *fluid* and *capacity* proving as fruitful in the science of heat as it did later in the science of electricity.

Black's other discovery concerning the capacity of bodies for acquiring heat was even more surprising. The plausible notion that when a solid (such as ice, or a metal) was brought to its melting point as shown by a thermometer, only a small amount of further heat was required to liquefy it had not been challenged. Similarly it was thought that a minute loss of heat was enough to cause water at 32° F. to congeal : no discontinuity between the solid and liquid states, in relation to heat lost or absorbed, was imagined. Black, however, observed that after a spell of cold weather masses of ice and snow would last for weeks without melting; were it not so 'torrents and inundations would be incomparably more irresistible and dreadful. They would tear up and sweep away everything, and that so suddenly that mankind should have great difficulty to escape their ravages.' A piece of melting ice, he reasoned, must still be capable of absorbing a great quantity

[1] e.g., mixing equal weights of water at 0° and gold at 103°, the resultant common temperature being 3°; the water has been heated through 3° by the quantity of heat that the gold lost in cooling through 100°. Hence the thermal capacities of water and gold are as 100 to 3, or the specific heat of gold is 0·03.

of heat, although the water running from it was at freezing-point. In other words, the ice was capable of taking up heat, without a corresponding increase in temperature being apparent; the only effect of this extra heat was the liquefying of the ice. It was *latent*, because it seemed to be 'absorbed or concealed within the water, so as not to be discoverable by the application of the thermometer.' Black then proceeded to measure experimentally the amount of latent heat taken up by ice on its conversion into water, and by water on its conversion into steam. Thus he attained the conception of a definite quantity of the "matter of heat"—insensible to the thermometer—being involved in any change of physical state; this "quantum" of heat was not a mere agent dissolving ice into water, but was a physical constituent of water differentiating it from ice, or alternatively, from steam.

Later, Laplace and Lavoisier were able to perfect Black's experiments with the ice-calorimeter which they invented.[1] They pointed out that such quantitative determinations could be carried out without theoretical preconceptions, but, like Black, they favoured the material theory, as is very evident in Lavoisier's *Traité Élementaire de Chimie*. Like Black, they concluded that any sample of matter (in a given physical state, at a given temperature) consists of substance and heat, the absolute degree of heat being incapable of registration on a thermometer, and residing in all bodies even at the lowest attainable temperatures. Their experiments went much further than Black's in studying the evolution or loss of heat in chemical operations, whence it appeared that the theory of chemistry would need to account not merely for matter-reactions (syntheses, analyses, exchanges and so forth) but for the heat-reactions with which these are integrally associated. In modern terms, they had realized that questions of energy are involved in chemical processes; as the concept *energy* was unknown to them, however, they naturally tended to make the heat-reaction cognate to the matter-reaction, by treating heat itself as a material entity, and measurements of "quantities of heat" as measurement of *something*, which would be conceived (at least in imagination) as existing apart from the matter whose quantity of heat was measured. The very obvious formulation

$$\text{ice} + \text{heat} \rightarrow \text{water}; \quad \text{water} + \text{heat} \rightarrow \text{steam}$$

[1] In this instrument, the quantity of heat lost by a body in falling to 0° C. was measured by the weight of water melted from a surrounding jacket of ice.

emphasized the dichotomy between the matter (water-corpuscles) identical in ice, water and steam, and the absorbed quantity of something measurable making the distinction between its three states. Attempts, throughout the eighteenth century, to measure the mass of heat by weighing heated and cooled bodies, which had a negative result, did not disturb those who held that heat was a weightless fluid.[1] Only in 1798 was attention called to the fact that bodies can evolve an indefinite amount of heat through friction by Benjamin Thompson's (Count Rumford) experiments. On the material theory the quantity of heat contained in a body at a given temperature was limited absolutely, but in spite of this difficulty, the kinetic view of heat (as it was less applicable, at this stage, to any other phenomena than those of friction) failed to displace the material theory for another half-century.

This fact is itself enough to provoke reflection among those who hold that quantitative experiments are infallible instruments of scientific progress. When the theory of heat was entirely qualitative, not to say speculative, in the seventeenth century, a more "correct" kinetic view prevailed. The material theory—which so successfully resisted, in the decade 1840–50, the attrition of Mayer and Joule—was decisively established by the quantitative experiments of Black, Laplace, Lavoisier and others. The "correct" kinetic view was in fact decisively obstructed by its failure to yield neat and easily comprehensible quantitative results, and for this reason it could not, perhaps, ever have been established save (as it eventually was) under the cloak of a generalized concept of energy.

In many of its aspects eighteenth-century physics represents the pre-history of this concept. Mechanical energy, as *vis viva*, and the power to do work, was very attentively considered. The first steps towards the study of chemical energy were taken by Laplace and Lavoisier. The very complex rôle of heat in physical and chemical changes was at least partially disclosed, and in a purely practical way the connection between heat energy and mechanical energy was of great interest to engineers, ever extending the utility and efficiency of the steam-engine as a prime mover. The major invention of this period—Watt's introduction of the separate con-

[1] It is very strange that the implausibility of the concept of matter without weight (which has been held by some to have inspired Lavoisier's attack on phlogiston) was one which he himself embraced in his own theory of caloric.

denser—is said to have been inspired by Black's discovery of latent heat, and certainly Watt carried on his own investigations into the physics of heat.[1] It was clearly his purpose to squeeze the maximum mechanical value from the "quantity of heat" contained in steam—and so cut down the coal bills of those who bought his engine—but the theoretical implications, in terms of a "perfectly efficient" engine, were only worked out by Sadi Carnot about 1824. Heat energy in the form of invisible radiation was also known, and the likeness in properties between this form of heat and light was recognized. Experiments to find out whether light itself has energy—for if a beam of light consisted of a stream of corpuscles travelling at a very high speed, their impact upon an opaque surface should be detectable, even though very minute—yielded some positive effects, but these were most probably due to other causes. Perhaps most important of all was the elucidation of a new form of energy, electricity. Certainly this was the most striking, the most original, and the most progressive branch of eighteenth-century physics. The spectacle of the erect hair, the nasal sparks, of an electrified youth hung in silk cords from the ceiling excited the rather coarse humour of the age; the mystery of lightning drawn off down an iron rod and confined, like a jinn, in a Leyden bottle, was witness to the strange power of nature and man's intellectual mastery of it; while, at a more serious and prosaic level, a new corpus of experimental and theoretical knowledge was taking shape, of incalculable importance for the future. No one could have foreseen, in Franklin's day, the extent to which physics was to become the science of electricity, yet already by 1800 almost every experimental physicist was to some degree an electrician.

All this grew from very humble origins. Gilbert had shown that the attractive property of rubbed amber was quite widespread in nature, and had coined such terms as "electric" (from Greek *elektron* = amber), and "charged body." Soon afterwards the

---

[1] It was not at all necessary, however, for Watt's purpose that the main cause of the wastefulness of contemporary steam-engines (the chilling of the cylinder and piston by the injection of cold water to produce a vacuum) should have been so scientifically diagnosed. An intuitive realization of the folly of alternately heating and cooling the same masses of metal without any productive purpose would have been amply sufficient. So that, whatever advice Watt may have received from Black, it can hardly be said that the separate condenser was an immediate fruit of the physics of heat.

mutual repulsion of similarly charged fragments of light materials was noticed for the first time. But the first hints of more remarkable effects followed upon the introduction of the first electrical machines, and thus anticipated the course of the later history of electricity: every important step was brought about by some new instrument or device. For electrical phenomena are not made manifest in nature; the few that occur (such as lightning) could never conceivably have been interpreted correctly in the light of reason. They were entirely hidden from artisans and other practical men skilled in nature's ways, no tools or instruments of science or art could be easily adapted to an inquiry into electricity, and the human body is very limited in its reactions to electrical stimuli from outside. There could, therefore, be no progress in electrical science without means of creating charges, currents etc., and means of revealing their various properties. Theoretical rationalization was bound to be, in the very early stages, of relatively little importance, and in any case subject to extremely rapid fluctuations.

The first electrical machine was a globe of sulphur or glass, mounted on an axle and rotated by a handle, which was rubbed against the hand until highly electrified. The glow, visible in the dark, produced by discharge between the globe and the hand was first noted by Otto von Guericke, who also succeeded in transmitting the electrical effect along a linen thread. Another curious phenomenon observed at about the same time was for long unrelated to electricity. This was the luminosity in the vacuum of a barometer when the mercury was shaken. Only in 1745 was it shown that under these conditions the glass tube became electrified, though Hawksbee, about 1710, had caused a similar glow to appear upon an electrical machine worked *in vacuo*, and inside an exhausted vessel rubbed externally. Hawksbee allowed a chain to hang against the globe of his electrical machine, so that the charge would be taken to a large "prime conductor," but the next improvement—the use of a soft rubber instead of the operator's hand—was only introduced a little before the middle of the century.

The study of conduction was taken further by Stephen Gray (1732), who found that charges could be transmitted along, or induced into, very long lines of thread when these were suitably supported. He was thus led to make the fundamental distinction between *insulators* and *conductors*, for silk filaments did not permit his charges to leak away, while equally fine copper wires did. Hair, resin and glass proved like silk non-conducting. A Frenchman, Charles Dufay, had the ingenuity to mount a variety of substances upon insulating supports in order to demonstrate that all—including the metals—could be electrified by friction when isolated from the earth; he saw that Gray's distinction between insulators and conductors was really more primary than the established one between electrics and non-electrics, to which it had seemed analogous. Dufay, moreover, discovered that a fragment of gold leaf, charged with an electrified glass rod, was not repelled (as he expected) by a piece of electrified amber, but strongly attracted to it. The lesson, from magnetism, was obvious enough; the two charges, the one "vitreous" and the other "resinous," were of opposite sign. In the neutral state all bodies contained equal quantities of both electricities; the action of friction was to remove a part of one or the other, leaving behind a superfluity of the second.

Any substance could be charged on a suitable stand. A number of experimenters tried to electrify water in insulating glass vessels: they discovered, to their distress, that if with one hand grasping the jar full of liquid, with the other hand they tried to take away the wire leading into it from the electrical machine, they experienced a frightful shock. They had, in fact, formed a condenser and discharged it through their bodies: it was an easy step to make the "Leyden jar" more convenient by lining it within and without with metal foil—and to learn to handle it with greater caution. With the aid of powerful frictional machines, and the large charge built up in a Leyden jar, it was possible by 1750 to produce very striking sparks, and discharges heavy enough to kill small animals or to be transmitted through long circuits of wire or water. Attempts were even made to estimate the velocity of the motion of a charge by the interval between two sparks across gaps separated by a long circuit: but of course without success. The heating effect of electricity (e.g. in melting fine conductors) was easily perceived as were some of the effects associated with discharge through a vacuum. Upon electrical theory the effect of the discovery of the condenser was profound. The dualistic hypothesis of Dufay was clearly susceptible of simplification: it was unnecessary to suppose that there were two kinds of electricity (comparable to the two poles of a magnet) for "oppositeness" could be taken, as in mathematics, as a difference in quantity rather than a difference in quality. On such a view, with a normal charge a body would be neutral, with an excess of electricity positively charged, and with a defect of electricity negatively charged. This view could be applied with particular success to the novel phenomena revealed by the condenser.

The man who so applied and developed the unitary theory of electricity was Benjamin Franklin (1706–90), retired printer of Philadelphia, popular philosopher, later hero of the American revolution and elder statesman of the young republic. With his plain common sense and distrust of subtlety, Franklin combined an active scientific imagination sharpened, perhaps, by his almost complete ignorance (during the creative stages of his work) of European theories. In his opinion, which strongly reflected the corpuscularian ideas of the age, electricity was a fluid, consisting of particles mutually repelling each other, but strongly attractive to other matter, which distributed itself uniformly as an "atmosphere" about a body or connected system of bodies. Electrification was a process whereby an excess of this fluid was collected upon a particular body, such as a glass rod, by friction or other means, so that it became positively charged; or removed from it so that it became negatively charged. Franklin believed (mistakenly) that a discharge was simply a transfer, often in the form of a unidirectional spark, from a body more highly charged with the fluid to one less charged; and the repulsion between two positively charged cork balls was readily attributed by him to the repulsion between the excess of electric particles. When Franklin became aware that *negatively* charged bodies also repel each other, he encountered a difficulty which his theory could not overcome. He fully realized the importance of the deduction, from his fluid theory, that within a closed system the quantity of electricity must be conserved: a person, placed on an insulating stand, could collect a positive charge upon a glass rod by rubbing it, but only by electrifying his own body negatively to an equal degree, that is, by forcing electricity from his body into the tube. By inverse reasoning from the same principle Franklin explained the action of the condenser, for every addition of electricity to one of its surfaces produced a corresponding loss (to earth) from the other, as shown by the fact that the Leyden jar could not be charged unless one plate was earthed. The total quantity of electricity in the jar was always constant, only its distribution being modified by electrification since a connection between the plates (under any conditions) ensured a return to the neutral state. Franklin regarded the condenser as fully charged when *all* the electric "atmosphere" had been driven off the earthed plate, for then no more could be added to the positive plate, since this would have increased the total quantity present.

This theory, and the experiments intended to confirm it, of which many were already familiar to European electricians, were warmly welcomed, as indeed were all contributions to science from the New World at this time. But at first Franklin's letters did not carry conviction, nor did they have any very dramatic effect. Strangely, it was a less creditable, but more showy, suggestion that brought about his lionization. Many electricians had noted the similarity between the electric spark and lightning and between the accompanying crackle and thunder, speculating on the possible identity of the two effects; there was therefore nothing very new in Franklin's similar speculation. He, however, had noticed particularly the powerful action of pointed conductors in "drawing off" a charge silently and conjectured that if thunder-clouds were, as he supposed, positively charged, the fact could be revealed by drawing electricity away to a high, pointed conductor. Having described the experiment, he failed to execute himself (for a variety of reasons, among which lack of courage was certainly not one). It was first successfully performed in France, soon after the translation of Franklin's early letters was made. The well-known "Kite experiment" at Philadelphia was carried out, with a like result, before news of the French attempts and of his own sudden fame as the tamer of lightning had reached Franklin in America. Richmann, at St. Petersburg, was the first "martyr" to the pursuit of this new branch of electrical science.

Until the moment of Franklin's intervention (1746–55) elec-tricians had been wholly preoccupied with qualitative effects. Most of the material published had dealt with the description of phenomena, which the experimenters had sought to explain in the light of *ad hoc* hypotheses, each of them jejune in varying degrees and inadequate to explain all the facts. Franklin himself had not sensibly modified this state of affairs, for though his single-fluid theory was more comprehensive than any other, it was not completely so, nor did it really rise above the phenomenalistic level of his work. The revisions he introduced himself were sufficiently serious for it to be classified (logically) as no more than a provisional working hypothesis. The investigation of new effects continued to be of some importance—as, for example, in the study of induction, of the rôle of the dielectric in condensers, and of pyro-electric phenomena—but a more rigorous inquiry into the *quantitative* aspects of electrical phenomena grew up alongside it during the second half of the century. In accordance with the general principle already mentioned (p. 237), the electrician's apparatus, which had formerly been limited to the revelation of the qualities of electricity, now began to be adapted to measure quantities. An interesting instance of this is the elaboration, from the pith-balls and gold-foils formerly employed to test for the existence of charges, of the torsion-balance of Coulomb (*c.* 1784) and the electrometer of Bennet (1787). Hitherto, though the achievement of solid additions to knowledge and the ingenuity of theorization must be duly recognized, experiment and thinking in electricity had been amateurish, for the former had often partaken of the nature of a parlour game, and the latter had included much extravagance. The rapid success of Franklin, starting almost from zero, is an indication as well of the superficiality of the subject, as of his own ability. Now, however, electrical science began to acquire a more serious status as a branch of physics.

Among the first to deny themselves the pleasure of declaring what electricity is, was Joseph Priestley. His precisely regulated experiments on conduction contain the seeds of later ideas of electrical resistance; like Æpinus a little earlier, he found that the distinction between conductors and insulators was far from absolute. Priestley used the length of a spark-gap, across which a discharge would jump instead of traversing a long circuit, as a measure of the circuit's resistance. He repeated Franklin's experiment to show that there is no charge inside an electrified hollow vessel of metal, and deduced from it (by analogy with a familiar theorem on gravitational attraction) the opinion that 'the attraction of electricity is subject to the same laws with that of gravitation' that is to say, it varied proportionately to the inverse square of the distance. This was perhaps the first proposition about

electricity that could be formulated mathematically, and moreover a prediction to which experimental verification could be
applied. This was done by Robison two years later (in 1769), by
Cavendish, and by the French engineer Coulomb. Cavendish
designed an ingenious apparatus by which one metal sphere could
be enclosed within another, with or without electrical contact
between the two, finding that the charge applied was invariably
confined to the outer sphere. He ascertained that the electric
force must vary inversely as the square of the distance within
limits of ± 1 per cent. Many other quantitative experiments
(partially anticipating the later work of Michael Faraday) were
performed by him about 1771–3, which with typical unconcern
he kept to himself. He was the first electrician to adopt a standard
of capacity (a metal-covered sphere, 12·1 inches in diameter), with
which he compared the capacities of other bodies, stating these
as "inches of electricity," that is as the diameters of spheres of
equal capacity, and to realize that the charge upon a conductor
is proportional to both its capacity and the "degree of electrification" applied. By "degree of electrification" Cavendish understood the extent to which the "electric fluid" was compressed
into a body, so that he approached very near to the later concept
of electrical potential. He also knew that the capacity of a condenser varies with the material of the dielectric, and made several
measurements of what Faraday was afterwards to name specific
inductive capacity. He measured the conductivity of a variety of
solutions in glass tubes, discovering that this was independent of
the size of the discharge passed through them. Considering that
he made use of crude pith-ball electrometers, and relied upon his
own senses to compare the violence of electric shocks, the numerical results he obtained were astonishingly good. All this work,
unfortunately, had no effect upon the subsequent progress of
electrostatics, as it was totally unknown. The inverse square law of
electric force was first demonstrated in print by Coulomb (1785),
in experiments one of which was similar to that of Cavendish,
while others made use of Coulomb's torsion-balance for the direct
measurement of forces. These experiments in turn served as the
foundation for Poisson's mathematical study of electrostatic forces
in the early nineteenth century.

The interval was marked by no important discoveries. This was
undoubtedly due in very large part to the sudden fascination of a

new set of phenomenalistic effects, wholly unsuspected, in which electricity appeared in another of its Protean forms. Hitherto the manifestations of electricity had been limited to two groups: (a) shocks and sparks, (b) repulsions and attractions. To the first group belonged, besides the laboratory effects, those of lightning and the torpedo or electric fish. No serious physiological study had been made of the first group of effects—it was merely known that a shock produced violent muscular contractions, and followed a more or less direct path through the body—though the administration of shocks was (like most other things) regarded as having a medical value. Hence experimentation in electricity had been practically confined to the exploitation of the attraction-repulsion effect in a variety of different ways. This in turn had its influence on theory. In the first place electricity was literally regarded as an effluvium, a particulate atmosphere surrounding the charged body. How, Newton had asked, can an electrified body 'emit an exhalation so rare and subtle, and yet so potent, as by its emission to cause no sensible diminution of the weight of the electrick body . . .?'[1] Then again, the Abbé Nollet (1700–70) had supposed repulsion and attraction to be the work of outflowing and inflowing streams of the electric fluid. Later in the century, the analogy between electrical and gravitational attraction becoming more obvious, the electric effluvium seemed as absurd as the gravitational effluvium of pre-Newtonian physicists, and "action at a distance" was accepted. Thus the theory of electricity passed through various phases of mechanical explanation, much as the theory of gravity had done in the seventeenth century, owing to the focusing of attention upon its mechanical manifestations. Just as the weightless fluid caloric was capable of mechanically expanding bodies, so the weightless electric fluid was capable of putting them in motion—even of causing continuous rotation in a light wheel. Mechanical analogies really justified the concept of electricity as a fluid, whose particles were capable of action at a distance, for electricity could flow along conductors, filling bodies to their "capacity," and yet be impeded by "impermeable" substances. Borrowings from the languages and ideas of hydraulics are indeed obvious; Cavendish's concept of electrical "pressure" showed how fertile they could be.

It is the nature of a fluid, even an elastic fluid, to flow, and the

[1] *Opticks*, Query 22.

study of "electrostatics" had actually embraced some investigation into the flow of electricity along conductors. But the effects produced by the flow of charge (under the prevailing conditions of experiment) were not striking as compared with the mechanically obvious effects of a static charge. The motion of electricity, revealed by a spark or a shock, was in any case a transient event, restoring the apparatus to a condition of electrical neutrality, so that the phenomenon of electricity had always appeared to be discontinuous, indeed so much so as to be almost adventitious. Charges were immediately annihilated by conduction to earth, and no perfectly insulating material was known. They were formed only by the chance electrification of a cloud, or by the discontinuous action of friction, which suggested that they were mechanically produced. The appearance of all the known effects depended upon the discontinuity in the movement of electricity, leading to the accumulation of large static charges.

The discovery of new manifestations of electricity was thus of absorbing interest for a variety of reasons. These were not the result of mechanical action, nor were they themselves mechanical in nature. They were continuous, and they required no effort. In the prevailing theory some segregating action was necessary to bring about the conditions of electrification—the fluid had to be impelled from one body to another, leaving the former unnaturally empty and making the latter unnaturally full—so creating an unbalanced state which nature herself sought to adjust at the first opportunity. The new effects predicated no such positive action; it was not necessary, in order to produce them, to build up mechanically an artificial "degree of electrification."

The differences between the new phenomena and the old were sufficient, at first, to obscure the connection between them. About 1780, in the laboratory of the Italian anatomist Luigi Galvani (1737–98) at Bologna, an assistant happened to notice that when he touched with his scalpel the crural nerve of a frog's leg which he was dissecting the muscles were violently contracted. It was remarked that this occurred while a spark was being drawn from an electrical machine placed on the same table. Galvani repeated the strange experiment, under the same conditions and with a like result. He introduced many variants, all of which proved that the experiment failed unless the operator was in electrical connection with the nerve, and the frog's leg effectively earthed. From this

Galvani suspected that some sort of electrical circuit was involved —for muscular contractions were known to occur when discharges were sent through dead animals—even though the frog was not directly linked to the electrical machine. Many experiments were performed at this stage to gain conviction that the stimulus was really electrical. The next major step was a successful demonstration that lightning flashes acted upon the limb in an identical fashion when similar electrical connections were made to it. In the course of these atmospheric experiments Galvani observed that frogs hung from an iron lattice in his garden by brass hooks penetrating into the spinal marrow gave occasional convulsions. Once, happening to press one of the hooks firmly against the iron, he saw immediate contractions. At first he thought that these were due to the escape to earth of some atmospheric electricity accumulated in the frog. To test this suspicion he re-created the same conditions indoors, placing the frog on an iron plate and pressing the brass hook firmly against it. Still the convulsions occurred (1786). Other combinations of metals, or even a circuit through his own body, or a homogeneous wire, with which he made a "conducting arc" between nerve and muscle, had the same effect, but not insulators. Again Galvani satisfied himself by elaborate experiments that the decisive circumstance was the existence of a path along which electricity could flow.

Had he been a more enthusiastic electrician, Galvani might have inquired more fully into the curious situation in which a spark from an electrical machine or Leyden jar could influence a frog's leg insulated from it. Instead, after his discovery that the spark was unnecessary provided that a direct connection was made between nerve and muscle, he abandoned that subject. At this point two other lines of inquiry suggested themselves. He could examine more carefully the nature of the electrical circuit, and this he did in some detail, finding that some metals were less effective than others in the conducting arc, that liquids could be used, and that a single conductor was less effective than one made up from two metals. But he did not attach great importance to the bimetallic circuit, because convulsions were obtained with a single conductor. He was therefore led to concentrate upon the second line of inquiry, convinced as he was that Leyden jars, machines, thunderstorms and other familiar sources of electricity could be excluded from his explanation of the phenomena, and that the conducting arc was simply an ordinary conductor of electricity. The physiology of the frog's reactions now drew his attention, since it appeared that the mere metallic connection of nerve and muscle was capable of causing the same contractions as the application of an electrical stimulus to these parts. In the former case the electricity moving along the conductor must have been supplied by the frog itself. To hypothesize further still, was not the stimulus given by a nerve to a muscle always electrical, for (as he said) 'the hidden nature of animal spirits, searched for in vain until now, appears at last as scarcely obscure'—it was electricity! Prepared in the brain, and distributed by the nerves, this "animal electricity" on entering the muscles caused their particles to attract each other more strongly, and the fibres correspondingly to contract. In the muscles also electricity was stored, as in a Leyden jar, so that from a communication between them and a nerve ensued the convulsions witnessed in his last series of experiments. Thus Galvani brought his electrical discoveries to a close by riding off on a physiological hobby-horse, abandoning physics in order to debate how his new knowledge of animal electricity might be applied to the cure of disease. Many other scientists eagerly followed his example.

An alternative and less dramatic interpretation would have made the frog's leg merely a sensitive detector of an electric current through the circuit linking nerve and muscle, as in Galvani's first experiments, the stimulus for the convulsions being always supplied by an external source of electricity. On this view, little compatible at first sight with his later discoveries, Galvani had not discovered a new example of animal electricity (already familiar in electric fish); he had invented a new electrical instrument, and a new source of electricity in motion—the bimetallic conductor of his last experiments. This was the interpretation of Galvani's work put forward by Alessandro Volta (1745–1827) of Pavia in 1792, about a year after the first account of them was published. At first Volta had given credence to Galvani's own explanation, hailing his discoveries as no less epoch-making than those of Franklin. Gradually, however, Volta uncovered facts which compelled him to differ from Galvani. The most delicate electrometer revealed no electricity in animal tissues; to cause the muscles to contract, it was only necessary to apply an electrical stimulus to the nerves, and not to the muscles themselves; and to create this stimulus a bimetallic junction was essential.[1] Finally, in a letter destined for the Royal Society, Volta asserted that the frog's leg as prepared by Galvani was nothing other than a delicate electrometer, and that most of the phenomena attributed to animal electricity were 'really the effects of a very feeble artificial electricity, which is generated in a way that is beyond doubt by the simple application of two different metals.'

> By this time [wrote Volta in November 1792] I am persuaded that the electric fluid is never excited and moved by the proper action of the organs, or by any vital force, or extended to be brought from one part of the animal to another, but that it is determined and constrained by virtue of an impulse which it receives in the place where the metals join.[2]

He had invented a new theory of contact electricity, and was compelled, in its defence, to criticize the physiological views of the wretched Galvani, who died in despair after some years of futile controversy. Meanwhile Volta continued his experiments. He found that the metals whose contact caused a flow of electric fluid could be arranged in a definite order, that other conductors such as carbon had the same effect, and that any moist conductor, such as water, served to bridge the different metals as well as animal tissues. By 1795 he had propounded the "law" that whenever two dissimilar metallic conductors were in contact with each other and a moist conductor, a flow of electricity took place. In 1796 he demonstrated that the mere contact of different metals produced equal and opposite charges upon them, made visible by the electroscope. This was the first proof of the identity of galvanic and frictional electricity, of which Volta had always been convinced. The charges, which appeared to be indefinitely procurable, were positive or negative according to the order of the metals used in the series which he had discovered earlier.[3] The intensity of the effects produced was still minute, and Volta realized that it could not be increased by multiplying the number of bimetallic junctions—so much was obvious from the distribution of charges. The case was different when two or more pairs of metal

---

[1] Volta ascribed the effects obtained by Galvani with a single conducting element to lack of homogeneity and other differences in the metal.

[2] *Opere* (Firenze, 1816), vol. II, pt. i, pp. 165–6.

[3] i.e. + Zn, Pb, Sn, Fe, Cu, Ag, Au, C−. These experiments were only made possible by Volta's improvement of the electroscope in earlier years.

plates in contact were joined together, not directly or by use of a third metal conductor, but by one of the moist conductors such as salt water, either placed in cups into which the metal plates were dipped, or soaked into cardboard discs inserted between each bimetallic pair. In this arrangement (described by Volta in 1800) the intensity of the effects produced was proportional to the number of the pairs of plates, so that Volta could charge condensers, produce sparks, and give severe shocks. This "artificial electrical organ" (which Volta compared to the natural electrical organs of certain fish), electromotor, or *pile*, as it was afterwards called, was as he said like a feebly charged Leyden jar of immense capacity, for it would yield electricity continuously. The continuity of the flow of electricity from terminal to terminal of the pile was particularly emphasized by Volta in his descriptive letter to the Royal Society; as there was no instrument suitable for the detection of continuous currents, he had to quote his physiological sensations in proof of their existence. The situation was paradoxical, but none the less real.

It seems clear that Volta, who was little interested in the chemical effects associated with the passage of an electric current, completely misunderstood the functioning of the single cells in his pile. He regarded the "electric force" as originating from the contact of two different metals, not from the reaction of these with the moist electrolyte between them. In modern theory the voltaic cell consists (for example) of copper, electrolyte and zinc; but to Volta himself the copper-zinc junction was the "cell," the source of electricity, and the electrolyte served merely to connect these together without neutralizing the charges collected on the metals. He was still thinking, essentially, in electrostatic terms:

> The action exciting and moving the electric fluid is not due, as is falsely believed, to the contact of the humid substance with the metal; or at any rate it is only due to that in a very small degree, which may be neglected in comparison with that due to the contact of two different metals, as all my experiments prove. In consequence the active element in my electromotive apparatus, in piles, or in cups, or in any other form that may be constructed in accordance with the same principles, is the mere metallic junction of two different metals, certainly not a humid substance applied to a metal, or included between two different metals, as the majority of physicists have claimed. The humid layers in this apparatus serve only to connect the metallic

junctions disposed in such a way as to impel the electric fluid in one direction, and to make this connection so that there shall be no action in a contrary direction.[1]

This theory did not long survive. As Nicholson pointed out in 1802, an electric current could be drawn from cells consisting of one metal and two electrolytes, and by this time already the work on the new form of electricity was assuming a markedly chemical character. Fabroni, in 1796, had observed the oxidation of one of a pair of plates of different metals joined together and immersed in water. Ritter had perceived that the order of the metals in Volta's series was a chemical order—that of their exchange in solutions of their salts. Very soon after Volta's letter of 1800 reached London, Nicholson and Carlisle, experimenting with the first of his piles to be constructed in England, had, by following up a chance observation, electrolysed water into oxygen and hydrogen. Solutions of metallic salts were rapidly decomposed by the same means. Even in the mid-eighteenth century chemical changes had been effected by electrostatic discharges, so that a way of proving yet more firmly the identity of frictional and galvanic electricity offered itself. It was certain that electrical forces could bring about a chemical change; was the converse also true? Humphry Davy asserted this boldly. In 1800 he pointed out that the voltaic pile could act only when the electrolyte was capable of oxidizing one of the metal elements, and that the intensity of its effect was directly related to the readiness of the electrolyte to react with the metal. Six years later, in a Bakerian lecture, Davy said:

In the present state of our knowledge, it would be useless to attempt to speculate on the remote cause of the electrical energy, or the reason why different bodies, after being brought into contact, should be found differently electrified; its relation to chemical affinity is, however, sufficiently evident. May it not be identical with it, and an essential property of matter?

For on Davy's view, in accord with his experiments on contact electricity, in the simplest types of electrochemical activity an

alkali which receives electricity from the metal would necessarily, on being separated from it, appear positive; whilst [an] acid under similar

[1] *Opere*, vol. II, pt. ii, p. 158; (written in 1801).

circumstances would be negative; and these bodies having respectively with regard to the metals, that which may be called a positive and a negative electrical energy, in their repellent and attractive functions seem to be governed by laws the same as the common laws of electrical attraction and repulsion.[1]

Thus Davy held that the concept of affinity, upon the basis of which the phenomena of chemical combination were explicable, was itself to be explained in terms of electrical forces or "energies." His prediction that electrolysis would prove to be a most valuable tool of chemical analysis was well borne out in the following year when, by this method, he isolated potassium from potash and sodium from common salt.

In the same lecture of 1806 Davy presented his electrochemical theory of the voltaic pile. He thought that, owing to the opposite "electrical energies" of the metals used, decomposition of the electrolyte occurred by the breaking down of its natural affinity, as in a solution of salt (for example) 'the oxygene and the acid are attracted by the zinc [plate], and the hydrogene and the alkali by the copper [plate].' The production of electricity, when the plates were joined in a circuit, was continuous because the chemical action, that is the solution of zinc in the electrolyte and the evolution of hydrogen from the copper plate, was continuous: 'the process of electromotion continues, as long as the chemical changes are capable of being carried on.' Davy also believed that owing to the tension created by the electrically opposed metals, the whole of the electrolyte was in a state of continual decomposition and recomposition. There were obvious defects in this account, but it had the great merit of integrating in one hypothesis the facts of contact electricity, discovered by Volta, and the facts of chemical change associated with the flow of electric current through compound bodies. Evidently Davy realized that the actions of electrolysis and of the voltaic pile are essentially the same, though the one requires an electric current to be applied, and the other yields a current. At the same time he recognized that electricity could be produced without chemical change, and chemical changes occur with which no electrical effects were associated; therefore chemical changes could not be 'the primary causes of the phænomena of GALVANISM.'

The scope of electrochemical theory was extended and defined

---

[1] *Philosophical Transactions*, 1807, pp. 33, 39.

by the Swedish chemist J. J. Berzelius, whose first memoir appeared in 1803. To pursue it farther would be to exceed the limits of this volume. What may be noticed is the significance of the sudden emergence of this young branch of science, electricity, as a bridge between physics and chemistry. A situation in which certain phenomena had been studied for their physiological significance, then reinterpreted by a physicist, and finally taken up by chemists, was absolutely unprecedented in science. The still-mysterious unity of nature was once more vindicated. For twenty years the chemical manifestations of electricity dominated research almost as completely as its mechanical manifestations had in the eighteenth century; it was not until about 1820, with the work of Oersted and Faraday, that the mechanical effects of current electricity (provided by the voltaic battery) attracted attention. Similarly, and inevitably, the mathematical theory of electric current was many years younger than that of electric charges, since in each case the mathematical theory was developed from the quantitative measurement of mechanical effects. All this was the fruit of the one crucial invention of the voltaic pile, the battery as it was soon to be called, to which Volta was led systematically from the first chance observations of Galvani. The chemical effects of transient electrostatic discharges had aroused little interest, whereas those of current electricity were spectacular. They raised the question of the relationship between the thing "electricity" or "electric fluid" and ordinary ponderable matter in a new and challenging form. Not only did electricity appear (in Davy's words) as 'an essential property of matter'—this, after all, in a different context, was the conclusion already drawn by Franklin and Dufay—it was rather that electricity was an active and determinant concomitant of matter without which, according to Davy and Berzelius, the chemical behaviour of the elementary forms of matter would be inconceivable. It was a logical consequence of the electrochemical theory that electricity was very far from being a kind of discontinuous abnormality, of concern only to the curious electrician who took pains to disturb bodies from their normal condition of comfortable neutrality. Electricity was not an adventitious atmosphere, or other circumstance, like humidity, which could be left out of account except for very special circumstances. From being casual, the rapid progress of ten years rendered it causal, having a function deeply involved in the differentiation of the various species of matter. Within a few more years the physicist was constrained to follow the chemist in making necessary adjustments so that he too could own in what sense electricity was "an essential property of matter," thus commencing his long ascent to the truth that the concepts "electricity" and "matter" are not complementary, but actually inseparable. In short, through the electricians' research, the link between the mathematical conception of matter, begun by Newton, and the empirical (or at least pragmatic) conception of matter proper to chemistry, was at last indicated although it is even yet far from being completely established.

CONCLUSION

MUCH more has been learnt about Nature, from the structure of matter to the physiology of man, in the last century and a half than in all preceding time. Of this there can be no doubt. But the scientific revolution ends when this vastly detailed exploration began, for it was that which made such investigation possible. At this point, in the early nineteenth century, a scientific paper on almost any topic is intelligible as a direct precursor to research which still continues. With infinitely feebler tools, but with the same insistence upon accuracy in observation, the same confidence in quantitative experimentation, the same enmeshing of theory, hypothesis and factual reporting which philosophers then and now found so resistant to logical analysis, men were tackling their problems as scientists today are tackling far more complex ones.

It has become almost a truism to assert that the development of natural science is the most pregnant feature of Western civilization. With technology—and this is hardly any longer an independent characteristic—it is the one product of the West that has had a decisive, probably permanent, impact upon other contemporary civilizations. Compared with modern science, capitalism, the nation-state, art and literature, Christianity and democracy, seem regional idiosyncrasies, whose past is full of vicissitudes and whose future is full of dark uncertainty. Each of these features of Western civilization has made its contribution to the genesis of science, to which perhaps their combination was essential, but one may imagine that science can flourish after one or all of these has lost its historic individuality. Indeed, that is already happening. Modern science was the offspring of a form of society which has lasted some four hundred years, playing a dominant rôle in world history; that form of society is now in dissolution, but it seems unlikely that science will necessarily disappear with it. For this, more than any other feature of Western society, has been the cause of the changes that we witness, and this also will be the most powerful influence on the moulding of a future society.

If this much—or even a fraction of it—be granted, it seems

unfortunate that we understand the genesis of modern science as little as we do. The modern study of Nature alone has had a determining effect upon the course of civilization, on history in its political, economic and intellectual totality. The science of Egypt and Babylonia, China and India, Greece and Islam and medieval Christendom had no such effect. The recovery of it by historians contributes little in a positive way to the understanding of the rise and decay of these societies, though such work does illuminate the origins and peculiarities of modern science. Among the challenges to which these societies responded successfully, or failed to meet, the ubiquitous challenge of Nature was certainly one; but the organized, conscious, rational response to it that we call science was of minor importance. Because of this, because some of these earlier strivings with Nature are continuously connected, and because all of them share certain common characters distinguishing them from modern science, they may be grouped together as intermediate between yet more primitive attempts to explain and master the mysteries of man's environment, and modern science. We know that modern science emerged from this intermediate stage, from which no other society than the recent western European was capable of escaping, and it is this emergence that we do not adequately comprehend.

Many questions have been cursorily handled in this book because satisfactory discussion would require a different and much more extensive treatment. Perhaps it is not possible yet. Scholars have discovered a great deal about the pre-history of the scientific revolution, and no simple account of its causation can be satisfactory. No one, for instance, would now deny that although it was a reaction against medieval scholasticism, scholasticism itself contributed to it. Galileo did not create a new science of motion out of nothing; while transforming that science and giving it unprecedented scope and certainty he used the idea of impetus and the technique of geometrical analysis that he inherited from the fourteenth century. On the other hand it would be absurd to suppose that the science of the seventeenth century was nothing more than a straightforward development of that of the fourteenth.

Among a multiplicity of factors a few suggest themselves as especially important in transforming the scientific outlook of early modern times. Technological and social progress formed a society that was relatively free, rich, and literate. The challenge its wealth of empirical skills offered to explanation and inventiveness alike was not confined to an isolated craftsman class, but was felt among the learned too; while the same skills promised means of investigating nature far superior to any available before. Technological progress implied the idea of intellectual progress, just as chance discoveries implied the possibility of systematic ones. If scientific knowledge was real it should be practicable; if practical, useful. Passive, contemplative learning was no longer enough.

This was not the sole, nor perhaps the most important, reason for growing dissatisfaction with the traditional science embalmed in books. Its sheer intellectual inadequacy was emphasized and re-emphasized from the fifteenth century onwards. First its astronomy, then its anatomy, were found insufficient; later the whole logic of its structure was pulled to pieces. Here the classical revival was most significant: negatively in revealing how much more extensive knowledge had been in antiquity than in the recent past; positively in introducing new ideas and methods of which the recent past had known little or nothing. Copernicus escaped from the exhausted, hopeless tyranny of Ptolemy to the long-rejected tenets of the Pythagorean school; others, like Galileo, found an alternative to the fruitless pursuit of Aristotle offered by Plato, Archimedes and Epicuros. The Greeks had had wider intellectual horizons than the schoolmen. Exploration of them produced apparently curious discrepancies, for the antithesis between old and new was never complete and no one (except Peter Ramus, who sought a *succès de scandale*) ever thought that Aristotle was mistaken on every count. No less than he, Copernicus and Galileo could not conceive of celestial motions that were not circular; Descartes could not accept the vacuum; nor Kepler the infinite. Intellectual freedom was not to be enjoyed wholesale, but piecemeal and even haphazard.

In the philosophy of the middle ages nothing now seemed more abstruse and sterile than discussion of forms and qualities in the Aristotelean manner. It was an exercise in intellectual subtlety that seemed to have minimal relation to things as they are. The reaction against it, nourished by the rediscovery of Greek atomism, resulted in the mechanistic philosophy of the seventeenth century, reducing the mystery of endless variety in nature to the interplay of two variables, matter and motion. Its necessary concomitant, the theory of primary and secondary qualities, yielded an immense simplification in the scientific appreciation of nature, since it transposed the mystery into the philosophic problem of perception. Now the physical universe, by definition, consisted only of moving particles, the ultimate reality of physical science; its mysteriousness resided only in the observer's varied, often distorted and misleading, registration of them in his consciousness. Whatever the sacrifice inherent in this transposition it was not made (immediately) by science. On the contrary, for the mechanistic universe that was now the object of scientific study was rendered susceptible of complete description, in principle at least; and as a complete, material, logically connected whole it could readily be seen as God's work, a divine artefact, the machine of nature. The Hebrew-Christian concept of creation suddenly crystallized into its perfect expression while the organismic Aristotelean view, always so awkward for scholastics, vanished for ever.

By a curious twist of irony the mathematical theory that the middle ages had applied vainly to its theory of forms suddenly became immensely serviceable when it was applied in a totally alien context. For the mechanistic universe could only be described properly in mathematical terms, and to this new language the scholastics had furnished a partial key. Yet the idea that the formal, logical structure of relationships (to which the facts of nature are fitted by science as they are disclosed by observation and experiment, and by which they are connected together in an orderly, explanatory system) should be a mathematical one was basically Platonic; or rather (in justice to the first of mathematical physicists) Archimedean. The notion of mathematical harmony, relationship, and order in nature came upon the sixteenth century with overwhelming force—far greater than it had ever exerted in antiquity. Plato had displayed its metaphysical charms; Archimedes had shown how to use it in weaving the texture of a theory. Galileo and Kepler learnt from them both. Mathematical reasoning could be built into a formal system (as Archimedes had demonstrated long before, and Galileo was to do in the *Discourses*) no less effectively, and with far more rigour, than the verbal logic of Aristotle. Given correct premises its conclusions (as Galileo proclaimed) *could not be false*. Thus, if the conclusions were verified in fact, the premises must be true; if both premises and conclusions were true, then the theory must be not only logically valid, but actually applicable to the physical world of experience. Mathematical physics was not merely a formal structure (as the middle ages, in its treatment of forms, and of astronomy, had taken it to be), it was an actual description of the real world.

This—for all its partial anticipation in the middle ages—was the core of the intellectual revolution in science. Not that it was the *whole* of the revolution in science, since other novel attitudes such as the emphasis on factual realism so significant in biological description, and empiricism in both physical and biological investigations, were independent of the mathematical approach. And it is obvious enough that the mathematical, theoretical attack would soon have spent its force had it not been continually reinforced by fresh resources drawn from observation and experiment. Inevitably, however, the sciences whose development was chiefly characterized during the sixteenth and seventeenth centuries by the extent and accuracy of their factual content suffered the least modification in their theoretical structure, which was slight at best. It was mathematical science above all that enabled the conceptual scheme of traditional science to be replaced; that offered substitutes for the rejected terms in which the latter had tried to explain nature. This had been recognized since the mid-sixteenth century as the principal desideratum, by the arch-empiricist Bacon no less than by Galileo or Kepler. Bacon could have written as well as Galileo that: "It is necessary first to teach the reform of the human mind, and to render it capable of distinguishing truth from falsehood," and he realized equally plainly that the faults of the old scheme of ideas could only be remedied by creating a new one.

Nevertheless, Bacon missed the supreme virtue of mathematical science in this task—so obvious once it is appreciated at all. Mathematical reasoning is simple because it avoids doubt and confusion. It is capable of infinite ramification without falling into inconsistency. Since it has the logic of number it has also, when applied to the physical universe in the manner of Archimedes and Galileo, the logic of quantity. It is capable of definite verification or refutation. It is totally indifferent to the observer and his problems of perception since it cannot extend beyond what the seventeenth century called the primary qualities. And, in its pre-twentieth century formulations at any rate, the mathematical universe converges with the mechanistic one; this truth, which was rather grasped intuitively than proved by Galileo and which Descartes attempted to elude (to the cost of his physical system), was amply demonstrated by Newton. Only with the abandonment of Newtonian physics could there be any question of renouncing it.

Beneath the positive accomplishments of the scientific revolution, beneath its adoption of new theories and new methods, there lies as Burtt pointed out long ago a change in metaphysics. The new notions about the content of a scientific explanation of events, about the properties of things and the laws governing them that must be taken for granted, about what is subjective and what objective in man's perception of nature, were not the products of the new scientific knowledge, though that knowledge could be taken to confirm them. Rather they stimulated Galileo's "reform of the human mind" through which acquisition of the new knowledge became possible. These new notions were indeed rather metaphysical than physical. They partly arose out of the critical enquiries of thirteenth and fourteenth century philosophers, partly they were learned from ancient authors read with new eyes, perhaps most of all they were produced by the intellectual frustration caused by the stagnation of late medieval science. To account in turn for *them*, however, would necessitate a philosophical and psychological research into the nature and expression at different periods of the capacity of the intellect for original thinking which cannot yet, perhaps, be produced at all and is certainly far outside the range of this book.

APPENDIX A

BOTANICAL ILLUSTRATION[1]

CERTAINLY naturalistic representations, both botanical and zoological, may be attributed to the medieval period—but only to its beginning and end. Such are to be found in the Greek *Codex Vindobonensis* (fifth century) or in the Latin herbal of Pseudo-Apuleius (seventh century); in both, however, there is already evidence of degeneration from the best Hellenistic models. They occur again in the works of miniaturists and other artists from the end of the fourteenth century onwards. Between these periods formalism and symbolism flourished and 'it is safe to assert that at the beginning of the thirteenth century scientific botanical illustration reached its nadir in the west' (Blunt). The herbal was then a mere uncritical catalogue, illustrated by figures which are purely conventional and heavily stylized. Botanical knowledge revived under the influence of the translators, as in the *De Plantis* of Albert the Great (*c.* 1200–80), the *Herbal* of Rufinus (*c.* 1290) and the *Buch der Natur* of Conrad von Megenburg (1309–74), all of which show occasional evidence of first-hand observation. Some originality is also shown in utilitarian writings of the thirteenth century and later dealing with hunting, falconry and agriculture. But the revival of naturalistic illustration was wholly the work of the artist, during the first stages of the Renaissance. Some of the best early representations of plants come from the brushes of Botticelli or the brothers van Eyck, as later from Dürer and Leonardo. Manuscripts equally prove that it was the artist, not the man of learning, who returned to the natural model. Some excellent figures, like those of Cybo of Hyères, are purely decorative and have no relation to the text they adorn. The almost contemporary Burgundian school of illuminators and Italian artists developed in the first years of the fifteenth century great technical and artistic skill in the representation of plants, producing such masterpieces as the herbal of Benedetto Rinio, prepared by Andrea Amadio, now preserved at Venice. The earliest printed herbals, however, still maintained the old stylized convention, seen for example in the English *Grete Herball* (1526). Only with the work of Brunfels (*Herbarum vivæ icones*, 1530–36, illustrated by Hans Weiditz) and of Fuchs (*Historia stirpium*, 1542, illustrated by Albrecht Meyer) was naturalistic interpretation transferred to the wood-cut block.

[1] Cf. Charles Singer: *From Magic to Science* (London, 1928); and "The herbal in antiquity and its transmission to later ages," *J. Hellenic Studies*, vol. XLVII, 1–52; Wilfrid Blunt: *The Art of Botanical Illustration* (London, 1950).

APPENDIX B

COMPARISON OF THE PTOLEMAIC AND COPERNICAN SYSTEMS

THE geometrical equivalence of the geostatic and heliostatic methods of representing the apparent motions of the celestial bodies, adopted by Ptolemy and Copernicus respectively, is not often clearly emphasized

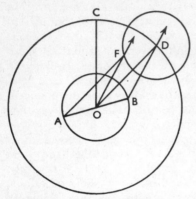

FIG. 11. The Upper Planets.

though it is vital to any discussion of the nature of the change in astronomical thought effected by Copernicus. In a simplified form it may be easily demonstrated.

Upper Planets (Fig. 11). On the Copernican hypothesis, O is the place of the sun, B that of the earth which has revolved through any angle COB, and D that of a planet which in the same time has revolved through the angle COD. The apparent position of the planet is on the line BD. Transposing this into Ptolemaic terms, O is the fixed earth, A

the sun, and D the centre of the planet's epicycle. If F is the place of the planet in the epicycle, then DF:DO = BO:DO. Moreover, AOB is a straight line, and $\angle FDO = \angle COB - \angle COD = \angle BOD$. These

relations follow from the form of the constants used by Ptolemy and Copernicus respectively (p. 15). Thus the apparent position of the planet is given as before, since the line FO is parallel to BD, and the triangles AFO, BDO, are identical.

Lower Planets (Fig. 12). Here the situation is slightly different. The Copernican representation is similar to that given above, with the sun at O, earth at B, planet at D, and the position given by the line BD. On the Ptolemaic representation the central earth is at O, the sun at A,

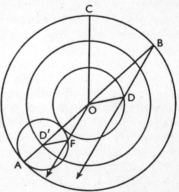

FIG. 12. The Lower Planets.

and the centre of the planet's epicycle is located on the radius AO. Values for the radii of the deferent and epicycle may be chosen so that D'F:D'O = DO:BO, provided that D'F + D'O <AO = BO. As be-

fore, these relations follow from the form of the constants adopted by Ptolemy and Copernicus. For the same reason the velocity of rotation of F about D′ is such that $\angle OD'F = \angle COD - \angle COB = \angle BOD$. Thus it follows (as with the upper planets) that FO and BD are parallel, and that the maximum elongation of the planet from the sun is the same on either hypothesis.

It may be noted that since *any* geometrical complexity added to one representation may be duplicated by a corresponding one in the other, the apparent positions of the planets are always the same on either hypothesis. In the case of the inferior planets, Venus and Mercury, however, the triangles OD′F, BOD, are similar but not identical. These planets cannot appear on the Ptolemaic hypothesis, as they do on the Copernican, on the remote side of the sun from the earth. On the latter hypothesis, but not on the former, Venus should exhibit phases similar to those of the moon. These were indeed observed by Galileo with the telescope. A simple adjustment to the Ptolemaic system (proposed long before) made it identical with the Copernican in this respect by centring the epicycles of Venus and Mercury upon the sun, i.e. drawing them about A.

APPENDIX C

SCIENTIFIC BOOKS BEFORE 1500

WITHOUT compiling any very elaborate statistics, it is apparent from the work of A. C. Klebs ("Incunabula Scientifica et Medica," *Osiris*, vol. IV, 1938) that the printed literature of science at the beginning of the sixteenth century was in the main dominated by its established traditions. Klebs lists more than 3,000 editions of 1,044 titles by about 650 authors. Among these occur 95 editions (and collections) of works attributed to Aristotle, 18 editions of the *Natural History* of Pliny the Elder, 7 of Ptolemy's *Cosmographia* (but none of the *Almagest*), and 5 of Lucretius. Of ancient medical writers, Dioscorides (*De materia medica*) was printed twice, and Galen (complete so far as known) once. But separate works of both Hippocrates and Galen were included in numerous collections of medical authorities. Celsus (*De medicina*) was printed four times. Collections of classical writers on military affairs, agriculture and astronomy were popular enough to justify more than one edition. Euclid was printed twice only.

Translations from the Arabic were often printed. They include Avicenna's *Canon of Medicine* (14 editions) and the works of other great Islamic physicians: al-Razi (15 editions besides titles in collections), Mesue (19 editions) and Serapion (4 editions). Averroes' commentaries on Aristotle were well known in print, but not Arabic astronomical writings.

Works deriving from the Latin West before 1400 form a very numerous group. The long-used *Etymologiæ* of Isidore of Seville was printed 8 times. An early product of the renaissance of learning in Europe, the *Quæstiones Naturales* of Adelard of Bath, merited 2 editions. Works by, or attributed to, Albert the Great were immensely popular, especially the *Secreta mulierum* and the *Liber aggregationis*, amounting to 150 editions. Thomas Aquinas appeared in 17 editions and numerous collections, Albert of Saxony in eleven. Two more sought-after books were the *Sphere* of Sacrobosco (31 editions) and the *Physiognomia* of Michael Scot (21 editions). Less well known medieval treatises, like those of Oresme (3), Thomas Bradwardine (3) or Walter Burley (5) all found publishers. The most widely read English author was undoubtedly Bartholomew (*De proprietatibus rerum*) with 12 editions in Latin, 8 in French, and others in English, Spanish and Flemish. Some medical writers of the middle ages were still in demand for study, to judge from the 31 editions of Arnald of Villanova, the 13 of Guy de Chauliac, and the 9 of Mondino. The most popular of all medical treatises was still the *Regimen sanitatis* of Salerno (41 editions). With these the 35 editions of printed herbals before 1500 may be linked.

The proportion of scientific books printed in the vernacular languages was small, but it was a microcosm of the whole. The German language was perhaps the best endowed in this way, and next French; in Italy, probably, the university tradition was too strong to render such vernacular printing profitable, and the literate class in fifteenth-century England did not at first demand many serious books. Many were of an ephemeral nature—prognostications and predictions, almanacks, and popular treatises on the maintenance of health. This last group includes the *Livre pour garder la sante* (1481), and the German *Regimen sanitatis zu Deutsch* (1472). Caxton printed *The Gouvernayle of helthe* in 1489. A book even more full of marvels than Bartholomew's, and equally widely available, was Mandeville's *Itinerarius* (in English, French, German, Italian and Flemish). Herbals were soon made accessible to the vernacular reader (German and French 1486, English 1525), as were treatises on anatomy deriving from Mondino (German, Hieronymus Brunschwig, 1497; Italian and Spanish, Johannes de Ketham, 1493 and 1494). The *Chirurgia* of Guy de Chauliac was represented in three languages before 1500. There were also some vernacular books on reckoning.

Here, then, is evidence drawn from two dissimilar sources enforcing the same conclusion. On the one hand, the great Italian printing-houses, supplying a highly literate and academic market, found it profitable to publish many of the "classics" of medieval science, often in several editions; on the other hand, vernacular printing shows the same intellectual content scaled down and vulgarized. There was no sudden craving for originality, no epidemic of criticism. This is also apparent from Mr. H. S. Bennett's study of early English publications dealing with science and medicine (*English Books and Readers*, 1475–1557, Ch. vi). The first surgical texts in English were based on Johannes de Vigo and Guy de Chauliac; the first anatomy text (by Thomas Vicary, 1548) on Mondino. Caxton's *Myrrour of the World* (1481) 'is a typical example of the encyclopædia beloved of the Middle Ages, and here made available to the ordinary reader with no attempt to bring it up to date.' Yet it was reprinted in 1490 and 1529. Such works, Mr. Bennett writes: 'contributed little or nothing that was new. Their compilers were content to reproduce knowledge that had been current for centuries, and the stationers traded in these wares confident that their customers would not be put off by their old-fashioned contents.' (But were the customers conscious that the contents were old-fashioned?) Caxton made no greater effort to revise the geographical or historical knowledge which he imparted in his books, much of it taken directly from the thirteenth and fourteenth centuries. In short, so far as the respective tastes of printers, students and the general public may be ascertained, they were conservative; those texts which had been most in demand before the invention of printing were the very ones that became most widely disseminated after it. Printing, therefore, had comparatively little immediate effect on the literature of science in the way of substituting novelties for traditions, though it did condemn to darker oblivion those texts which were already half-forgotten by the mid-fifteenth century, and, by reason of this, it reveals the strength of the conservative forces extant in the early stages of the renaissance.

BIBLIOGRAPHICAL NOTES

STUDIES on the history of science are already so numerous that it is only possible in these notes to give some indications of indebtedness, and of some possible lines of exploration. I restrict myself here mainly to the English and French languages, though these are (obviously) far from containing everything of importance.

General : The journals *Isis* and *Osiris, Annals of Science, Archives Internationales d'histoire des Sciences* (continuing *Archeion*), and *Revue d'histoire des Sciences*, are indispensable. The bibliographies in *Isis* analyse work in the history of science during the last forty years; cf. also G. Sarton, *A Guide to the History of Science* (Waltham, Mass., 1952) and the handlist published by the Historical Association (*Helps for Students of History, No. 52*). The period covered by this volume is treated in all the general histories of science (Singer, Dampier, Wightman, Mason etc.) and especially by Professor H. Butterfield, *The Origins of Modern Science* (London, 1949), and H. T. Pledge, *Science since 1500* (London, 1939). A. Wolf, *History of Science, Technology and Philosophy in the Sixteenth, Seventeenth and Eighteenth Centuries* gives many valuable scientific details but is very deficient in interpretation.

CHAPTER I

For reference and bibliography, G. Sarton, *Introduction to the History of Science* (Baltimore, 1927–48, 5 vols.) and Lynn Thorndike, *History of Magic and Experimental Science* (New York, 1923–58) are essential. There is no fully adequate short history for the medieval period. For Islam, cf. H. J. J. Winter, *Eastern Science* (London, 1952), A. Mieli, *La Science Arabe* (Leiden, 1938). Some of the older books, e.g. C. Singer, *From Magic to Science* (London, 1928) are still useful. A. C. Crombie, *From Augustine to Galileo* (London, 1952; rev. ed., *Medieval and Early Modern Science*, 1959) and *Robert Grosseteste and the Origins of Experimental Science* (Oxford, 1953), takes a favourable view of the medieval contribution to the scientific revolution, with much bibliographical information. C. H. Haskins, *The Renaissance of the Twelfth Century* (Cambridge, Mass., 1928), and *Studies in the History of Medieval Science* (Cambridge, Mass., 1924); J. Huizinga, *The Waning of the Middle Ages* (London, 1924); H. Rashdall, *Universities of Europe in the Middle Ages* (ed. Oxford, 1936), are useful for background. Editions of texts being published in quantity, e.g. M. Clagett, *The Science of Mechanics in the Middle Ages* (Madison, 1959); L. Thorndike, *The Sphere of Sacrosbosco* (Chicago, 1949), and *The Herbal of Rufinus* (Chicago, 1946). On renaissance learning see George Sarton, *Six Wings* (Bloomington, 1957).

CHAPTER II

(*a*) There are general histories of medicine by F. H. Garrison (Philadelphia, 1929) and A. Castiglioni (New York, 1947)—the latter has the more synthetic treatment. A valuable survey is C. Singer, *The Evolution of Anatomy* (London, 1926). J. B. de C. M. Saunders and C. D. O'Malley, *Andreas Vesalius* (New York,

1950) give the anatomical illustrations with short notes; *idem, Leonardo da Vinci on the Human Body* (New York, 1952). Cf. also C. Singer, *Vesalius on the Human Brain* (London, 1952)—the only translation of a fair portion of *De Fabrica*; H. Cushing, *A Bio-bibliography of Andreas Vesalius* (New York, 1943); S. W. Lambert *et al.*, *Three Vesalian Essays* (New York, 1952); Saunders and O'Malley, and Singer, in *Studies and Essays . . . offered to George Sarton* (New York, 1946), articles in the *Bull. Hist. Med.*, vol. XIV, 1943; J. P. McMurrich, *Leonardo da Vinci the Anatomist* (London, 1930), Vittorio Putti, *Berengario da Carpi* (Bologna, 1937); L. R. Lind has translated Berengario's *Short Introduction to Anatomy* (Chicago, 1959); G. Keynes, *The Apologie and Treatise of Ambroise Paré* (London, 1951); Sir C. Sherrington, *The Endeavour of Jean Fernel* (Cambridge, 1946).

(*b*) The edition of Ptolemy which I have found most accessible is that of the Abbé Halma (with French translation, Paris, 1813–16), and for Copernicus the original printing of 1543. Secondary works are: A. Armitage, *Copernicus, the Founder of Modern Astronomy* (London, 1938), H. Dingle in *The Scientific Adventure* (London, 1952), J. L. E. Dreyer, *History of Planetary Systems from Thales to Kepler* (Cambridge, 1906) (repr. as *History of Astronomy*, New York, 1953), F. R. Johnson, *Astronomical Thought in Renaissance England* (Baltimore, 1937), T. S. Kuhn, *The Copernican Revolution* (Harvard, 1957), E. Rosen, *Three Copernican Treatises* (New York, 1939), D. W. Singer, *Giordano Bruno, his Life and Thought* (London, 1950), D. Stimson, *The Gradual Acceptance of the Copernican Theory* (New York, 1917).

(*c*) Metallurgy and industrial chemistry are dealt with in G. Agricola, *De re metallica* (trans. H. C. and L. H. Hoover, New York, 1950); C. S. Smith and M. Gnudi, *The Pirotechnia of Vannoccio Biringuccio* (New York, 1943), C. S. Smith and A. Sisco, *Treatise on Ores and Assaying of Lazarus Ercker* (Chicago, 1951). On history of pharmacology, E. Kremers and G. Urdang, *History of Pharmacy* (Philadelphia, 1940).

CHAPTER III

There is no wholly satisfactory English work on Galileo. J. J. Fahie, *Galileo, his Life and Works* (London, 1903) is uncritical and out of date. F. Sherwood Taylor, *Galileo and the Freedom of Thought* (London, 1938) is better but limited. In translation there are *Dialogues concerning Two New Sciences* (H. Crew and A. de Salvio, New York, 1914, 1952), Stillman Drake, *Dialogue concerning the Two Chief World Systems* (Berkeley, 1953); *idem, Discoveries and Opinions of Galileo* (New York, 1957); *idem* and I. E. Drabkin, *Galileo on Motion and on Mechanics* (Madison, 1960). The best book on Galileo and the church is Giorgio de Santillana, *The Crime of Galileo* (Chicago, 1955); he has also revised Salusbury's translation of the *Dialogue on the Great World Systems* (Chicago, 1953). Pierre Duhem's classic *Études sur Léonard de Vinci* (Paris, 1906–13) and *Origines de la Statique* (Paris, 1905–6) are invaluable. Cf. also L. Cooper, *Aristotle, Galileo, and the Leaning Tower of Pisa* (Ithaca, 1935), René Descartes, *Œuvres*, ed. Ch. Adam and P. Tannery (Paris, 1897–1913), R. Dugas, *Histoire de la Mécanique* (Neuchatel, 1950), G. Galilei, *Opere*, ed. A. Favaro (Firenze, 1890–1909), A. Koyré, *Études Galiléennes* (Paris, 1939; *Actualités Scientifiques et Industrielles* Nos. 852–4—most important), E. Mach, *Science of Mechanics* (Chicago, 1907), A. Maier, *Die*

Vorläufer Galileis in 14. Jahrhundert (Rome, 1949) and *Die Impetustheorie* (Rome, 1951), R. Marcolongo, "Lo sviluppo della meccanico sino ai discepoli di Galileo" in *Atti d. R. Acc. dei Lincei* (Physical Series) vol. XIII (Rome, 1920), A. Mieli, "Il Tricentario dei 'Discorsi e Dimostrazioni Matematiche' di Galileo Galilei" in *Archeion*, vol. XXI (Rome, 1938)—critical of Duhem etc., N. Oresme, "Le Livre du Ciel et du Monde" in *Medieval Studies*, vols. III–V (1941–3), G. Sarton, "Simon Stevin of Bruges" in *Isis*, vol. XXI (1934).

CHAPTER IV

In addition to works already mentioned, A. Armitage, "The Deviation of Falling Bodies," *Ann. Sci.*, vol. V (1947), I. Boulliau, *Astronomia Philolaica* (Paris, 1645), M. Caspar (trs. C. D. Hellman), *Johannes Kepler* (London, 1959); J. L. E. Dreyer, *Tycho Brahe* (Edinburgh, 1890), Johann Kepler, *Gesammelte Werke* (Munich, 1938–), A. Koyré, "A documentary history of the problem of fall from Kepler to Newton," *Trans. Amer. Phil. Soc.*, N.S. xlv, 1955; S. I. Mintz, "Galileo, Hobbes, and the Circle of Perfection," *Isis*, vol. 43 (1952), D. Shapley, "Pre-Huygenian Observations of Saturn's Rings," *Isis*, vol. 40 (1949).

CHAPTER V

The best study is H. P. Bayon, "William Harvey, Physician and Biologist," *Ann· Sci.*, vols. III, IV (1938–9); cf. also L. Chauvois, *William Harvey* (London, 1957). General books are F. J. Cole, *Early Theories of Sexual Generation* (Oxford, 1930) and *History of Comparative Anatomy* (London, 1949), M. Foster, *History of Physiology* (Cambridge, 1924), J. Needham, *History of Embryology* (Cambridge, 1934), E. Nordenskiold, *History of Biology* (New York, 1946), C. Singer, *History of Biology* (New York, 1950). Cf. also H. Brown, "John Denis and Transfusion of Blood, Paris, 1667–8," *Isis*, vol. 39 (1938), L. D. Cohen, "Descartes and More on the Beast-Machine," *Ann. Sci.*, vol. I (1936), C. Dobell, *Antony van Leeuwenhoek and his "Little Animals"* (London, 1932), G. Keynes, "The History of Blood Transfusion," *Science News*, vol. III (1947), A. van Leeuwenhoek, *Collected Letters* (Amsterdam, 1939–), W. Pagel, "William Harvey and the Purpose of the Circulation," *Isis*, vol. 42 (1951), C. E. Raven, *John Ray* (Cambridge, 1950), F. Redi, *Opere* (Napoli, 1778, Milano, 1809–11), J. Trueta, "Michael Servetus and the Discovery of the lesser Circulation," *Yale Jo. of Biol. and Med.*, vol. XXI (1948), R. Willis, *Works of William Harvey* (London, 1847).

CHAPTER VI

There is to my knowledge no complete history of scientific method. Useful contemporary studies are R. B. Braithwaite, *Scientific Explanation* (Cambridge, 1953), M. R. Cohen and E. Nagel, *Introduction to Logic and Scientific Method* (London, 1934), N. R. Hanson, *Patterns of Discovery* (Cambridge, 1958); S. Toulmin, *Philosophy of Science* (London, 1953). Cf. also F. H. Anderson, *Philosophy of Francis Bacon* (Chicago, 1948), A. G. A. Balz, *Cartesian Studies* (New York, 1951), E. A. Burtt, *Metaphysical Foundations of Modern Physical Science* (London, 1925), A. C. Crombie and H. Dingle, *op. cit.* for Chs. I and II, A.

Gewirtz, "Experience and the non-mathematical in Descartes," *Jo. Hist. Ideas*, vol. II, 1941, E. Gilson, *La Philosophie au Moyen Age* (Paris, 1944), A. Koyré, "An Experiment in Measurement," *Proc. Amer. Phil. Soc.*, XCVII, 1953, J. H. Randall, "The Development of the Scientific Method in the School of Padua," *Jo. Hist. Ideas*, vol. I (1940), B. Russell, *History of Western Philosophy* (London, 1946).

CHAPTER VII

(a) An excellent survey is M. Ornstein, *The Rôle of Scientific Societies in the 17th Century* (Chicago, 1938). Cf. also J. Bertrand, *L'Académie des Sciences et les Académiciens de 1666 à 1793* (Paris, 1869), T. Birch, *History of the Royal Society* (London, 1756), H. Brown, *Scientific Organization in 17th Century France* (Baltimore, 1934). A. Favaro, "Documenti per la Storia dell'Accademia dei Lincei," *Bullettino di Bibliografia e di Storia delle Scienze*, vol. XX (Rome, 1887), A. J. George, "The Genesis of the Académie des Sciences," *Ann. Sci.*, vol. III (1938), F. R. Johnson, "Gresham College: Precursor of the Royal Society," *Jo. Hist. Ideas*, vol. I (1940), R. F. Jones, *Ancients and Moderns* (St. Louis, 1936), R. Lenoble, *Mersenne ou la Naissance du Mécanisme* (Paris, 1943), Sir H. Lyons, *The Royal Society* (London, 1944), *Notes and Records of the Royal Society, passim*.

(b) A very important essay, with full references, is M. Boas, "The Establishment of the Mechanical Philosophy," *Osiris*, vol. X, 1952. Also, H. Brown, "The Utilitarian Motive in the Age of Descartes," *Ann. Sci.*, vol. I (1936), A. R. Hall, "The Scholar and the Craftsman in the Scientific Revolution," in *Critical Problems in the History of Science* (Madison, 1959); R. K. Merton, "Science, Technology and Society in 17th Century England," *Osiris*, vol. IV (1938), G. Milhaud, *Descartes Savant* (Paris, 1921), J. F. Scott, *The Scientific Work of René Descartes* (London, 1952), P. Mouy, *La Développement de la Physique Cartésienne* (Paris, 1934), J. R. Partington, "Origins of the Atomic Theory," *Ann. Sci.*, vol. IV (1939).

CHAPTER VIII

M. Cantor, *Vorlesungen über Geschichte der Mathematik* (Leipzig, 1880–1908), is still essential for reference. J. E. Montucla, *Histoire des Mathématiques* (Paris, 1799–1802) is well worth reading on the seventeenth century. Cf. also C. B. Boyer, *History of the Calculus* (repr. New York, 1959); D. E. Smith, *History of Mathematics* (New York, 1923), and more popular accounts by E. T. Bell, A. Hooper and others. M. Daumas, *Les Instruments Scientifiques aux 17e et 18e Siècles* is excellent. The histories of single instruments are R. S. Clay and T. H. Court, *History of the Microscope* (London, 1932) and H. C. King, *History of the Telescope* (London, 1955). Further, I. B. Cohen, "Roemer and Fahrenheit," *Isis*, vol. 39 (1948), J. W. Olmsted, "The Application of Telescopes to Astronomical Instruments," *Isis*, vol. 40 (1949), L. D. Patterson, "The Royal Society's Standard Thermometer," *Isis*, vol. 44 (1953), F. Sherwood Taylor, "The Origins of the Thermometer," *Ann. Sci.*, vol. V (1947).

CHAPTER IX

(i) *Sources:* R. T. Gunther, *Early Science in Oxford* (esp. vols. VI, VII, VIII); Christiaan Huygens, *Œuvres Complètes* (La Haye, 1888–1950); Isaac Newton, *Correspondence* (ed. H. W. Turnbull, Cambridge, 1959–); *Opticks* (London, 1931; New York, 1952); *Principia* (ed. F. Cajori, Berkeley, 1946). Cf. also I. B. Cohen *et al.*, *Isaac Newton's Papers and Letters on Natural Philosophy* (Cambridge, Mass., 1958); A. R. and M. B. Hall, *Unpublished Scientific Papers of Sir Isaac Newton* (Cambridge, 1962).

(ii) *Biographies:* A. E. Bell, *Christiaan Huygens and the Development of Science in the Seventeenth Century* (London, 1947); Sir D. Brewster, *Memoirs of Sir Isaac Newton* (Edinburgh, 1855); M. 'Espinasse, *Robert Hooke* (London, 1956); L. T. More, *Isaac Newton* (London, 1934).

(iii) *Studies:* H. G. Alexander, *The Leibniz-Clarke Correspondence* (Manchester, 1956); W. W. R. Ball, *An Essay on Newton's Principia* (London, 1893); I. B. Cohen, *Franklin and Newton* (Philadelphia, 1956); W. J. Greenstreet (ed.), *Isaac Newton Memorial Volume* (London, 1927); History of Science Society, *Isaac Newton* (London, 1928); A. Koyré, *From the Closed World to the Infinite Universe* (Baltimore, 1957); A. J. Snow, *Matter and Gravity in Newton's Physical Philosophy* (London, 1926).

(iv) *Articles:* E. N. da C. Andrade, "Robert Hooke," *Proc. Royal Society* A, CCI, 1950; A. Armitage, " 'Borel's hypothesis' and the Rise of Celestial Mechanics," *Ann. Sci.*, VI, 1948–50; A. R. Hall, "Sir Isaac Newton's Notebook," *Cam. Hist. Journ.*, IX, 1948, "Further Optical Experiments of Isaac Newton," *Ann. Sci.*, XI, 1955, "Newton on the Calculation of Central Forces," *Ann. Sci.*, XIII, 1957, "Correcting the *Principia*," *Osiris*, XIII, 1958; A. R. and M. B. Hall, "Newton's Mechanical Principles," *Journ. Hist. Ideas*, XX, 1959, "Newton's Chemical Experiments," *Arch. Int. d'Hist. Sci.*, XI, 1948, "Newton's Theory of Matter," *Isis*, LI, 1960. A. Koyré, "La Mécanique Céleste de J. A. Borelli," *Rev. d'Hist. Sci.*, V, 1952, "Significance of the Newtonian Synthesis," *Arch. Int. d'Hist. Sci.*, XI, 1950, "Unpublished Letter of Robert Hooke to Isaac Newton," *Isis*, XLIII, 1952, "La Gravitation Universelle de Kepler à Newton," *Actes VI Cong. Int. d'Hist. Sci.*, 1953, "L'Œuvre astronomique de Kepler," *XVIIᵉ Siècle*, 1956, "L'hypothèse et l'expérience chez Newton," *Bull. Soc. franc. Philosophie*, I, 1956. L. D. Patterson, "Hooke's Gravitation Theory and its Influence on Newton," *Isis*, XL, 1949; E. W. Strong, "Newton and God," *Journ. Hist. Ideas*, XIII, 1952; R. S. Westfall, "Isaac Newton, Religious Rationalist or Mystic?", *Rev. of Religion*, XXII, 1958.

CHAPTER X

To the books listed for Ch. V may be added: A. Arber, *Herbals* (Cambridge, 1950), P. G. Fothergill, *Historical Aspects of Organic Evolution* (London, 1952, with full bibliography), Knut Hagberg, *Carl Linnæus* (London, 1952), A. Hughes, *History of Cytology* (London, 1959), W. P. Jones, "The Vogue of Natural History in England, 1750–70," *Ann. Sci.*, vol. II, 1937, A. O. Lovejoy, *The Great Chain of Being* (Cambridge, Mass., 1948), C. E. Raven, *English Naturalists from Neckam to Ray* (Cambridge, 1947), and *Natural Religion and Christian Theology* (Cambridge, 1953), C. Singer, "The Dawn of Microscopical Discovery," *Jo. Roy. Microscopical Soc.* (1915).

CHAPTER XI

There is no modern history of chemistry on the large scale. H. M. Leicester, *The Historical Background of Chemistry* (London, 1956) and J. R. Partington, *Short History of Chemistry* (London, 1948) are good introductions. On 18th century chemical industry A. and N. Clow, *The Chemical Revolution* (London, 1952)—mainly Scotland—and H. Guerlac, "Some French Antecedents of the Chemical Revolution," *Chymia*, 1958, are excellent. Further, M. Boas, *Robert Boyle and 17th Century Chemistry* (Cambridge, 1958); idem, "Acid and Alkali in 17th Century Chemistry," *Arch. Int. d'Hist. Sci.*, IX, 1956, "Structure of Matter and Chemical Theory in the 17th and 18th Centuries," *Critical Problems in the History of Science*, (Madison, 1959); L. J. M. Coleby, *Chemical Studies of P. J. Macquer* (London, 1938); M. Daumas, *Lavoisier* (Paris, 1941), *Lavoisier, théoricien et expérimentateur* (Paris, 1955); H. Guerlac, "Joseph Black and Fixed Air," *Isis*, XLVIII, 1957, "John Mayow and the Aerial Nitre," *Actes VII Cong. Int. d'Hist. Sci.*, 1953, "The Origin of Lavoisier's Work on Combustion," *Arch. Int. d'Hist Sci.*, XII, 1959; D. McKie, *Antoine Lavoisier* (London, 1935), and *Essays of Jean Rey* (London, 1951), and "Black's Chemical Lectures," *Ann. Sci.*, vol. I (1936), H. Metzger, *Les Doctrines Chimiques en France du début du 17e à la fin du 18e Siècle* (Paris, 1923) and *Newton, Stahl, Boerhaave et la Doctrine Chimique* (Paris, 1930), J. R. Partington, "Joan Baptista van Helmont," *Ann. Sci.*, vol. I (1936), and with D. McKie, "Historical Studies in the Phlogiston Theory," *Ann. Sci.*, vols. II–IV (1937-9), also, "Life and Works of John Mayow," *Isis*, XLVII, 1956; T. S. Patterson, "Jean Beguin and his Tyrocinium Chemicum," *Ann. Sci.*, vol. II (1937), and "John Mayow in Contemporary Setting," *Isis*, vol. 15 (1931), J. M. Stillman, *The Story of Early Chemistry* (New York, 1924), J. H. White, *History of the Phlogiston Theory* (London, 1932).

CHAPTER XII

P. Brunet, *L'Introduction des Théories de Newton en France au 18e Siècle* (Paris, 1931), I. B. Cohen, *Benjamin Franklin's Experiments* (Cambridge, Mass., 1941—with useful historical introduction) and *Franklin and Newton* (Philadelphia, 1956), E. S. Cornell, "The Radiant Heat Spectrum," *Ann. Sci.*, vol. III (1938), D. Fleming, "Latent Heat and the Invention of the Watt Engine," *Isis*, vol. 43 (1952), L. Galvani, *Opere edite e inedite* (Bologna, 1841), D. McKie and N. H. de V. Heathcote, *The Discovery of Specific and Latent Heats* (London, 1935), J. C. Maxwell, *Electrical Researches of Henry Cavendish* (Cambridge, 1879), *Philosophical Magazine*, 'Natural Philosophy through the 18th Century' (Commemoration Number, 1948), C. Truesdell, introductions to *Euleri Opera Omnia*, Series II, vols. XI, XII (Zürich, 1954, 1956). A. Volta, *Collezione del Opere* (Firenze, 1816), W. Cameron Walker, "The Detection and Measurement of Electric Charges in the 18th Century," *Ann. Sci.*, vol. I (1936), Sir E. Whittaker, *History of Theories of Aether and Electricity*, vol. I (Cambridge, 1951).

INDEX

Watt, James (1736–1819), engineer, 350

Wesley, John (1703–91), 341

Whiston, William (1667–1752), divine, 299

White, Gilbert (1720–93), naturalist, 295, 301

Wilkins, John (1614–72), divine, 103, 193, 194

William of Moerbeke (c. 1215–86), translator, 39

William of Ockham (c. 1295–1349), philosopher, 56, 163

Willughby, Francis (1635–72), naturalist, 286

Wilson, George (1631–1711), chemist, 320

Wolff, Caspar (1738–94), biologist, 295

Wordsworth, William (1770–1850), 295

Wren, Christopher (1632–1723), architect, 153, 194

Wyclif, John (d. 1384), 2

Young, Thomas (1773–1829), physicist, 245

Zaluzian, Adam Zaluzanski of (c. 1500–1610?), botanist, 283